AP①

ENCYCLOPAEDIA OF
NANO CHEMISTRY

ENCYCLOPAEDIA OF
NANO CHEMISTRY

R. Thomson

ANMOL PUBLICATIONS PVT. LTD.
NEW DELHI - 110 002 (INDIA)

ANMOL PUBLICATIONS PVT. LTD.
H.O.: 4374/4B, Ansari Road, Darya Ganj,
New Delhi-110 002 (India)
Ph.: 23278000, 23261597

B.O.: No. 1015, Ist Main Road, BSK IIIrd Stage
IIIrd Phase, IIIrd Block,
Bangalore - 560 085 (India)
Visit us at: www.anmolpublications.com

Encyclopaedia of Nano Chemistry

First Published, 2006

ISBN 81-261-2984-0

PRINTED IN INDIA

Printed at Mehra Offset Press, Delhi.

Content

Preface

This book entitled "Encyclopaedia of Nano Chemistry" provides readers with an elaborate introduction to molecular nanotechnology, nanostructured materials, nanoparticles and nanodevices. An overview of relationship between nanotechnology and nano-veliro is presented herein. The modern global Issues regarding nanoparticles, nanostructured materials and polymer nanotechnology are discussed in detail. The subject areas of nanocomposites and nanostructured materials are covered detailibg about their respective properties and applications. The inter-relatioship among today's nanoworld, nanochemistry and the all important process of self-assembly is highlighted. An overview of nanotech products is presented.

There are so many other applications of nano chemistry ranging from mundane to incredibly profound. Other applications include energy storage materials, photovoltaic cells, fuel cells, polymer nanocomposites, super-tough nanocoatings, landmine detectors, and thousands more. Although it is a field that has already seen its share of hype, there is a general consensus that over the long term, nano chemistry is likely to reconfigure everything. Nanostructured materials do not represent a new phenomenon. However, the ability to probe, manipulate, understand and engineer matter at atomic scales has only recently come within our grasp. Developments such as the invention of the Scanning Tunneling Microscope in 1981 have since made nanoscale science a reality. Today, the term nanotechnology is used to describe specific construction and manipulation performed on an atomic and molecular scale. Generally, systems with a maximum size of less than a hundred nanometers are classified as nanotechnology. Nanotechnology is a completely new production process in a molecular world, opening up possibilities far beyond what we have known before. Nature provides the inspiration for this new technology – from time immemorial, nanomachines haye been at work in all organic cells, building our world of plants, animals and people from a store of chemical elements.

Experts world over believe nanotechnology will have a more far-reaching impact than the silicon integrated chip with many more applications than electronics. Nanotechnology is expected to revolutionize certain areas, such as semiconductors, pharmaceuticals and materials. The total societal impact of nanotechnology is expected to be much greater than that of the silicon integrated circuit because it is applicable in many more fields than electronics. Nanotechnology experts believe that because of nano chemistry, we would see more changes in the next 30 years than we saw in the entire last century.

This book tends to cover the aforementioned subject areas in their broadest possible terms. All useful supplementary research and reference tools have been included for readers' further investigations in the subject area of Nano chemistry. This book mainly focuses on the following areas: Nanotechnology; Nanoparticles, Nanomaterials and Nanostructures. This book seeks to explain this complex field of study. It also seeks to provide an understanding of what is happening now in the field of nano chemistry, as well as what to expect in future by generations to come.

—Editor

1 Introduction to Molecular Nanotechnology, Nanostructured Materials, Nanoparticles and Nanodevices

Molecular Nanotechnology: An Introduction

Manufactured products are made from atoms. The properties of those products depend on how those atoms are arranged. If we rearrange the atoms in coal we can make diamond. If we rearrange the atoms in sand (and add a few other trace elements) we can make computer chips. If we rearrange the atoms in dirt, water and air we can make potatoes. Todays manufacturing methods are very crude at the molecular level. Casting, grinding, milling and even lithography move atoms in great thundering statistical herds. It's like trying to make things out of LEGO blocks with boxing gloves on your hands. Yes, you can push the LEGO blocks into great heaps and pile them up, but you can't really snap them together the way you'd like. In the future, nanotechnology will let us take off the boxing gloves. We'll be able to snap together the fundamental building blocks of nature easily, inexpensively and in most of the ways permitted by the laws of physics. This will be essential if we are to continue the revolution in computer hardware beyond about the next decade, and will also let us fabricate an entire new generation of products that are cleaner, stronger, lighter, and more precise.

It's worth pointing out that the word "nanotechnology" has become very popular and is used to describe many types of research where the characteristic dimensions are less than about 1,000 nanometers. For example, continued improvements in lithography have resulted in line widths that are less than one micron: this work is often called "nanotechnology." Sub-micron lithography is clearly very valuable (ask anyone who uses a computer!) but it is equally clear that conventional lithography will not let us build semiconductor devices in which individual dopant atoms are located at specific lattice sites. Many of the exponentially improving trends in computer hardware capability have remained steady for the last 50 years. There is fairly widespread belief that these trends are likely to continue for at least another several years, but then conventional lithography starts to reach its limits. If we are to continue these trends we will have to develop a new manufacturing technology which will let us inexpensively build computer systems with mole quantities of logic elements that are molecular in both size and precision and are interconnected in complex and highly idiosyncratic patterns. Nanotechnology will let us do this. When it's unclear from the context whether we're using the specific definition of "nanotechnology" (given here) or the broader and more inclusive definition (often used in the literature), we'll use the terms "molecular nanotechnology" or "molecular manufacturing." Whatever we call it, it should let us

- Get essentially every atom in the right place.
- Make almost any structure consistent with the laws of physics that we can specify in molecular detail.
- Have manufacturing costs not greatly exceeding the cost of the required raw materials and energy.

There are two more concepts commonly associated with nanotechnology:

- Positional assembly.
- Massive parallelism.

Clearly, we would be happy with any method that simultaneously achieved the first three objectives. However, this seems difficult without using some form of positional assembly (to get the right molecular parts in the right places) and some form of massive parallelism (to keep the costs down). The need for positional assembly implies an interest in molecular robotics, e.g., robotic devices that are molecular both in their size and precision. These molecular scale positional devices are likely to resemble very small versions of their everyday macroscopic counterparts. Positional assembly is frequently used in normal macroscopic manufacturing today, and provides tremendous advantages. Imagine trying to build a bicycle with both hands tied behind your back! The idea of manipulating and positioning individual atoms and molecules is still new and takes some getting used to. However, as Feynman said in a classic talk in 1959: "The principles of physics, as far as I can see, do not speak against the possibility of maneuvering things atom by atom." We need to apply at the molecular scale the concept that has demonstrated its effectiveness at the macroscopic scale: making parts go where we want by *putting* them where we want!

One robotic arm assembling molecular parts is going to take a long time to assemble anything large — so we need *lots* of robotic arms: this is what we mean by massive parallelism. While earlier proposals achieved massive parallelism through self replication, today's "best guess" is that future molecular manufacturing systems will use some form of convergent assembly. In this process vast numbers of small parts are assembled by vast numbers of small robotic arms into larger parts, those larger parts are assembled by larger robotic arms into still larger parts, and so forth. If the size of the parts doubles at each iteration, we can go from one nanometer parts (a few atoms in size) to one meter parts (almost as big as a person) in only 30 steps. A very important tool. Supramolecular chemistry is the chemistry beyond the molecule, and molecules are being designed to self-assemble into larger structures. In this case, biology is a place to find inspiration: cells and their pieces are made from self-assembling biopolymers such as proteins and protein complexes. One of the things being explored is synthesis of organic molecules by adding them to the ends of complementary DNA strands such as A and B, with molecules A and B attached to the end; when these are put together, the complementary DNA strands hydrogen bonds into a double helix, AB, and the DNA molecule can be removed to isolate the product AB. Natural or man-made particles or artifacts often have qualities and capabilities quite different from their macroscopic counterparts, for example, which is chemically inert at normal scales, can serve as a potent chemical catalyst at nanoscales.

Nanoparticles, Nanomaterials and Nanostructures: An Overview

Manipulation of structure in the *nanoscale* regime is an area of growing demand due to the unique physical properties observed in the *nanoscale* size regime. Well defined *nanoscale* materials provide potential applications in biotechnology, microelectronics, data storage, functional devices, communication and technology. Due to these vast technologies of importance, bottom-up assembly is currently a widely investigated methodology in meeting the material demand for the miniaturization of electronic devices. A current limitation in nanotechnology is the lack of robust and versatile routes to template *nanoscale* building blocks into controllable structures Numeral strategies in assembling nanoparticles have been reported such as lithography templated surfaces molecular recognition processes, interfacial assembly block copolymers templates and field induced dipolar association ix into complex structures. In and co-workers demonstrated the assembly of silica particles and polystyrene beads via template assisted self assembly on surfaces. Although this method provides a wide range of one-dimensional (1D) nanostructures, there is still a need for synthetic methodologies to generate hierarchically self assembled systems in solution. Here, we investigate the feasibility of inverse ferrofluids as template for *nanoscale* building blocks because its solution morphology can be actuated by an external field. Inverse ferrofluids consist of a suspension of non magnetic particle (100-500nm) and magnetic colloids (6-10nm).

In the presence of an external magnetic field, Fe colloids phase separated into extended chains, which concurrently segregates nonmagnetic colloids into their own domains. Phase separation occurred in this system is driven by depletion demixing forces, which increases the net entropy gain in the system. De Gans et al, demonstrated the assembling of silica particles into chains that extend into micron in length in ferrofluids. It is of our interest to permanently stitch the bundles of silica shown by De Gans. It is possible to crosslink the methacrylate coated silica particles using gluteraldehyde, 2,2-azo-bis-isobtyronitrile (AIBN) and benzoyl peroxide. Subsequently, magnetic colloids can be removed via simple centrifugation, or magnetic fractionation. We are also interested in studying the interfacial effects and/or properties of 1D structure (wire) compared to its bulk and individual nanocrystal counterpart. This template serves as an important model system in studying co-operative phenomena and complex processes. The synthetic method of preparing inverse ferrofluid template is reported. Monodisperse silica particle (100-500nm) was synthesized using the Stober method and provides a well defined model system in obtaining a wide range of uniform particle sizes.

Monodispersed Co ferrofluid will be synthesized via rapid decomposition of $Co_2(CO)_8$ (dicobalt octacarbonyl) in the presence of suitable surfactants such as trioctylphosphine oxide and oleic acid. Co ferrofluid and SiO_2 nanoparticles will be synthesized as reported. The Stober process provides a well defined model system to prepare silica colloids processing a wide range of particle sizes with narrow size distribution. The surface of the silica beads will be functionalized with methacrylate groups which will served as a steric stabilizer to prevent silica particles from aggregation and also provide polymerizable moieties. In the blend of dispersion of inverse ferrofluids, 1-D chains like structure are obtained by applying an external field. The assembled silica chains will be preserved by polymerizing the methacrylate group with neighboring silica particles by the addition of radical sources such as 2,2-azo-bis- isobtyronitrile (AIBN) and benzoyl peroxide. The conventional C-shaped bar magnets are used to supply the magnetic field while observing the formation of gap spanning chains using an optical microscope. In addition to designing template for nanoscale building blocks, we are also interested in magnetic materials in the nanoscale regime. Magnetic nanostructure is worth exploring because of their application high density magnetic storage media.

We are interested in investigating ferromagnetic Co nanoparticles that exhibit particle interaction stronger than thermal energy; aggregating into chains or columnar bundles depending on the concentration. Herein the mechanism of chaining behavior in ferromagnetic Co nanoparticles stabilized with functional polymers is being studied. Preliminary observation in the retention of chaining behavior after spin and drop casting on carbon coated grid suggested that these chains are covalently bonded. Additionally, blending of these particles with methacrylate coated silica particle exhibits unusual morphologies suggesting that magnetic nanoparticles are linked into chains. In this blending experiment, transmission electron microscopy (TEM) images shows that the Co chains preferentially wrap around the surface of silica particles. The suggested mechanism of this interaction is the coupling of radical functional polymer with the reactive surface group on the silica particle. Free radicals generated at the particle periphery enable covalent connectivity with silica particles. This finding demonstrates that Hawker-type alkoxyamine end groups on polymer surfactants have the potential as crosslinker agents to bound nanoparticles into chains. To investigate this hypothesis, we will conduct several control experiments. Our first approach is to synthesize ferromagnetic Co nanoparticles stabilized by non reactive groups such as oleic acid and TOPO. The chaining behavior and interactions of these Co particles with silica nanoparticles will be compared to the functionalized polymer coated Co nanopartic

Nanomaterial technology is receiving a great deal of attention in the coatings industry today. Several approaches within this technology can be used to achieve organic-inorganic nanocomposite or nanostructured coatings. These approaches include incorporation of preformed nanoparticles in organic resin systems, in-situ generation of nanoparticles or nanophases, and other nanostructuring mechanisms. Following an overview of benefits and critical issues of nanomaterial technology, this article reviews advancements in the application of nanomaterial technology to coatings. Several commercial applications of the technology are described.

Conventional materials have grains sizes ranging from microns to several millimeters and contain several billion atoms each. Nanometer sized grains contain only about 900 atoms each. As the grain size decreases, there is a significant increase in the volume fraction of grain boundaries or interfaces. This characteristic strongly influences the chemical and physical properties of the material. For example, nanostructured ceramics are tougher and stronger than the coarser grained ceramics. Nanophase metals exhibit significant increases in yield strength and elastic modulus. It has also been shown that other properties(electrical, optical, magnetic...) are influenced by the fine grained structure of these materials. Using a variety of synthesis methods, it is possible to produce nanostructured materials in the following forms: thin films, coatings, powders and as a bulk material. There is also considerable interest in the generation of carbon nanostructures, which are related to the famous Buckyball. In addition, the use of nano-sized materials as fillers for composite materials is generating interest. specifically in the case of polymer nanocomposites.

There are several different types of nanostructured materials. These range from zero dimensional atom clusters to three dimensional equiaxed grain structure. Each class has at least one dimension in the nanometer range. Atom clusters and filaments are defined as zero modulation dimensionality and can have any aspect ratio from 1 to ¥2. Any multilayered material with layer thickness in the nanometer range is classified as one-dimensionally modulated. Layers in the nanometer thickness range consisting of ultrafine grains (nanometer range diameter) are two-dimensionally modulated. This class includes coatings, buried layers and thin films. The last class is that consisting of three dimensionally modulated microstructures or nanophase materials.

The methods employed to produce nanostructured materials are numerous, with each method having advantages and disadvantages depending on the desired properties or application. Atom clusters are typically synthesized via a vapor condensation route which is essentially evaporation of a solid metal followed by rapid condensation to form nano-sized clusters. These resulting powders are essentially agglomerates of the nano-sized atom clusters. The powders can be then be used as fillers for composite materials or consolidated into bulk material. The cluster assembly method can be used to produce ceramic or metal nanostructured powders. The main advantage of this method is the extremely low contamination levels and the ability to control the final cluster size by manipulation of the process parameters such as temperature, gas environment and evaporation rate. Several different methods have been developed using the vapor condensation concept. These include inert gas condensation, chemical vapor condensation, laser ablation, electron beam deposition, arc evaporation, etc.

Both metals and ceramics can be produced using a variety of chemical approaches such as sol-gel or thermal decomposition. These methods provide large quantities of nano-sized agglomerates at a low cost. Chemical processes also allow effective control of the stoichiometry of the final product. However, the precursor chemicals required may leave residual contamination on the particle surfaces, which leads to difficulties in compaction and sintering. Additionally, powders produced through wet chemical techniques often have difficulties with agglomeration. Other methods which implement chemical reactions include Examples: A common method of producing nanostructured powders is through mechanical deformation, i.e. milling or shock deformation. These processes produce nanostructured materials through severe mechanical deformation of a coarse grained precursor material. The nanometer sized grains nucleate within the dislocation cell structures located in the shear bands. The final grain size is a function of the amount of energy input during the milling as well as time, temperature during milling and milling atmosphere. The precursor materials can be crystalline or amorphous materials in the form of powders or tapes. Through mechanical milling, it is possible to produce nanostructured powders of systems which are otherwise immicsible. A major disadvantage of this method is the possibility of contamination from the milling media due to the large forces and energies involved in the milling process. Other synthesis methods based on *mechanical deformation*: friciton/sliding wear. Examples: Three dimensional nanostructured materials are also synthesized through thermal crystallization of an amorphous material. By controlling the nucleation and growth during annealing of an amorphous material, one can produce a bulk material with average grain sizes of less than 20 nm without the need for consolidation and sintering steps. However, this process is limited to material composition which are readily available in the form of a metallic glass, which has an amorphous microstructure.

The atomic structure of nanostructured materials is unlike that seen in glasses or crystals because of the large volume fraction (greater than 50 per cent) of grain boundaries or interfaces. Coarse-grained materials have less than 3 per cent of all atoms associated with grain boundaries or interfaces. However, in materials with nanosized grains between 5 per cent and 50 per cent of all atoms are associated with grain boundaries or interfaces. The grain boundary structure in these nanostructured materials is similar to that of the conventional coarse grained materials.

The material properties of the nanostructured materials show remarkable improvement or deviation from the properties exhibited by the coarser grained material. These unique properties are attributed to the significant increase in grain boundary area due to the small grain size. In terms of the mechanical properties, nanostructured metals have shown increases in hardness values (ultimate and yield strengths). Specifically, pure nanophase metals have shown a clear increase in hardness with decreasing grain size, following the well known Hall-Petch equation. Nanostructured ceramics have exhibited superplastic properties at low temperatures. This is significant as ceramics are conventionally brittle materials. Further deviations from trends observed in conventional materials are observed in creep of nanostructured metals.

The common theme of this WTEC study is the engineering of materials with novel (i.e., improved) properties through the controlled synthesis and assembly of the material at the nanoscale level. The range of applications is extremely broad, and these will be described in further detail in subsequent chapters in this report. The corresponding means of synthesis and assembly are similarly wide-ranging. But however multifaceted the synthesis approaches and the ultimate applications, there are common issues and unique defining features of these nanostructured materials. First, there is the recognition of critical scale lengths that define the material structure and organization, generally in the nanometer range, and that ultimately determine the fundamental macroscopic properties of the material. Research in nanostructured materials is motivated by the belief that ability to control the building blocks or nanostructure of the materials can result in enhanced properties at the macroscale: increased hardness, ductility, magnetic coupling, catalytic enhancement, selective absorption, or higher efficiency electronic or optical behavior. Synthesis and assembly strategies accommodate precursors from liquid, solid, or gas phase; employ chemical or physical deposition approaches; and similarly rely on either chemical reactivity or physical compaction to integrate nanostructure building blocks within the final material structure. The variety of techniques is shown schematically

Schematic of variety of nanostructure synthesis and assembly approaches. The "bottom-up" approach first forms the nanostructured building blocks and then assembles them into the final material. An example of this approach is the formation of powder components through aerosol techniques (Wu et al. 1993) and then the compaction of the components into the final material. These techniques have been used extensively in the formation of structural composite materials. One "top-down" approach begins with a suitable starting material and then "sculpts" the functionality from the material. This technique is similar to the approach used by the semiconductor industry in forming devices out of an electronic substrate (silicon), utilizing pattern formation (such as electron beam lithography) and pattern transfer processes (such as reactive ion etching) that have the requisite spatial resolution to achieve creation of structures at the nanoscale. This particular area of nanostructure formation has tremendous scope, warranting its own separate study, is a driving issue for the electronics industry, and will not be a principal theme of this study. Another top-down approach is "ball-milling," the formation of nanostructure building blocks through controlled, mechanical attrition of the bulk starting material (Koch 1989). Those nano building blocks are then subsequently assembled into a new bulk material. In fact, many current strategies for material synthesis integrate both synthesis and assembly into a single process, such as characterizes chemical synthesis of nanostructured materials (Murray et al. 1993; Katari et al. 1994). The degree of control required over the sizes of the nanostructure components, and the nature of their distribution and bonding within the fully formed material varies greatly, depending on the ultimate materials application. Achieving selective optical absorption in a material (e.g., UVblocking dispersions) may allow a wide range of sizes of the component nanostructure building blocks, while quantum dot lasers or single electron transistors require a far tighter distribution of size of the nanostructure components. Compaction methods may provide excellent adhesion for nanocomposite materials of improved structural performance (e.g., ductility), but such interfaces may be unsatisfactory for electronic materials. The intention of this chapter of the report is not to recapitulate in detail the various synthesis and assembly techniques that have been and are being employed in the fabrication of nanostructured materials; that detail can be found in succeeding chapters as well as in excellent summary descriptions provided in the May 8-9, 1997 WTEC workshop proceedings (Siegel, Hu, and Roco 1998). Rather, in attempting to capture the salient features of a new impetus for and interest in a field of nanostructure science and technology, it is more useful to identify the emerging commonalities than the differences among synthesis and assembly approaches.

However broad the range of synthesis approaches, the critical control points fall into two categories:

1. control of the size and composition of the nanocluster components, whether they are aerosol particles, powders, semiconductor quantum dots, or other nanocomponents

2. control of the interfaces and distributions of the nanocomponents within the fully formed materials These two aspects of nanostructure formation are inextricably linked; nevertheless, it is important to understand how to exercise separate control over the nucleation of the nanostructure building blocks and the growth (for example, minimizing coagulation or agglomeration) of those components throughout the synthesis and assembly process. This latter issue is related to the importance of the following:

- the chemical, thermal, and temporal stability of such formed nanostructures
- the ability to scale-up synthesis and assembly strategies for low-cost, large-scale production of nanostructured materials, while at the same

time maintaining control of critical feature size and quality of interfaces (economic viability is a compelling issue for any nanostructure technology) All researchers in this area are addressing these issues.

Nanostructure Synthesis and Assembly

There has been steady technological progress in all fields of nanostructure synthesis and assembly, in no small part because of the more general availability of characterization tools having higher spatial, energy, and time resolution to clearly distinguish and trace the process of nanostructure formation. As transmission electron microscopy and X-ray diffraction techniques helped in an earlier period to relate the improved properties of "age-hardened" aluminum alloys to their nanostructure (Koch 1998), today's technological advances in materials characterization are providing new insights into the role of the nanostructure in determining macroscopic properties. The tightly-coupled iteration between characterizing the nanostructure, understanding the relationship between nanostructure and macroscopic material properties, and improved sophistication and control in determining nanostructure size and placement have accelerated the rate of progress and helped to define the critical components of this "new" field of nanostructure science and technology. Tightly focused (1-2 μm), high brightness synchrotron X-ray sources provide detailed structural information on colloids, polymers, alloys, and other material structures, highlighting the inhomogeneities of the material with suitable spatial resolution (Hellemans 1998).

Improved characterization - higher spatial resolution - higher sensitivity leads t o motivat es Better control of size and placement Understanding of - structure-property link - influence of interfaces stimulates.

Interactive cycle of characterization, understanding and enhanced control in the synthesis and assembly of nanostructures. Another important enabling technology has been the now widely available scanning probe technology, including scanning tunneling microscopy and atomic force microscopy. The power of these techniques has provided impetus for developing even higher performance scanning probe tips, fabricated through microfabrication techniques. Development of different tip structures in various materials has given rise to an entire family of powerful scanning probe techniques that encompass such a wide range of characterization capabilities that one can envision "a laboratory on a tip" (Berger et al. 1996). The development of a tip technology also impacts the synthesis and assembly processes themselves: scanning probe technologies have been used as the basis of materials patterning and processing at nanometer scales (Held et al. 1997; Snow et al. 1997; and Wilder et al. 1997) and have provided information on the mechanical and thermal properties of materials at the nanoscale (Nakabeppu et al. 1995; Tighe et al. 1997; and Zhang et al. 1996). More sophisticated in situ monitoring strategies have provided greater understanding and control in the synthesis of nanostructured building blocks, particularly those formed in vacuum environments. Molecular beam epitaxy (MBE) represents a physical vapor (gas phase) deposition technique where sub-monolayer control can be imposed on the formation of two-dimensional and, more recently, three-dimensional nanostructured materials (Leonard et al. 1993). A great deal of the understanding and control derives from the ability to carry out sensitive monitoring of the growth process in situ: reflection high energy electron diffraction (RHEED) details the nature of the surface and surface bonding, and oscillations of the RHEED intensity provide information on the growth rate (Neave et al. 1983). The improvements brought about by these advances in technology have been substantial, but perhaps of greater importance for this nascent field of nanostructure science and technology has been the development of strategies and technologies that have been formed across the former disciplines. More reliable means of controlling nanostructure size and placement, with an end view of being able to scale up the production of such materials while maintaining the control over the nanostructure, have given impetus to a common search for novel synthesis and assembly strategies. In that search, it is apparent that the naturally occurring synthesis and assembly of biological materials can provide us with some critical insights.

Major efforts in nanoparticle synthesis can be grouped into two broad areas: gas phase synthesis and sol-gel processing. Nanoparticles with diameters ranging from 1 to 10 nm with consistent crystal structure, surface derivatization, and a high degree of monodispersity have been processed by both gas-phase and sol-gel techniques. Typical size variances are about 20%; however, for measurable enhancement of the quantum effect, this must be reduced to less than 5% (Murray et al. 1993). Initial development of new crystalline materials was based on nanoparticles generated by evaporation and condensation (nucleation and growth) in a subatmospheric inert-gas environment (Gleiter 1989; Siegel 1991, 1994). Various aerosol processing techniques have been reported to improve the production yield of nanoparticles (Uyeda 1991, Friedlander 1998). These include synthesis by combustion flame (Zachariah 1994, Calcote and Keil 1997, Axelbaum 1997, Pratsinis 1997); plasma (Rao et al. 1997); laser ablation (Becker et al. 1997); chemical vapor condensation (Kear et al. 1997); spray pyrolysis (Messing et al. 1994); electrospray (de la Mora et al. 1994); and plasma spray (Berndt et al. 1997). Sol-gel processing is a wet chemical synthesis approach that can be used to generate nanoparticles by gelation, precipitation, and hydrothermal treatment (Kung and Ko 1996). Size distribution of semiconductor, metal, and metal oxide nanoparticles can be manipulated by either dopant introduction (Kyprianidou-Leodidou et al. 1994) or heat treatment (Wang et al. 1997). Better size and stability control of quantum-confined semiconductor nanoparticles can be achieved through the use of inverted micelles (Gacoin 1997), polymer matrix architecture based on block copolymers (Sankaran et al. 1993) or polymer blends (Yuan et al. 1992), porous glasses (Justus et al. 1992), and ex-situ particle-capping techniques (Majetich and Canter 1993; Olshavsky and Allcock 1997).

Additional nanoparticle synthesis techniques include sonochemical processing, cavitation processing, microemulsion processing, and highenergy ball milling. In sonochemistry, an acoustic cavitation process can generate a transient localized hot zone with extremely high temperature gradient and pressure (Suslick et al. 1996). Such sudden changes in temperature and pressure assist the destruction of the sonochemical precursor (e.g., organometallic solution) and the formation of nanoparticles. The technique can be used to produce a large volume of material for industrial applications. In hydrodynamic cavitation, nanoparticles are generated through creation and release of gas bubbles inside the sol-gel solution (Sunstrom et al. 1996). By rapidly pressurizing in a supercritical drying chamber and exposing to cavitational disturbance and high temperature heating, the sol-gel solution is mixed. The erupted hydrodynamic bubbles are responsible for nucleation, growth, and quenching of the nanoparticles. Particle size can be controlled by adjusting the pressure and the solution retention time in the cavitation chamber. Microemulsions have been used for synthesis of metallic (Kishida et al. 1995), semiconductor (Kortan et al. 1990; Pileni et al. 1992), silica (Arriagada and Osseo-Assave 1995), barium sulfate (Hopwood and Mann 1997), magnetic, and superconductor (Pillai et al. 1995) nanoparticles. By controlling the very low interfacial tension (~10-3 mN/m) through the addition of a cosurfactant (e.g., an alcohol of intermediate chain length), these microemulsions are produced spontaneously without the need for significant mechanical agitation. The technique is useful for large-scale production of nanoparticles using relatively simple and inexpensive hardware (Higgins 1997). Finally, high energy ball milling, the only top-down approach for nanoparticle synthesis, has been used for the generation of magnetic (Leslie-Pelecky and Reike 1996), catalytic (Ying and Sun 1997), and structural (Koch 1989) nanoparticles. The technique, which is already a commercial technology, has been considered dirty because of contamination problems from ball-milling processes. However, the availability of tungsten carbide components and the use of inert atmosphere and/or high vacuum processes have reduced impurities to acceptable levels for many industrial applications. Common drawbacks include the low surface area, the highly polydisperse size distributions, and the partially amorphous state of the asprepared powders.

Field-Controlled Molecular Switching Devices

A second category of molecular electronics device that appears promising for the development of molecular nanoelectronic digital computers is the electric-field controlled molecular switching devices, including molecular quantum-effect devices. These are most closely descended from the solid-state microelectronics and nanoelectronic devices described earlier, and promise to be the fastest and most densely integrated among all of the current alternatives. A group at Purdue University has used self-assembly to fabricate and demonstrate functioning arrays of molecular electronic quantum confinement structures connected by molecular wires. As another example, J.M. Tour proposes embedding a quantum well in a molecular wire by inserting pairs of barrier groups that break the sequence of conjugated porbitals, forming a two-terminal molecular RTD; other structures for three-terminal molecules have been suggested. (In February

1999, Tour reported the demonstration of RTD effects in the manner proposed above; at press time, the paper was still in review [J.M. Tour, personal communication, 1999].) Nondissipative (adiabatic) Thouless electron pumping has also been described at 0.33 K.

Merkle and Drexler consider a hypothetical "helical logic" device in which a single electron can switch another single electron, with 1 or 0 represented as the presence or absence of individual electrons. The electrons are constrained to move along helical paths, driven by a rotating immersive electric field (normal to the helical axis) that confines each charge carrier to a fraction of a turn of a single helical loop, moving it like water in an Archimedean screw. A logic operation involves two helices, one of which splits into two "descendant" helices. At the point of divergence, differences in the electrostatic potential resulting from the presence or absence of a carrier in the adjacent helix control the direction taken by a carrier in the splitting helix; the sequence is reversible, allowing two initially distinct helical paths to merge into a single outgoing helical path without forcing a dissipative transition. Energy dissipation at ~10 GHz is ~10^{-6} zJ/gate-cycle at an operating temperature of 1 K.

Other classes of molecular electronic devices have been studied, including picosecond photoactive or photochromic molecular switching devices using light, and electrochemical molecular devices using electrochemical reactions, to change the shape, orientation, or electron configuration of a molecule in order to switch a current. However, photoactive devices in a dense network would require an NSOM-like emitter to switch individually, since optical photon pathways are not easily confined at length scales very much below optical wavelengths of ~500-1000 nm; electrochemical molecular devices might require immersion in solvent to operate.

Fullerenes exhibit a wide range of quantized conductivities that may be useful in nanoelectronic computers. For example, chemisorption alters properties: 3 alkali dopant atoms per C_{60} buckyball gives high conductivity, while none or 6 dopant atoms gives an insulator. Buckytube conductivity also varies with mechanical stress and torsion, tube diameter, and other geometrical parameters. Depending upon the direction of the nanotube chiral vector, carbon nanotubes may be either metallic or semiconducting. Suitably-sized, doped, flexed, stressed, or chiralized fullerenes could be nested to make insulated wires or joined to make nanoelectronic computer components including resistors with quantized "staircase" resistance of $nh/2e^2$, n = 1, 2,... (e is electronic charge, h is Planck's constant), two-terminal devices such as diodes, three- or four-terminal devices such as heterojunction, TUBEFET, or thin-film transistors, and gates that introduce electric fields. A single-molecule transistor using semiconducting buckytubes has been demonstrated experimentally at room temperature.

Carbon nanotubes have been cut to length, size-sorted by field flow fractionation, their open ends derivatized with thiol groups and then tethered to 10-nm gold particle "junction boxes" which AFM images show can connect together at least two separate tubes. If the tethering process can be spatially controlled using DNA complementarity or by other means, then fullerene nanoelectronic circuit elements could be assembled into three-dimensional CPU-like structures. P. Collins expects an all-carbon 8-bit adder with nanometer dimensions by the year 2002, and J.C. Ellenbogen of MITRE Corp. has designed a molecular half-adder that measures ~10 nm x 10 nm in size. This research area was extremely active in 1998.

Fullerenes may also allow compact memory devices. For example, it has long been speculated that a ~1 nm buckyball encapsulated in a ~1 nm diameter nanotube segment may glide freely back and forth, trapped weakly in each end cap by van der Waals forces. An external voltage could nudge a charged C_{60} molecule to one end of the other, creating a two-state "buckyshuttle RAM" storage device. Such a device could store one bit of information in a ~10 nm^3 volume, a storage density of ~10^8 bits/micron^3; the C_{60} thermal velocity of ~100 m/sec at 310 K inside a ~10 nm memory element gives a typical read/write time of ~0.1 nanosec. In early 1999, Kwon and colleagues reported that finely dispersed 4-6 nm diamond powder thermally annealed by heating in a graphite crucible under argon at 1800°C for 1 hour produced oblong multiwall carbon nanostructures which in some cases could move around inside each other, like C_{60} trapped within C_{480}. Kwon noted that there are potential energy minima at either end, so a charged buckyball (e.g., an enclosed ion) could be shuttled with an applied electric field, which could be used for memory storage. Computational simulation of writing a bit showed that the interior buckyball neatly shuttled to the correct end, then gently bounced to rest in ~0.01 nanosec (i.e., ~100 GHz operation) dissipating only ~10 kT in the C_{480}. Kwon proposed a high density memory board using aligned buckyshuttles in a hexagonal lattice with addressing wires above and below, not unlike a ferrite core memory; the memory would likely be nonvolatile at room temperature, with mass production based on self-assembly.

Is Molecular Manufacturing Possible?

Most contemporary industrial fabrication processes are based on "top-down" technologies, wherein small objects are sawn or machined from larger objects, or small features are imposed on larger objects, in either case by removing unwanted matter. The results of such processes may be small, such as micron-featured integrated circuits, or very large, such as jet aircraft, but in most cases the material is being processed in chunks far larger than molecular scale.

Molecular manufacturing, on the other hand, represents a "bottom-up" technology. Desired products will be built directly by "assembler" machines, molecule by molecule, making larger and larger objects with atomic precision. The results of such processes may also be very small or very large, much as biology builds both micron-sized bacteria and 100-meter tall sequoia trees. However, since assemblers add matter only where it is intended, little need be removed and hence there may be minimal waste during the process. By guiding with precision the assembly of molecules and supramolecular structures, such a manufacturing system could construct an extraordinarily wide range of products of unprecedented quality and performance.

In 1959, the Nobel physicist Richard P. Feynman observed that:

"The principles of physics, as far as I can see, do not speak against the possibility of maneuvering things atom by atom. It is not an attempt to violate any laws; it is something, in principle, that can be done; but, in practice, it has not been done because we are too big . . . Ultimately, we can do chemical synthesis."

"A chemist comes to us and says, 'Look, I want a molecule that has the atoms arranged thus and so; make me that molecule. The chemist does a mysterious thing when he wants to make a molecule. He sees that it has got that ring, so he mixes this and that, and he shakes it, and he fiddles around. And, at the end of a difficult process, he usually does succeed in synthesizing what he wants.

"But it is interesting that it would be, in principle, possible (I think) for a physicist to synthesize any chemical substance that the chemist writes down. Give the orders and the physicist synthesizes it. How? Put the atoms down where the chemist says, and so you make the substance. The problems of chemistry and biology can be greatly helped if our ability to see what we are doing, and to do things on an atomic level, is ultimately developed a development which I think cannot be avoided."

Nearly 40 years after Feynman's famous "Plenty of Room at the Bottom" speech, and a decade after Drexler's original proposal for a bottom-up approach to machinebuilding using molecular assemblers, Nobel chemist Richard Smalley also largely agreed that this objective should prove feasible. Noted Smalley: "On a length scale of more than one nanometer, the mechanical robot assembler metaphor envisioned by Drexler almost certainly will work..."

Many skeptical questions arise when one first encounters the ideas of molecular nanotechnology and molecular assemblers. It is useful to keep in mind the proven feasibility of such systems in the biological tradition over billions of years of natural evolution. In one sense, molecular nanotechnology will be a refinement and expansion upon how nature works at the molecular scale. Nature's examples, such as human beings, certainly answer the most basic skeptical questions, such as: Can macroscopic objects be built from molecular scale processes? (Yes, thanks to cellular replication.) Are molecular objects stable? (Of course; the human population alone contains $>10^{29}$ reasonably stable and quite functional ribosomal "nanomachines"). Observes Nobel chemist Jean-Marie Lehn: "The chemist finds illustration, inspiration and stimulation in natural processes, as well as confidence and reassurance since they are proof that such highly complex systems can indeed be achieved on the basis of molecular components."

What about quantum effects? The uncertainty principle makes electron positions somewhat fuzzy, but the atom as a whole has a comparatively definite position set by the relatively great mass of the atomic nucleus. The quantum probability function of electrons in atoms tends to drop off exponentially with distance outside the atom, giving atoms a moderately sharp "edge". Mathematically, the positional uncertainty of a single carbon atom of mass $m_C = 2 \times 10^{26}$ kg bound in a single C-C bond of stiffness $k_C = 440$ N/m may be crudely estimated from the classical vibrational frequency $n_C = (k_C/m_C)^{1/2} = 1.5 \times 10^{14}$ Hz. This sets the zero-point vibrational bond energy $E_C = h\, n_C / 2 = 4.9 \times 10^{20}$ J $= k_C\, DX_C^2 / 2$ where $h = 6.63 \times 10^{34}$ J-sec (Planck's constant) and $DX_C \sim 0.015$ nm is the maximum classical amplitude of the bound carbon atom (roughly the same as the 3 dB point for the gaussian wavefunction, notes J. Soreff). Thus DX_C is just ~5% of the typical atomic electron cloud diameter of ~0.3 nm, imposing only a modest additional constraint on the fabrication and stability of nanomechanical structures. (Even in most liquids at their boiling points, each molecule is free to move only ~0.07 nm from its average position.)

How about the effects of Brownian bombardment on nano-machines? Describing a nanomechanical component as a harmonic oscillator embedded in a gas, Drexler notes: "At equilibrium, an impinging gas molecule is as likely to absorb energy as to deliver it, and so molecular bombardment has no net effect on the amplitude of vibration. How a system is coupled to a thermal bath can affect its detailed dynamics, but not the statistical distribution of dynamical quantities." Nanomachines must also obey the laws of thermodynamics. Nanorobots cannot be used as a "Maxwell's demon" — energy must be expended to do useful work; there can be no free lunch.

Will high-energy radiation damage nanomachines? Radiation can break chemical bonds and disrupt molecular machines. The annual failure rate for a properly designed ~1 micron^3 molecular machine (containing many billions of atoms) can be made as low as several percent. This estimate is derived from a consideration of all known thermal, photochemical, radiation, and other damage mechanisms, but the dominant error mechanism that appears difficult to substantially reduce is damage caused by background radiation. Failure rates are not zero, but are nonetheless remarkable by today's standards, especially given error detection and correction at the module level.

During nanodevice fabrication, the problems of crossbonding and reactive intermediates pose additional constraints that must be dealt with using appropriate assembler designs. Specifically, assemblers with rigid positioning components operating in vacuo can restrict reactive intermediates and molecular tool tips to desired locations with a precision of at least ~0.1 nm, or slightly less than an atomic diameter. Within the confines of an evacuated or "eutactic" assembler workspace, the location of every structural atom may be known to within the uncertainty created by thermal noise. Internal structural components may possess relatively high stiffness, so positional uncertainty caused by thermal noise may be small. In the classical analysis, positional variance $DX^2 = kT / k_s$, where $k = 1.381 \times 10^{21}$ J/kelvin (Boltzmann's constant), T is temperature in kelvins, and k_s is component stiffness, typically 10 N/m for nanometer-scale diamondoid components, hence the standard deviation of component position is ~0.02 nm at room temperature, or ~0.01 nm at cryogenic (e.g., LN_2) temperatures. By employing sufficiently stiff mechanisms the position of every structural atom in the system can be known to within a fraction of an atomic diameter with high reliability and without the need for explicit positional sensing. Steric tool hindrance — in molecular manufacturing, the conical volume occupied by a tool tip or manipulator platform — is a technical design problem that may be overcome by using sufficiently stiff tools or multi-tip tools, working in vacuo. Additionally, while some synthetic methods might involve the manipulation of individual atoms by appropriate tools, such as the hydrogen abstraction tool which removes a single selected hydrogen atom from a diamondoid surface, other reactions will involve small clusters of atoms or larger molecular components.

What about friction and wear among nanomechanical components? Properly designed molecular machines lack wear mechanisms, although other damage mechanisms remain. Dry bearings with atomically precise surfaces can have negligible static friction and wear, and very low dynamic friction or drag. Nanomechanical components are better viewed as moving smoothly in a force field than as sliding subject to friction. For example, total drag power dissipated by an isolated molecular sleeve bearing of radius 2 nm spinning at 1 MHz is dominated by band-stiffness scattering, amounting to 0.000004 picowatts (pW), very small compared to the 1-1000 pW power draw that will be typical of micron-size medical nanorobots. The strong precise surfaces of nanomachines experience no change during a typical operational cycle, hence zero wear. Within the single-point failure model, the first step in a wear process (e.g., a dislocated atom) is regarded as fatal, hence cumulative wear plays no role in determining device lifetimes. In eutactic nanomachines, contaminants of all kinds are rigorously excluded. The equivalent of wear particles cannot appear until the device has failed. On this scale, repulsive fields provide "lubrication" and an oil molecule would be a contaminant object, not a lubricant.

The general conclusion is that molecular nanotechnology violates no physical laws. In 1998, it was generally accepted that molecular nanotechnology will be developed, although there remains some disagreement about how long it will take. To progress from today's limited capabilities to the complex nanorobots described later in this trilogy will clearly require a great deal of research and development effort. Still, the proper question is no longer "if" but "when." This Chapter presents a few of the many possible technical paths leading to a working molecular assembler that were being discussed in 1998. The reader is strongly urged to consult the most current literature for the latest results. The conventional top-down approaches to nanotechnology. Several possible direct bottom-up pathways to molecular manufacturing including biotechnology, supramolecular chemistry, and scanning probes. Gimzewski and Joachim note that "bottom-up approaches to nanofabrication may one day compete with conventional top-down approaches in providing nanotechnologies for the next millenium. Top-down refers to increasing miniaturization through extension of existing microfabrication schemes, [whereas] the bottom-up scenario is one of ever-increasing complexity at the molecular level while maintaining control on an atom-by-atom basis". Surveys molecular mechanical components and design work on molecular assemblers and molecular manufacturing systems from the perspective of 1998.

Other Synthesis Issues

One of the most challenging problems in synthesis is the controlled generation of monodispersed nanoparticles with size variance so small that size selection by centrifugal precipitation or mobility classification is not necessary. Among all the synthesis techniques discussed above, gas-phase synthesis is one of the best techniques with respect to size monodispersity, typically achieved by using a combination of rigorous control of nucleationcondensation growth and avoidance of coagulation by diffusion and turbulence as well as by the effective collection of nanoparticles and their handling afterwards. The stability of the collected nanoparticle powders against agglomeration, sintering, and compositional changes can be ensured by collecting the nanoparticles in liquid suspension. For semiconducting particles, stabilization of the liquid suspension has been demonstrated by the addition of polar solvent (Murray et al. 1993); surfactant molecules have been used to stabilize the liquid suspension of metallic nanoparticles.

Alternatively, inert silica encapsulation of nanoparticles by gas-phase reaction and by oxidation in colloidal solution has been shown to be effective for metallic nanoparticles (Andres et al. 1998). New approaches need to be developed for the generation of monodisperse nanoparticles that do not require the use of a size classification procedure. An example of this is a process developed in Japan where very monodispersed gold colloidal nanoparticles with diameters of about 1 nm have been prepared by reduction of metallic salt with UV irradiation in the presence of dendrimers (Esumi et al. 1998). Poly(amidoamine) dendrimers with surface amino groups of higher generations have spherical 3-D structures, which may have an effective protective action for the formation of gold nanoparticles. Although the specific role of dendrimers for the formation of monodispersed nanoparticles has yet to be defined, good monodispersity is thought to come from the complex reaction accompanying the decomposition of dendrimers, which eventually leads to the conversion of solution ions to gold nanoparticles.

Scaleup production is of great interest for nanoparticle synthesis. High energy ball milling, already a commercial high-volume process, as mentioned above, has been instrumental in generating nanoparticles for the preparation of magnetic, structural, and catalytic materials. However, the process produces polydispersed amorphous powder, which requires subsequent partial recrystallization before the powder is consolidated into nanostructured materials. Although gas-phase synthesis is generally a low production rate process (typically in the 100 milligrams per hour range) in research laboratories, higher rates of production (about 20 grams per hour) are being demonstrated at Ångström Laboratory at Uppsala University in Sweden. Even higher production rates (about 1 kg per hour) are now being achieved commercially. For sol-gel processing, the development of continuous processing techniques based on present knowledge of batch processing has yet to be addressed for economical scaleup production of nanoparticles. Other related sol-gel issues concern the cost of precursors and the recycling of solvent. Overall, sol-gel processing is attractive for commercial scale-up production.

A recent paradigm shift envisioned for optoelectronics and computational devices involves the assembly of molecular or quantum wires (Chidsey and Murray 1986). Large polymeric molecules have been used as nano building blocks for nanoporous molecular sieves, biocompatible materials, optical switching, data processing, and other nonlinear optical components. Chain aggregates of nanoparticles can be considered as polymer-like units with their primary particles composed of a few hundred to a few thousand molecules. Thus, these chain aggregates can be considered "heavy" quantum wires. In fact, nanoparticle chain aggregates have been studied extensively as magnetic materials (Zhang and Manthiram 1997), as reinforced elastomers (Pu et al. 1997), and as additives in concrete (Sabir 1997). These aggregates have been shown to have chemical and mechanical properties different from those of individual primary particles (Friedlander et al. 1998). Depending on the particle size and its compositional material, the bonding force responsible for holding the aggregates together varies from weak van der Waals force for micrometer particles to strong chemical bonds for nanometer particles to very strong magnetic dipolar bonds for nanosized magnetic particles. The mechanical, optical, and electronic transport properties can be varied by controlling the diameter and the monodispersity of the primary particles, the crystalline structure and morphology, aggregate length, interfacial properties, and material purity. These chain aggregates can be formed by allowing agglomeration of nanoparticles generated by any of the synthesis techniques discussed above, with the exception of high energy ball milling, which generates particles with low surface area and high anisotropic morphologies, both of which are detrimental for the formation of chain aggregates. Depending on the magnetic and electric charging properties of the nanoparticles, an external applied magnetic or electric field can be used to control the fractal dimension of aggregates. For optical applications of chain aggregates, lower fractal dimensions (i.e., relatively straight chain aggregates with few branches) are desirable.

Recent advances in the fabrication of nanometer fibers or tubes offer another form of building blocks for nanostructured materials. An effective way to generate nanometer fibers (or tubes) is based on the use of membrane-template techniques (Martin 1994). Membranes, with nanochannels generated by fission-fragment tracks or by electrochemical etching of aluminum metal, are used as templates for either chemical or electrochemical deposition of conductive polymers (Pathasarathy and Martin 1994), metal (van de Zande et al. 1997), semiconductor (Klein et al. 1993), and other materials for the generation of nanofibers or tubes. Since the nanochannels on membranes are very uniform in size, the diameter and the aspect ratio of the nanofibers (or tubes) synthesized by the membranetemplate technique can be precisely controlled. This greatly facilitates the interpretation of optical data and the processing of these fibers (or tubes) into 2-D nanostructured materials (de Heer et al. 1995). Single-crystal semiconductor nanofibers can also be grown catalytically by metalorganic vapor phase epitaxy and laser ablation vapor-liquid-solid techniques (Hiruma et al. 1995; Morales and Lieber 1998). The synthesis of these onedimensional structures with diameters in the range of 3 to 15 nm holds considerable technological promise for optoelectronic device applications, such as the p-n junctions for light emission at Hitachi Central Research Laboratory in Japan. The advent of carbon-based nanotubes has created yet another way to fabricate nanometer fibers and tubes. These nanotubes have been used as templates for the fabrication of carbide and oxide nanotubes (Dai et al. 1995; Kasuga et al. 1998). Synthesis of nanotubes based on BN, BC3 and BC2N have also been reported (Chopra et al. 1995; Miyamoto et al. 1994). These nanotubes potentially possess large third-order optical non-linearity and other unusual properties (Xie and Jiang 1998). Metallic nanofibers synthesized by carbon-nanotube-template techniques are useful in the design of infrared absorption materials. The carbon nanotubes can now be catalytically produced in large quantities and have been used for reinforcement of nanostructural composite materials and concrete (Peigney et al. 1997). The elegant complexity of biological materials represents the achievement of structural order over many length scales, with the full structure developed from the "nested levels of structural hierarchy" (Aksay et al. 1992), in which self-assembled organic materials can form templates or scaffolding for inorganic components. These notions of a multilevel material structure with strong interactions among levels and an interplay of perfection and imperfection forming the final material was discussed earlier by Cyril Stanley Smith (1981). Along with characteristic length scales, there are characteristic relaxation times of the material, bringing in the consideration of the temporal stability of the structured materials (Zener 1948). A more detailed discussion of the role of biological materials as both paradigm and tool for the fabrication of nanostructured materials is given. It is interesting how many of the synthesis and assembly approaches seek out and adapt two key features of biogenic fabrication: that of "self-assembly," and the use of natural nanoscaled templates.

Self-Assembly

"Self-assembly" is a term that by now figures prominently in the literature of nanostructured materials and nanofabrication. The term therefore carries a variety of implicit or explicit meanings, and we cite the definition given by Kuhn and Ulman: "Self-assembly is a process in which supermolecular hierarchical organization is established in a complex system of interlocking components." The mechanisms that produce the hierarchical organization are determined by competing molecular interactions (e.g., interactions between hydrophobic versus hydrophilic components, van der Waals, Coulombic, or hydrogen bonding), resulting in particular microphase separation or surface segregation of the component materials. Thus, the use of a hierarchy of bond strengths and/or chemical specificity can produce a hierarchy of lengths in the final nanostructured material (Muthukumar et al. 1998; Stupp and Braun 1998). As one example of such self-assembled or self-organized materials, McGehee et al. (1994) have mixed silica precursors with surfactants that have self-ordered to form various surfactant-water liquid crystals, producing various structures built from walls of amorphous silica, organized about a repetitive arrangement of pores up to a hundred angstroms in diameter. In the "natural" formation of inorganic nanostructures, the addition of organic molecules can strongly influence the resulting structure of inorganic components. Such strategies have been adopted in synthetic formation of nanostructures, such as in the formation of networks of gold clusters (Andres et al. 1998). Gold clusters 3.7 nm in diameter, formed in the vapor phase, are encapsulated in organic surfactants, such as dodecanethiol, forming a colloidal suspension. The surfactants prevent the agglomeration of the gold clusters. Addition of small amount of dithiol precipitates out a 3-D cluster network, which can in turn be deposited onto another solid substrate. A transmission electron microscope (TEM) image of a cluster array spin-cast onto MoS2. The methodology of self-assembly has even been extended to physical vapor deposition processes where it would seem more difficult to control the nucleation and growth of three-dimensional nanostructures. Utilizing the strain inherent in the epitaxial growth of lattice-mismatched materials, and the expected strain-induced transition from two-dimensional (layered) to three-dimensional (islanded) growth, together with careful monitoring of the growth process through RHEED analysis, researchers have been able to form arrays of semiconductor quantum dots, ~ 200-300 Å in diameter, ~ 1011 cm-2 in density, and with a size variation of about ±7% (Leonard et al. 1993). An example of such "self-assembled" semiconductor quantum dots.

The achievement of arrays of several billions of quantum dots of these dimensions with such a size variation is beyond the capability of standard high resolution lithographic and pattern transfer processes. Moreover, the controlled formation of critical surfaces and interfaces without the intercession of ion-assisted processing that can introduce potential defects into the materials has produced a rich source of optically and electronically efficient quantum structures. A number of researchers have already incorporated such self-assembled dots into laser structures (Bimberg et al. 1998). Chemical specificity may provide the most robust means of ensuring control of size and placement of nanostructured building blocks, and recent work in the synthesis of compound semiconductor quantum dots from chemical precursors have provided even tighter distributions of size variation (±5%) than those shown in the strain-induced self-assembled dots (Katari et al. 1994; Murray et al. 1993).

Natural Templates or "Scaffolds"

Biogenic systems employ natural templates or scaffolding in the construction of nanoscaled materials, where the templates and scaffolding can help set the proper, critical dimensions that are characteristic of the final material. The templates can be formed "artificially," through the use of lithographically defined patterning and processing of the substrate material in order to achieve selective nucleation and growth of the nanostructured materials. However, there are also a host of "natural" templating materials that researchers can avail themselves of. The zeolites have for a long time been utilized as the basis of high surface area materials that enable catalysis. Recent work has taken these natural cage structures and formed large cage structures with varying pore sizes (Estermann et al. 1991; Ying and Sun 1997; Bu et al. 1998). When zeolite cages are "loaded" with various materials, the controlled proximity enforced by the cage structure can result in various magnetically active or nonlinear optical materials. More recently, the C-based fullerenes and C-nanotubes have provided a tremendously rich basis for nanostructure research; the impact of such materials is further described. The by now "easy" formation of such structures and ready availability of such materials provide a rich source of nanostructures with nearly immediate applications for electronic devices, storage, and enhancement of structural materials. Further, these materials can serve as the nanostructured precursors of nanostructure synthesis approaches (such as the starting material for cluster or aerosol deposition), or can provide an avenue for the synthesis of templates of varying sizes. Such templates can be electronically "doped" and filled with other materials. Recent work reported the synthesis of "nanorods" of GaN through confined reactions in carbon nanotubes (Han et al. 1998).

Opportunities and Challenges

The broad explorations of this WTEC study have shown a great diversity of impressive work on nanostructured materials. One could argue that many aspects of the work on nanostructured materials have been long-established efforts, with well-developed techniques that have been brought forth into the manufacturing arena. How do we then explain the current excitement and interest in nanostructured materials as a promising new endeavor? Part of the answer lies in the recognition of common scientific issues and common enabling technologies that link this group of researchers together. Ready availability of ever more sophisticated characterization methods that allow us to visualize and probe materials at the nanoscale has accelerated the pace of activities in this field; at the same time, recognition of the common critical issues of control over nanostructure size and placement motivates sharing of solutions over the boundaries of conventional disciplines. Thus, some of the most exciting findings of this study manifest the crossfertilization of techniques and ideas; for example, aerosol particles are being integrated with more traditionally fabricated electronic nanostructures, with placement achieved through the manipulation of STM tips, in order to explore ideas of electronic device enhancement at the nanostructure level (Junno et al. 1998). The ordered assembly of diblock copolymers define three-dimensional nanostructures, and those structures are transferred into electronic (semiconductor) substrates through high resolution pattern transfer processes (Möller 1998). Recognizing the need for effective utilization of ideas across disciplines brings the responsibility of establishing the educational infrastructure that will adequately train young scientists to more fully develop the concepts and applications of nanostructured materials in the future. Such an infrastructure will require researchers and educators who are familiar with the properties of a broad range of materials, including polymers, biomolecular materials, metals, ceramics, and semiconductors. It will benefit from an informed perspective on critical applications, and must provide access to a wide range of synthesis and assembly techniques and characterization methods that currently reside separately in the disciplines of physics, chemistry, electrical engineering, biology, etc. Finally, a critical enabler for the future of this field and for its educational infrastructure is further development of computational tools that encompass the full range of atomistic calculations to macroscopic materials properties. This will require a "systems approach" (Olson 1997) that will span a variety of different computational models, addressing the different critical length scales, starting with solutions of the Schrodinger and Poisson equations, solving interatomic forces, and scaling all the way up to simulations of macroscopic properties and behavior (Goddard 1998). Increased appreciation of and access to the diverse means of nanoparticle synthesis and assembly that have been developed within many different disciplines, and a common development of enabling tools and technologies, will enhance the pace of accomplishments in this new area of nanoscale synthesis and assembly.

Dispersions and Coatings

The object of the controlled assembly of nanoparticulates is to make materials with new properties and assemble them with practical applications. A unique value of nanoparticulates is their extremely high particle surface area; having many more sites for achieving property enhancements makes them ideal for a wide variety of applications as dispersions and coatings. Dispersive and coating applications of nanoparticles include optical, thermal, and diffusion barriers. Significant work on nanoscale dispersions and coatings is underway worldwide in the areas of ceramics, cosmetics, biosensors, colorants, and abrasion-resistant polymers. Other applications include imaging ink jet materials, electrophotography, pharmaceuticals, flavor enhancers, pesticides, lubricants, and other proprietary applications specific to industry. Still another application is in a new, post-silicon generation of electronic devices that includes nanotubes and fullerenes as constituent units of carbon nanoelectronic devices; here, dispersion takes on a more quantum consideration in which the number of atoms in a cluster is compared to the number of surface atoms to determine its dispersion function. Also in the semiconductor industry, a monolayer or thin film coating of atoms or molecules is deposited on foils, metal sheets, or glass to enhance storage capacity and accelerate responses from the electronic component. All these applications deal with dispersions or coatings of particles that enhance specific features. Also, ability to manufacture smaller functional systems enhances performance, cost, and efficiency. These considerations power the drive towards miniaturization and precision finishing.

Property Amplification

What's particularly exciting in the study of nanoparticles, nanostructured materials, and nanodevices is their ability to also add value to materials and products through enhancement of specific properties, such as the following:

- Mechanical strength. Nanostructured powders have been produced by plasma processing where the reactor vaporizes coarse metal particles (Froes 1998, 105-6); by combustion synthesis where redox reactions take place at elevated temperatures, followed by quenching; and by mechanical alloying with gas atomization. When nanocrystalline powders achieved by these means are compacted and applied as a coating, they lend significant strength and ductility to a variety of conventional materials such as ceramic, composites, and metal alloys.
- Superconductivity. Brus (1996) describes the nature of the superconductivity effect in detail. It involves a deposition of nanocrystals on substrates, leading to improved optical and electrical properties.
- Covering power. Because nanostructuring increases the number of active sites—there are many more atoms per grain boundary—the enhanced surface area leads to a reduced material requirement, which in turn can lower cost (Solomaon and Hawthorne 1983).
- Ability to incorporate high cost materials. Expensive materials such as colorants and drugs may be effectively dispersed in small and controlled quantities through nanostructuring (Schnur 1994).
- Environmental value. Improvements in environmental impact are achieved by utilizing nanostructure particulates in coatings and thus eliminating the requirement for toxic solvents. By eliminating hazardous wastes, nanocoatings can both reduce a company's disposal costs and improve its environmental position. Thus, nanostructured dispersions and coatings can significantly reduce material costs and improve performance and functionality in a large variety of applications. In all of the sites the WTEC panel visited in Europe, Russia, and Japan, research groups were interested in achieving one or more of the properties listed above.

Enablers

For nanoparticulate dispersions and coatings, certain enablers must be present to achieve success: (1) effective particle preparation, (2) stabilization of the dispersed phase, (3) scaleup and control of the process, and (4) the existence of excellent analytical capabilities. The main methods or issues for each are described below.

1. Particle preparation. Wet chemical methods such as liquid phase

precipitation or sol-gel methods are of high interest in particle preparation (Friedlander 1993). In the area of hybrid methods, both spray pyrolysis and flame hydrolysis are utilized. Numerous physical methods such as mechanical size reduction are also often employed.

2. Stabilization of the dispersed phase. For stabilization of the dispersed phase, it is necessary to understand how particles can be kept as distinct entities. Either a charged stabilized system or a sterically stabilized approach is required. A successful preparation for use as a dispersion should be free from agglomeration in the liquid state so as to maintain particle integrity. The dry-coated format of nanoparticles should also minimize any presence of particle aggregation
3. Scaleup and control of the process. The scaleup and control of nanoparticle processing is well described by Kear (1998). Here, the issue in achieving a high rate of production of powder is the effective pyrolysis of the gas stream containing the precursor. Issues such as scaling and reproducibility from one run of nanoparticle material or dispersions to another are all inherent in the concept of precision engineering or invariant process control.
4. Analytical capabilities. Analytical capabilities are absolutely essential for characterizing dispersions and coatings (Angstrom 1995). Particle size determinations, assay analysis, and interfacial properties are all important. Transmission electron microscopes, atomic force microscopy, nuclear magnetic resonance, and scanning tunneling microscopy are just some of the tools utilized in characterizing nanoparticle dispersions and coatings, particularly at the very small end of the nanoscale.

Applications

Below are some general examples of the vast array of applications of nanoparticulate dispersions and coatings. There are a number of applications in all of the categories below that are already in the public domain; however, most are still highly proprietary, especially drug delivery applications. Cosmetics. An area of nanoparticle technology that has incredible commercial potential is the cosmetic industry (Crandall 1996, 251-267). Here there is a great demonstrated demand, and the technology can be made simple, since properties of color and light fastness are achieved by component mixing in the cosmetic preparation. A survey in 1990 indicated a worldwide gross volume of $14-18 billion for toiletries (Crandall 1996, 61), i.e., traditional hygiene products such as powders, sprays, perfumes, and deodorants. The large markets for sunscreens and skin rejuvenation preparations promise additional revenues. The diet industry is said to gross $33 billion annually (Crandall 1996, 61). One way that nanoparticle technology is addressing this market is through introducing nanoparticulate taste enhancers into low-calorie substrates. Medicine/Pharmacology. In the area of medical applications, finely dispersed pharmaceuticals offer rapid drug delivery and reduced dosages for patients (POST 1995). Dispersions of strong and resilient biocompatible materials suggest opportunities for artificial joints. These generally are ceramic materials containing nanoparticulates. Overall, much of the demand for nanoparticulate dispersions and coatings comes from the cosmetic and pharmaceutical industries; in particular, liquid dispersion preparations will be widely used to apply topical coatings to the human epidermis because they can be absorbed faster and more completely than conventional coatings. Microelectromechanical systems (MEMS). Although MEMS technologies will support the semiconductor industry in particular, there are many other applications being explored, such as in medicine, ceramics, thin films, metal alloys, and other proprietary applications. In the United States a particular focus is applying sputtering coatings to achieve MEMS technology in concert with these applications. Printing. In the areas of image capture/image output addressed by ink jet technology, nanoscience can help control the properties of the inks themselves. The production and use of nanoengineered ink products benefits from such complimentary technology as laser-assist delivery of the ink jet droplet to maintain an accurate deposit of the ink on its target (POST 1996). Another application in this field is using nanoscale properties to tailor the inks to achieve ideal absorption and drying times for desired color properties and permanency. Semiconductors. One form of "bottom up" technology that is receiving

1996). Here single atoms or molecules are deposited by physical vapor deposition, which could be achieved through sputtering, molecular beam epitaxy, or chemical vapor deposition. Sputtering is used on a large scale to coat metal sheets, glass, polymer substrates and other receptive materials in order to produce enhanced electronic properties for information storage and processing speed. Sensors. Chemical or physical sensors often use nanoparticles because they provide high surface area for detecting the state of chemical reactions, because the quality of detection signals is improved, and because earlier and more accurate determination of leakage reduces waste. Some commercial sensors and actuators composed of thin films are already used for environmental vapor monitoring in reactors, for example. Other likely applications of nanotechnology involving dispersions and coatings include nanofabricated surface coatings for keeping windows and surfaces clean (POST 1996). Here, the transparent nanocoating on a surface prevents fog and dirt particles from depositing on the substrate. Commercial products (achieved through gas phase condensation) also include aluminum oxide/ epoxy dispersions yielding 19 times more wear resistance than conventional products.

Molecular Technology Today

One dictionary definition of a machine is "any system, usually of rigid bodies, formed and connected to alter, transmit, and direct applied forces in a predetermined manner to accomplish a specific objective, such as the performance of useful work." Molecular machines fit this definition quite well. To imagine these machines, one must first picture molecules. We can picture atoms as beads and molecules as clumps of beads, like a child's beads linked by snaps. In fact, chemists do sometimes visualize molecules by building models from plastic beads (some of which link in several directions, like the hubs in a Tinkertoy set). Atoms are rounded like beads, and although molecular bonds are not snaps, our picture at least captures the essential notion that bonds can be broken and reformed. If an atom were the size of a small marble, a fairly complex *molecule* would be the size of your fist. This makes a useful mental image, but atoms are really about 1/10,000 the size of *bacteria*, and bacteria are about 1/10,000 the size of mosquitoes. (An atomic *nucleus*, however, is about 1/100,000 the size of the atom itself; the difference between an atom and its nucleus is the difference between a fire and a nuclear reaction.)

The things around us act as they do because of the way their molecules behave. Air holds neither its shape nor its volume because its molecules move freely, bumping and ricocheting through open space. Water molecules stick together as they move about, so water holds a constant volume as it changes shape. Copper holds its shape because its atoms stick together in regular patterns; we can bend it and hammer it because its atoms can slip over one another while remaining bound together. Glass shatters when we hammer it because its atoms separate before they slip. Rubber consists of networks of kinked molecules, like a tangle of springs. When stretched and released, its molecules straighten and then coil again. These simple molecular patterns make up passive substances. More complex patterns make up the active nanomachines of living *cells*.

Biochemists already work with these machines, which are chiefly made of protein, the main engineering material of living cells. These molecular machines have relatively few atoms, and so they have lumpy surfaces, like objects made by gluing together a handful of small marbles. Also, many pairs of atoms are linked by bonds that can bend or rotate, and so protein machines are unusually flexible. But like all machines, they have parts of different shapes and sizes that do useful work. All machines use clumps of atoms as parts. Protein machines simply use very small clumps.

Biochemists dream of designing and building such devices, but there are difficulties to be overcome. Engineers use beams of light to project patterns onto silicon chips, but chemists must build much more indirectly than that. When they combine molecules in various sequences, they have only limited control over how the molecules join. When biochemists need complex molecular machines, they still have to borrow them from cells. Nevertheless, advanced molecular machines will eventually let them build nanocircuits and nanomachines as easily and directly as engineers now build microcircuits or washing machines. Then progress will become swift and dramatic.

Genetic engineers are already showing the way. Ordinarily, when chemists make molecular chains—called "polymers"—they dump molecules into a vessel where they bump and snap together haphazardly in a liquid. The resulting chains have varying lengths, and the molecules are strung together in no particular order.

But in *modern gene synthesis machines*, genetic engineers build more orderly polymers—specific *DNA* molecules—by combining molecules in a particular order. These molecules are the nucleotides of DNA (the letters of the genetic alphabet) and genetic engineers don't dump them all in together. Instead, they direct the machine to add different nucleotides in a particular sequence to spell out a particular message. They first bond one kind of *nucleotide* to the chain ends, then wash away the leftover material and add chemicals to prepare the chain ends to bond the next nucleotide. They grow chains as they bond on nucleotides, one at a time, in a programmed sequence. They anchor the very first nucleotide in each chain to a solid surface to keep the chain from washing away with its chemical bathwater. In this way, they have a big clumsy machine in a cabinet assemble specific molecular structures from parts a hundred million times smaller than itself.

But this blind assembly process accidentally omits nucleotides from some chains. The likelihood of mistakes grows as chains grow longer. Like workers discarding bad parts before assembling a car, genetic engineers reduce errors by discarding bad chains. Then, to join these short chains into working genes (typically thousands of nucleotides long), they turn to molecular machines found in bacteria.

These protein machines, called restriction enzymes, "read" certain DNA sequences as "cut here." They read these genetic patterns by touch, by sticking to them, and they cut the chain by rearranging a few atoms. Other *enzymes* splice pieces together, reading matching parts as "glue here"—likewise "reading" chains by selective stickiness and splicing chains by rearranging a few atoms. By using gene machines to write, and restriction enzymes to cut and paste, genetic engineers can write and edit whatever DNA messages they choose.

But by itself, DNA is a fairly worthless molecule. It is neither strong like Kevlar, nor colorful like a dye, nor active like an enzyme, yet it has something that industry is prepared to spend millions of dollars to use: the ability to direct molecular machines called ribosomes. In cells, molecular machines first transcribe DNA, copying its information to make *RNA* "tapes." Then, much as old numerically controlled machines shape metal based on instructions stored on tape, ribosomes build proteins based on instructions stored on RNA strands. And proteins are useful.

Proteins, like DNA, resemble strings of lumpy beads. But unlike DNA, protein molecules fold up to form small objects able to do things. Some are enzymes, machines that build up and tear down molecules (and copy DNA, transcribe it, and build other proteins in the cycle of life). Other proteins are hormones, binding to yet other proteins to signal cells to change their behavior. Genetic engineers can produce these objects cheaply by directing the cheap and efficient molecular machinery inside living organisms to do the work. Whereas engineers running a chemical plant must work with vats of reacting chemicals (which often misarrange atoms and make noxious byproducts), engineers working with bacteria can make them absorb chemicals, carefully rearrange the atoms, and store a product or release it into the fluid around them.

Genetic engineers have now programmed bacteria to make proteins ranging from human growth hormone to rennin, an enzyme used in making cheese. The pharmaceutical company *Eli Lilly* (*Indianapolis*) is now marketing Humulin, human insulin molecules made by bacteria.

Existing Protein Machines

These protein hormones and enzymes selectively stick to other molecules. An enzyme changes its target's structure, then moves on; a hormone affects its target's behavior only so long as both remain stuck together. Enzymes and hormones can be described in mechanical terms, but their behavior is more often described in chemical terms.

But *other proteins serve basic mechanical functions*. Some push and pull, some act as cords or struts, and parts of some molecules make excellent bearings. The machinery of muscle, for instance, has gangs of proteins that reach, grab a "rope" (also made of protein), pull it, then reach out again for a fresh grip; whenever you move, you use these machines. Amoebas and human cells move and change shape by using fibers and rods that act as molecular muscles and bones. A reversible, variable-speed motor drives bacteria through water by turning a corkscrew-shaped propeller. If a hobbyist could build tiny cars around such motors, several billions of billions would fit in a pocket, and 150-lane freeways could be built through your finest *capillaries*.

Simple molecular devices combine to form systems resembling industrial machines. In the 1950s engineers developed machine tools that cut metal under the control of a punched paper tape. A century and a half earlier, Joseph-Marie Jacquard had built a loom that wove complex patterns under the control of a chain of punched cards. Yet over three billion years before Jacquard, cells had developed the machinery of the *ribosome*. Ribosomes are proof that nanomachines built of protein and RNA can be programmed to build complex molecules.

Then consider viruses. One kind, the *T4 phage*, acts like a spring-loaded syringe and looks like something out of an industrial parts catalog. It can stick to a bacterium, punch a hole, and inject viral DNA (yes, even bacteria suffer infections). Like a conqueror seizing factories to build more tanks, this DNA then directs the cell's machines to build more viral DNA and syringes. Like all organisms, these viruses exist because they are fairly stable and are good at getting copies of themselves made.

Whether in cells or not, nanomachines obey the universal laws of nature. Ordinary chemical bonds hold their atoms together, and ordinary chemical reactions (guided by other nanomachines) assemble them. Protein molecules can even join to form machines without special help, driven only by thermal agitation and chemical forces. By mixing viral proteins (and the DNA they serve) in a test tube, molecular biologists have assembled working T_4 viruses. This ability is surprising: imagine putting automotive parts in a large box, shaking it, and finding an assembled car when you look inside! Yet the T_4 *virus* is but one of many *self-assembling structures*. Molecular biologists have taken the machinery of the ribosome apart into over fifty separate protein and RNA molecules, and then combined them in test tubes to form working ribosomes again.

To see how this happens, imagine different T_4 protein chains floating around in water. Each kind folds up to form a lump with distinctive bumps and hollows, covered by distinctive patterns of oiliness, wetness, and electric charge. Picture them wandering and tumbling, jostled by the thermal vibrations of the surrounding water molecules. From time to time two bounce together, then bounce apart. Sometimes, though, two bounce together and fit, bumps in hollows, with sticky patches matching; they then pull together and stick. In this way protein adds to protein to make sections of the virus, and sections assemble to form the whole.

Protein engineers will not need nanoarms and nanohands to assemble complex nanomachines. Still, tiny manipulators will be useful and they will be built. Just as today's engineers build machinery as complex as player pianos and robot arms from ordinary motors, bearings, and moving parts, so tomorrow's biochemists will be able to use protein molecules as motors, bearings, and moving parts to build robot arms which will themselves be able to handle individual molecules.

Designing With Protein

How far off is such an ability? Steps have been taken, but much work remains to be done. Biochemists have already mapped the structures of many proteins. With gene machines to help write DNA tapes, they can direct cells to build *any protein they can design*. But they still don't know how to design chains that will fold up to make proteins of the right shape and function. The forces that fold proteins are weak, and the number of plausible ways a protein might fold is astronomical, so designing a large protein from scratch isn't easy.

The forces that stick proteins together to form complex machines are the same ones that fold the protein chains in the first place. The differing shapes and kinds of stickiness of *amino acids*—the lumpy molecular "beads" forming protein chains—make each protein chain fold up in a specific way to form an object of a particular shape. Biochemists have learned rules that suggest how an amino acid chain might fold, but the rules aren't very firm. Trying to predict how a chain will fold is like trying to work a jigsaw puzzle, but a puzzle with no pattern printed on its pieces to show when the fit is correct, and with pieces that seem to fit together about as well (or as badly) in many different ways, all but one of them wrong. False starts could consume many lifetimes, and a correct answer might not even be recognized. Biochemists using the best computer programs now available still cannot predict how a long, natural protein chain will actually fold, and some of them have despaired of designing protein molecules soon.

Yet most biochemists work as scientists, not as engineers. They work at predicting how natural proteins will fold, not at *designing* proteins that will fold predictably. *These tasks may sound similar*, but they differ greatly: the first is a scientific challenge, the second is an *engineering* challenge. Why should natural proteins fold in a way that scientists will find easy to predict? All that nature requires is that they in fact fold correctly, not that they fold in a way obvious to people.

Proteins *could* be designed from the start with the goal of making their folding more predictable. Carl Pabo, *writing in the journal Nature*, has suggested a design strategy based on this insight, and some biochemical engineers have designed and built *short chains of a few dozen pieces* that fold and nestle onto the surfaces of other molecules as planned. *They have designed from scratch a protein* with properties like those of melittin, a toxin in bee venom. They have modified existing enzymes, *changing their behaviors in predictable ways*. Our understanding of proteins is growing daily.

In 1959, *according to biologist Garrett Hardin*, some geneticists called genetic engineering impossible; today, it is an industry. Biochemistry and computer-aided design are now exploding fields, and as *Frederick Blattner wrote in the journal Science*, "computer chess programs have already reached the level below the grand master. Perhaps the solution to the protein-folding problem is nearer than we think." *William Rastetter* of *Genentech*, writing *in Applied Biochemistry and Biotechnology* asks, "How far off is *de novo* enzyme design and synthesis? Ten, fifteen years?" He answers, "Perhaps not that long."

Forrest Carter of the *U.S. Naval Research Laboratory*, Ari Aviram and Philip Seiden of *IBM*, Kevin Ulmer of Genex Corporation, and other researchers in university and industrial laboratories around the globe have already begun theoretical work and experiments aimed at developing molecular switches, memory devices, and other structures that could be incorporated into a protein-based computer. The U.S. Naval Research Laboratory has held *two international workshops on molecular electronic devices*, and a meeting sponsored by the *U.S. National Science Foundation* has *recommended support for basic research* aimed at developing molecular computers. Japan has reportedly begun a multimillion-dollar program aimed at developing self-assembling molecular motors and computers, and *VLSI Research Inc.*, of San Jose, *reports* that "It looks like the race to bio-chips [another term for molecular electronic systems] has already started. *NEC*, *Hitachi*, *Toshiba*, *Matsushita*, *Fujitsu*, *Sanyo-Denki* and *Sharp* have commenced full-scale research efforts on bio-chips for bio-computers."

Biochemists have other reasons to want to learn the art of protein design. New enzymes promise to perform dirty, expensive chemical processes more cheaply and cleanly, and novel proteins will offer a whole new spectrum of tools to biotechnologists. We are already on the road to protein engineering, and as Kevin Ulmer notes in the quote from *Science* that heads this chapter, this road leads "toward a more general capability for molecular engineering which would allow us to structure matter atom by atom."

New Replicators

History shows us that hardware evolves. Test tube RNA, viruses, and dogs all show how evolution proceeds by the modification and testing of replicators. But hardware (today) cannot reproduce itself—so where are the replicators behind the evolution of technology? What are the machine genes?

Of course, we need not actually identify replicators in order to recognize evolution. Darwin described evolution before Mendel discovered genes, and geneticists learned much about heredity before *Watson* and *Crick* discovered the structure of DNA. Darwin needed no knowledge of molecular genetics to see that organisms varied and that some left more descendants. A replicator is a pattern that can get copies of itself made. It may need help; without protein machines to copy it, DNA could not replicate. But by this standard, some *machines* are replicators! Companies often make machines that fall into the hands of a competitor; the competitor then learns their secrets and builds copies. Just as genes "use" protein machines to replicate, so such machines "use" human minds and hands to replicate. With nanocomputers directing assemblers and *disassemblers*, the replication of hardware could even be automated.

The human mind, though, is a far subtler engine of imitation than any mere protein machine or assembler. Voice, writing, and drawing can transmit designs from mind to mind before they take form as hardware. The ideas behind methods of design are subtler yet: more abstract than hardware, they replicate and function exclusively in the world of minds and symbol systems.

Where genes have evolved over generations and eons, mental replicators now evolve over days and decades. Like genes, ideas split, combine, and take multiple forms (genes can be transcribed from DNA to RNA and back again; ideas can be translated from language to language). *Science* cannot yet describe the neural patterns that embody ideas in brains, but anyone can see that ideas mutate, replicate, and compete. Ideas evolve.

Richard Dawkins calls bits of replicating mental patterns "memes" (*meme* rhymes with *cream*). He says "examples of memes are tunes, ideas, catch-phrases, clothes fashions, ways of making pots or of building arches. Just as genes propagate themselves in the gene pool by leaping from body to body [generation to generation] via sperms or eggs, so memes propagate themselves in the meme pool by leaping from brain to brain via a process which, in the broad sense, can be called imitation."

The Creatures of the Mind

Memes replicate because people both learn and teach. They vary because people create the new and misunderstand the old. They are selected (in part) because people don't believe or repeat everything they hear. As test tube RNA molecules compete for scarce copying machines and subunits, so memes must compete for a scarce resource—human attention and effort. Since memes shape behavior, their success or failure is a deadly serious matter.

Since ancient times, mental models and patterns of behavior have passed from parent to child. Meme patterns that aid survival and reproduction have tended to spread. (Eat this root only after cooking; don't eat those berries, their evil spirits will twist your guts.) Year by year, people varied their actions with varying results. Year by year, some died while others found new tricks of survival and passed them on. Genes built brains skilled at imitation because the patterns imitated were, on the whole, of value—their bearers, after all, had survived to spread them.

Memes themselves, though, face their own matters of "life" and "death": as replicators, they evolve solely to survive and spread. Like viruses, they can replicate without aiding their host's survival or well-being. Indeed, the meme for martyrdom-in-a-cause can spread itself through the very act of killing its host.

Genes, like memes, survive by many strategies. Some duck genes have spread themselves by encouraging ducks to pair off to care for their gene-bearing eggs and young. Some duck genes have spread themselves (when in male ducks) by encouraging rape, and some (when in female ducks) by encouraging the planting of eggs in other ducks' nests. Still other genes found in ducks are virus genes, able to spread without making more ducks. Protecting eggs helps the duck species (and the individual duck genes) survive; rape helps one set of duck genes at the expense of others; infection helps viral genes at the expense of duck genes in general. As Richard Dawkins points out, genes "care" only about their own replication: they appear selfish.

But *selfish motives can encourage cooperation*. People seeking money and recognition for themselves cooperate to build corporations that serve other people's wants. Selfish genes cooperate to build organisms that themselves often cooperate. Even so, to imagine that genes *automatically* serve some greater good (—of their chromosome?—their cell?—their body?—their species?) is to mistake a common effect for an underlying cause. To ignore the selfishness of replicators is to be lulled by a dangerous illusion.

Some genes in cells are out-and-out parasites. Like herpes genes inserted in human chromosomes, they exploit cells and harm their hosts. Yet if genes can be parasites, why not memes as well?

In *The Extended Phenotype*, Richard Dawkins describes a worm that parasitizes bees and completes its life cycle in water. It gets from bee to water by making the host bee dive to its death. Similarly, ant brainworms must enter a sheep to complete their life cycle. To accomplish this, they burrow into the host ant's brain, somehow causing changes that make the ant "want" to climb to the top of a grass stem and wait, eventually to be eaten by a sheep. As worms enter other organisms and use them to survive and replicate, so do memes. Indeed, the *absence* of memes exploiting people for their own selfish ends would be amazing, a sign of some powerful—indeed, nearly perfect—mental immune system. But parasitic memes clearly do exist. Just as viruses evolve to stimulate cells to make viruses, so rumors evolve to sound plausible and juicy, stimulating repetition. Ask not whether a rumor is true, ask instead how it spreads. Experience shows that ideas evolved to be successful replicators need have little to do with the truth.

At best, chain letters, spurious rumors, fashionable lunacies, and other mental parasites harm people by wasting their time. At worst, they implant deadly misconceptions. These meme systems exploit human ignorance and vulnerability. Spreading them is like having a cold and sneezing on a friend. Though some memes act much like viruses, infectiousness isn't necessarily bad (think of an infectious grin, or infectious good nature). If a package of ideas has merit, then its infectiousness simply increases its merit—and indeed, the best ethical teachings also teach us to teach ethics. Good publications may entertain, enrich understanding, aid judgment—and advertise gift subscriptions. Spreading useful meme systems is like offering useful seeds to a friend with a garden.

Selecting Ideas

Parasites have forced organisms to evolve immune systems, such as the enzymes that bacteria use to cut up invading viruses, or the roving white blood cells our bodies use to destroy bacteria. Parasitic memes have forced minds down a similar path, evolving meme systems that serve as mental immune systems.

The oldest and simplest mental immune system simply commands "believe the old, reject the new." Something like this system generally kept tribes from abandoning old, tested ways in favor of wild new notions—such as the notion that obeying alleged ghostly orders to destroy all the tribe's cattle and grain would somehow bring forth a miraculous abundance of food and armies of ancestors to drive out foreigners. (*This meme package infected the Xhosa people* of southern Africa in 1856; by the next year 68,000 had died, chiefly of starvation.)

Your body's immune system follows a similar rule: it generally accepts all the cell types present in early life and rejects new types such as potential cancer cells and invading bacteria, as foreign and dangerous. This simple reject-the-new system once worked well, yet in this era of organ transplantation it can kill. Similarly, in an era when science and technology regularly present facts that are both new and trustworthy, a rigid mental immune system becomes a dangerous handicap.

For all its shortcomings, though, the reject-the-new principle is simple and offers real advantages. Tradition holds much that is tried and true (or if not true, then at least workable). Change is risky: just as most mutations are bad, so most new ideas are wrong. Even reason can be dangerous: if a tradition links sound practices to a fear of ghosts, then overconfident rational thought may throw out the good with the bogus. Unfortunately, traditions evolved to *be* good may have less appeal than ideas evolved to *sound* good—when first questioned, the soundest tradition may be displaced by worse ideas that better appeal to the rational mind.

Yet memes that seal the mind against new ideas protect themselves in a suspiciously self-serving way. While protecting valuable traditions from clumsy editing, they may also shield parasitic claptrap from the test of truth. In times of swift change they can make minds dangerously rigid.

Much of the history of philosophy and science may be seen as a search for better mental immune systems, for better ways to reject the false, the worthless, and the damaging. The best systems respect tradition, yet encourage experiment. They suggest standards for judging memes, helping the mind distinguish between parasites and tools.

The principles of evolution provide a way to view change, whether in molecules, organisms, technologies, minds, or cultures. The same basic questions keep arising: What are the replicators? How do they vary? What determines their success? How do they defend against invaders? These questions will arise again when we consider the consequences of the assembler revolution, and yet again when we consider how society might deal with those consequences.

The deep-rooted principles of evolutionary change will shape the development of nanotechnology, even as the distinction between hardware and life begins to blur. These principles show much about what we can and cannot hope to achieve, and they can help us focus our efforts to shape the future. They also tell us much about what we can and cannot foresee, because they guide the evolution not only of hardware, but of knowledge itself.

Molecular Replicators

Cells replicate. Their machines copy their DNA, which directs their ribosomal machinery to build other machines from simpler molecules. These machines and molecules are held in a fluid-filled bag. Its membrane lets in fuel molecules and parts for more nanomachines, DNA, membrane, and so forth; it lets out spent fuel and scrapped components. A cell replicates by copying the parts inside its membrane bag, sorting them into two clumps, and then pinching the bag in two. Artificial replicators could be built to work in a similar way, but using assemblers instead of ribosomes. In this way, we could build cell-like replicators that are not limited to molecular machinery made from the soft, moist folds of protein molecules.

But engineers seem more likely to develop other approaches to replication. *Evolution* had no easy way to alter the fundamental pattern of the cell, and this pattern has shortcomings. In synapses, for example, the cells of the brain signal their neighbors by emptying bladders of chemical molecules. The molecules then jostle around until they bind to sensor molecules on the neighboring cell, sometimes triggering a neural impulse. A chemical *synapse* makes a slow switch, and neural impulses move slower than sound. With assemblers, molecular engineers will build entire computers smaller than a synapse and a millionfold faster.

Mutation and selection could no more make a synapse into a mechanical *nanocomputer* than a breeder could make a horse into a car. Nonetheless, engineers have built cars, and will also learn to build computers faster than brains, and replicators more capable than existing cells.

Some of these replicators will not resemble cells at all, but *will instead resemble factories* shrunk to cellular size. They will contain nanomachines mounted on a molecular framework and conveyor belts to move parts from machine to machine. Outside, they will have a set of assembler arms for building replicas of themselves, an atom or a section at a time.

How fast these replicators can replicate will depend on their assembly speed and their size. Imagine an advanced assembler that contains a million atoms: it can have as many as ten thousand moving parts, each containing an average of one hundred atoms—enough parts to make up a rather complex machine. In fact, the assembler itself looks like a box supporting a stubby robot arm a hundred atoms long. The box and arm contain devices that move the arm from position to position, and others that change the molecular tools at its tip.

Behind the box sits a device that reads a tape and provides mechanical signals that trigger arm motions and tool changes. In front of the arm sits an unfinished structure. Conveyors bring molecules to the assembler system. Some supply energy to motors that drive the tape reader and arm, and others supply groups of atoms for assembly. Atom by atom (or group by group), the arm moves pieces into place as directed by the tape; chemical reactions bond them to the structure on contact.

These assemblers will work fast. *A fast enzyme*, such as carbonic anhydrase or ketosteroid isomerase, can process almost a million molecules per second, even without conveyors and power-driven mechanisms to slap a new *molecule* into place as soon as an old one is released. It might seem too much to expect an assembler to grab a molecule, move it, and jam it into place in a mere millionth of a second. But small appendages can move to and fro very swiftly. A human arm can flap up and down several times per second, fingers can tap more rapidly, a fly can wave its wings fast enough to buzz, and a mosquito makes a maddening whine. Insects can wave their wings at about a thousand times the frequency of a human arm because an insect's wing is about a thousand times shorter.

An assembler arm will be about fifty million times shorter than a human arm, and so (as it turns out) it will be able to move back and forth *about fifty million times more rapidly*. For an assembler arm to move a mere million times per second would be like a human arm moving about once per minute: sluggish. So it seems a very reasonable goal.

The speed of replication will depend also on the total size of the system to be built. Assemblers will not replicate by themselves; they will need materials and energy, and instructions on how to use them. Ordinary chemicals can supply materials and energy, but nanomachinery must be available to process them. Bumpy *polymer* molecules can code information like a punched paper tape, but a reader must be available to translate the patterns of bumps into patterns of arm motion. Together, these parts form the essentials of a replicator: the tape supplies instructions for assembling *a copy of the assembler, of the reader, of the other nanomachines, and of the tape itself.*

A reasonable design for this sort of replicator will likely include several assembler arms and several more arms to hold and move workpieces. Each of these arms will add another million atoms or so. The other parts—tape readers, chemical processors, and so forth-may also be as complicated as assemblers. Finally, a flexible replicator system will probably include a simple computer; following the mechanical approach that I mentioned in this encyclopaedia, this will add roughly 100 million atoms. Altogether, these parts will total less than 150 million atoms. Assume instead a total of one billion, to leave a wide margin for error. Ignore the added capability of the additional assembler arms, leaving a still wider margin. Working at one million atoms per second, the system will still copy itself in one thousand seconds, or a bit over fifteen minutes—about the time a bacterium takes to replicate under good conditions.

Imagine such a replicator floating in a bottle of chemicals, making copies of itself. It builds one copy in one thousand seconds, thirty-six in ten hours. In a week, it stacks up enough copies to fill the volume of a human cell. In a century, it stacks up enough to make a respectable speck. If this were all that replicators could do, we could perhaps ignore them in safety.

Each copy, though, will build yet more copies. Thus the first replicator assembles a copy in one thousand seconds, the two replicators then build two more in the next thousand seconds, the four build another four, and the eight build another eight. At the end of ten hours, there are not thirty-six new replicators, but over 68 billion. In less than a day, they would weigh a ton; in less than two days, they would outweigh the Earth; in another four hours, they would exceed the mass of the Sun and all the planets combined—if the bottle of chemicals hadn't run dry long before. Regular doubling means *exponential growth*. Replicators multiply exponentially unless restrained, as by lack of room or resources. Bacteria do it, and at about the same rate as the replicators just described. People replicate far more slowly, yet given time enough they, too, could overshoot any finite resource supply. Concern about population growth will never lose its importance. Concern about controlling rapid new replicators will soon become important indeed.

Molecules & Skyscrapers

Machines able to grasp and position individual atoms will be able to build almost anything by bonding the right atoms together in the right patterns. To be sure, building large objects one atom at a time will be slow. A fly, after all, contains about a million atoms for every second since the dinosaurs were young. Molecular machines can nonetheless build objects of substantial size—they build whales, after all.

To make large objects rapidly, a vast number of assemblers must cooperate, but replicators will produce assemblers by the ton. Indeed, with correct design, the difference between an assembler system and a replicator will lie entirely in the assembler's programming.

If a replicating assembler can copy itself in one thousand seconds, then it can be programmed to build something else its own size just as fast. Similarly, a ton of replicators can swiftly build a ton of something else—and the product will have all its billions of billions of billions of atoms in the right place, with *only a minute fraction misplaced*. To see the abilities and limits of one method for assembling large objects, imagine a flat sheet covered with small assembly arms-perhaps an army of replicators reprogrammed for construction work and arrayed in orderly ranks. Conveyors and communication channels behind them supply reactive molecules, energy, and assembly instructions. If each arm occupies an area 100 atomic diameters wide, then behind each assembler will be room for conveyors and channels totaling about 10,000 atoms in cross sectional area.

This seems room enough. A space ten or twenty atoms wide can hold a conveyor (perhaps based on molecular belts and pulleys). A channel a few atoms wide can hold a molecular rod which, like those in the mechanical computer mentioned in this encyclopaedia, will be pushed and pulled to transmit signals. All the arms will work together to build a broad, solid structure layer by layer. Each arm will be responsible for its own area, handling about 10,000 atoms per layer. A sheet of assemblers handling 1,000,000 atoms per second per arm will complete about one hundred atomic layers per second. This may sound fast, but at this rate piling up a paper-sheet thickness will take about an hour, and making a meter-thick slab will take over a year.

Faster arms might raise the assembly speed to over a meter per day, but they would produce more waste heat. If they could build a meter-thick layer in a day, the heat from one square meter could cook hundreds of steaks simultaneously, and might fry the machinery. At some size and speed, cooling problems will become a limiting factor, but there are other ways of assembling objects faster without overheating the machinery.

Imagine trying to build a house by gluing together individual grains of sand. Adding a layer of grains might take grain-gluing machines so long that raising the walls would take decades. Now imagine that machines in a factory first glue the grains together to make bricks. The factory can work on many bricks at once. With enough grain-gluing machines, bricks would pour out fast; wall assemblers could then build walls swiftly by stacking the preassembled bricks. Similarly, molecular assemblers will team up with larger assemblers to build big things quickly—machines can be any size from molecular to gigantic. With this approach, most of the assembly heat will be dissipated far from the work site, in making the parts.

Skyscraper construction and the architecture of life suggest a related way to construct large objects. Large plants and animals have vascular systems, intricate channels that carry materials to molecular machinery working throughout their tissues. Similarly, after riggers and riveters finish the frame of a skyscraper, the building's "vascular system"—its elevators and corridors, aided by cranes—carry construction materials to workers throughout the interior. Assembly systems could also employ this strategy, first putting up a scaffold and then working throughout its volume, incorporating materials brought through channels from the outside.

Imagine this approach being used to "grow" a large rocket engine, working inside a vat in an industrial plant. The vat—made of shiny steel, with a glass window for the benefit of visitors—stands taller than a person, since it must hold the completed engine. Pipes and pumps link it to other equipment and to water-cooled heat exchangers. This arrangement lets the operator circulate various fluids through the vat.

To begin the process, the operator swings back the top of the vat and lowers into it a base plate on which the engine will be built. The top is then resealed. At the touch of a button, pumps flood the chamber with a thick, milky fluid which submerges the plate and then obscures the window. This fluid flows from another vat in which replicating assemblers have been raised and then reprogrammed by making them copy and spread a new instruction tape (a bit like infecting bacteria with a *virus*). These new assembler systems, smaller than bacteria, scatter light and make the fluid look milky. Their sheer abundance makes it viscous.

At the center of the base plate, deep in the swirling, assembler-laden fluid, sits a "seed." It contains a nanocomputer with stored engine plans, and its surface sports patches to which assemblers stick. When an assembler sticks to it, they plug themselves together and the seed computer transfers instructions to the assembler computer. This new programming tells it where it is in relation to the seed, and directs it to extend its manipulator arms to snag more assemblers. These then plug in and are similarly programmed. Obeying these instructions from the seed (which spread through the expanding network of communicating assemblers) a sort of assembler-*crystal* grows from the chaos of the liquid. Since each assembler knows its location in the plan, it snags more assemblers only where more are needed. This forms a pattern less regular and more complex than that of any natural crystal. In the course of a few hours, the assembler scaffolding grows to match the final shape of the planned rocket engine.

Then the vat's pumps return to life, replacing the milky fluid of unattached assemblers with a clear mixture of organic solvents and dissolved substances—including aluminum compounds, oxygen-rich compounds, and compounds to serve as assembler fuel. As the fluid clears, the shape of the rocket engine grows visible through the window, looking like a full-scale model sculpted in translucent white plastic. Next, a message spreading from the seed directs designated assemblers to release their neighbors and fold their arms. They wash out of the structure in sudden streamers of white, leaving a spongy lattice of attached assemblers, now with room enough to work. The engine shape in the vat grows almost transparent, with a hint of iridescence.

Each remaining assembler, though still linked to its neighbors, is now surrounded by tiny fluid-filled channels. Special arms on the assemblers work like flagella, whipping the fluid along to circulate it through the channels. These motions, like all the others performed by the assemblers, are powered by molecular engines fueled by molecules in the fluid. As dissolved sugar powers yeast, so these dissolved chemicals power assemblers. The flowing fluid brings fresh fuel and dissolved raw materials for construction; as it flows out it carries off waste heat. The communications network spreads instructions to each assembler.

The assemblers are now ready to start construction. They are to build a rocket engine, consisting mostly of pipes and pumps. This means building strong, light structures in intricate shapes, some able to stand intense heat, some full of tubes to carry cooling fluid. Where great strength is needed, the assemblers set to work constructing rods of interlocked fibers of carbon, in its diamond form. From these, they build a lattice tailored to stand up to the expected pattern of stress. Where resistance to heat and corrosion is essential (as on many surfaces), they build similar structures of aluminum oxide, in its sapphire form. In places where stress will be low, the assemblers save mass by leaving wider spaces in the lattice. In places where stress will be high, the assemblers reinforce the structure until the remaining passages are barely wide enough for the assemblers to move. Elsewhere the assemblers lay down other materials to make sensors, computers, motors, solenoids, and whatever else is needed.

To finish their jobs, they build walls to divide the remaining channel spaces into almost sealed cells, then withdraw to the last openings and pump out the fluid inside. Sealing the empty cells, they withdraw completely and float away in the circulating fluid. Finally, the vat drains, a spray rinses the engine, the lid lifts, and the finished engine is hoisted out to dry. Its creation has required less than a day and almost no human attention.

What is the engine like? Rather than being a massive piece of welded and bolted metal, it is a seamless thing, gemlike. Its empty internal cells, patterned in arrays about a wavelength of light apart, have a side effect: like the pits on a laser disk they diffract light, producing a varied iridescence like that of a fire opal. These empty spaces lighten a structure already made from some of the lightest, strongest materials known. Compared to a modern metal engine, this advanced engine has over 90 percent less mass.

Tap it, and it rings like a bell of surprisingly high pitch for its size. Mounted in a spacecraft of similar construction, it flies from a runway to space and back again with ease. It stands long, hard use because its strong materials have let designers include large safety margins. Because assemblers have let designers pattern its structure to yield before breaking (blunting cracks and halting their spread), the engine is not only strong but tough.

For all its excellence, this engine is fundamentally quite conventional. It has merely replaced dense metal with carefully tailored structures of light, tightly bonded atoms. The final product contains no nanomachinery.

More advanced designs will exploit *nanotechnology* more deeply. They could leave a vascular system in place to supply assembler and *disassembler systems*; these can be programmed to mend worn parts. So long as users supply such an engine with energy and raw materials, it will renew its own structure. More advanced engines can also be literally more flexible. Rocket engines work best if they can take different shapes under different operating conditions, but engineers cannot make bulk metal strong, light, and limber. With nanotechnology, though, a structure stronger than steel and lighter than wood could change shape like muscle (*working, like muscle*, on the sliding fiber principle), An engine could then expand, contract, and bend at the base to provide the desired thrust in the desired direction under varying conditions. With properly programmed assemblers and disassemblers, it could even remodel its fundamental structure long after leaving the vat.

In short, replicating assemblers will copy themselves by the ton, then make other products such as computers, rocket engines, chairs, and so forth. They will make disassemblers able to break down rock to supply raw material. They will make solar collectors to supply energy. Though tiny, they will build big. Teams of nanomachines in nature build whales, and seeds replicate machinery and organize atoms into vast structures of cellulose, building redwood trees. There is nothing too startling about growing a rocket engine in a specially prepared vat. Indeed, foresters given suitable assembler "seeds" could grow spaceships from soil, air, and sunlight.

Assemblers will be able to make virtually anything from common materials without labor, replacing smoking factories with systems as clean as forests. They will transform technology and the economy at their roots, opening a new world of possibilities. They will indeed be engines of abundance.

It's a Small, Small, Small, Small World

Manufactured products are made from atoms. The properties of those products depend on how those atoms are arranged. If we rearrange the atoms in coal, we get diamonds. If we rearrange the atoms in sand (and add a pinch of impurities) we get computer chips. If we rearrange the atoms in dirt, water and air we get grass.

Since we first made stone tools and flint knives we have been arranging atoms in great thundering statistical heards by casting, milling, grinding, chipping and the like. We've gotten better at it: we can make more things at lower cost and greater precision than ever before. But at the molecular scale we're still making great ungainly heaps and untidy piles of atoms.

That's changing. In special cases we can already arrange *atoms* and *molecules* exactly as we want. *Theoretical analyses make it clear we can do a lot more.* Eventually, we should be able to arrange and rearrange atoms and molecules much as we might arrange LEGO blocks. In not too many decades we should have a manufacturing technology able to:

Build products with almost every atom in the right place.

Do so inexpensively.

Make most arrangements of atoms consistent with physical law.

Often called *nanotechnology, molecular nanotechnology or molecular manufacturing*, it will let us make most products lighter, stronger, smarter, cheaper, cleaner and more precise.

One warning: in contrast to the useage in this article *some researchers use the word "nanotechnology" to refer to high resolution lithographic technology while others use it to refer to almost any research where some critical size is less than a micron (1,000 nanometers).* When there is risk of confusion, the more specific terms "molecular nanotechnology" or "molecular manufacturing" should be used.

There are two main issues in nanotechnology:

- What might molecular manufacturing systems look like?
- How could we build such systems given our current technology?

As molecular manufacturing systems do not yet exist, and as it will likely be a few decades before we can build them, the answer to the first question must be based on theoretical and computational models. Such models serve several purposes. First and most obvious, some might argue that the goal is itself inherently impossible, e.g., that such systems cannot be made within the framework of existing physical law. Theoretical and computational models provide an inexpensive way to examine this question and provide assurances that proposed systems are possible.

Second, such models give us a feel for what molecular manufacturing systems might look like. The better you understand the goal, the better your chances of actually achieving it. To illustrate this point, we consider a historical example based on the single most important invention of the 20th century: the computer. The basic idea of the relay was known in the 1820's, and *the concept of a mechanical stored program computer was understood by Babbage by the mid 1800's*. Practical computers could have been built using relays by the 1860's if there had been an adequate theoretical understanding of "computation." An adequate theoretical understanding of molecular manufacturing should likewise ease the problems of practical development by opening for consideration approaches and designs that might not otherwise be developed for decades or — as happened in the case of the computer — almost a century.

In this paper we'll first address the question of what molecular manufacturing systems could look like, and then consider how to build them given our current technology. This second question involves not only a consideration of current experimental triumphs and how they might be extended, but also the definition of intermediate goals and systems.

If you'd like to see what's available on the web about nanotechnology (and there's quite a lot), a good starting place is *http: / /www.zyvex.com/nano*.

The advantages of nanotechnology

What would it mean if we could inexpensively make things with every atom in the right place? For starters, we could continue the revolution in computer hardware right down to molecular gates and wires — something that today's lithographic methods (used to make computer chips) could never hope to do. We could inexpensively make very strong and very light materials: shatterproof diamond in precisely the shapes we want, by the ton, and over fifty times lighter than steel of the same strength. We could make a Cadillac that weighed fifty kilograms, or a full-sized sofa you could pick up with one hand. We could make surgical instruments of such precision and deftness that they could operate on the cells and even molecules from which we are made — something well beyond today's medical technology. The list goes on — almost any manufactured product could be improved, often by orders of magnitude.

The advantages of positional control

One of the basic principles of nanotechnology is positional control. At the macroscopic scale, the idea that we can hold parts in our hands and assemble them by properly positioning them with respect to each other goes back to prehistory: we celebrate ourselves as the tool using species. Our wisdom and our knowledge would have done us scant good without an opposable thumb: we'd still be shivering in the bushes, unable to start a fire.

At the molecular scale, the idea of holding and positioning molecules is new and almost shocking. However, as long ago as 1959 *Richard Feynman, the Nobel prize winning physicist, said that nothing in the laws of physics prevented us from arranging atoms the way we want: "...it is something, in principle, that can be done; but in practice, it has not been done because we are too big."*

Before discussing the advantages of positional control at the molecular scale, it's helpful to look at some of the methods that have been developed by chemists — methods that don't use positional control, but still let chemists synthesize a remarkably wide range of molecules and molecular structures.

Self Assembly

The ability of chemists to synthesize what they want by stirring things together is truly remarkable. Imagine building a radio by putting all the parts in a bag, shaking, and pulling out the radio — fully assembled and ready to work! Self assembly — the art and science of arranging conditions so that the parts themselves spontaneously assemble into the desired structure — is a well established and powerful method of synthesizing complex molecular structures. A basic principle in self assembly is selective stickiness: if two molecular parts have complementary shapes and charge patterns — one part has a hollow where the other part has a bump, and one part has a positive charge where the other part has a negative charge — then they will tend to stick together in one particular way. By shaking these parts around — something which thermal noise does for us quite naturally if the parts are floating in solution — the parts will eventually, purely by chance, be brought together in just the right way and combine into a bigger part. This bigger part can combine in the same way with other parts, letting us gradually build a complex whole from molecular pieces by stirring them together and shaking.

Many viruses use this approach to make more viruses — if you stir the parts of the T4 bacteriophage together in a test tube, they will self assemble into fully functional viruses.

Positional devices and positionally controlled reactions

While self assembly is a *path* to nanotechnology, by itself it would be hard pressed to make the very wide range of products promised by nanotechnology. We don't know how to self assemble shatterproof diamond, for example. (We'll later discuss a way of making diamond using positional control). During self assembly the parts bounce around and bump into each other in all kinds of ways, and if they stick together when we don't *want* them to stick together, we'll get unwanted globs of random parts. Many types of parts have this problem, so self assembly won't work for them. To make diamond, it seems as though we need to use indiscriminately sticky parts (such as radicals, carbenes and the like). These parts can't be allowed to randomly bump into each other (or much of anything else, for that matter) because they'd stick together when we didn't want them to stick together and form messy blobs instead of precise molecular machines.

We can avoid this problem if we can hold and position the parts. Even though the molecular parts that are used to make diamond are both indiscriminately and *very* sticky (more technically, the barriers to bond formation are low and the resulting covalent bonds are quite strong), if we can position them we can prevent them from bumping into each other in the wrong way. When two sticky parts do come into contact with each other, they'll do so in the right orientation because we're *holding* them in the right orientation. In short, positional control at the molecular scale should let us make things which would be difficult or impossible to make without it. Given our macroscopic intuition, this shouldn't be surprising. If we couldn't use our hands to hold and position parts, there are lots of things we'd have a very hard time making! If we are to position molecular parts we must develop the molecular equivalent of "arms" and "hands." We'll need to learn what it means to "pick up" such parts and "snap them together." We'll have to understand the precise *chemical reactions that such a device would use*.

One of the first questions we'll need to answer is: what does a molecular-scale positional device look like? Current proposals are similar to macroscopic robotic devices but on a much smaller scale. The illustrations (from *Nanosystems, the best technical introduction to nanotechnology*) show a design for a molecular-scale robotic arm proposed by *Eric Drexler*, a pioneering researcher in the field. Only 100 nanometers high and 30 nanometers in diameter, this rather squat design has a few million atoms and roughly a hundred moving parts. It uses no lubricants, for at this scale a lubricant molecule is more like a piece of grit. Instead, the bearings are "run dry" (following a suggestion by *Feynman*) as described in the following paragraph.

Running bearings dry should work both because the diamond surface is very slippery (see the coefficient of friction for diamond in the table) and because we can make the surface very smooth — so smooth that there wouldn't even be molecular-sized asperities or imperfections that might catch or grind against each other. Computer models support our intuition: *analysis of the bearings shown here using computational chemistry programs* shows they should rotate easily.

Stiffness

Our molecular arms will be buffeted by something we don't worry about at the macroscopic scale: thermal noise. This makes molecular-scale objects wiggle and jiggle, just as Brownian motion makes small dust particles bounce around at random. *Can our molecular robotic arm maintain its position in the face of thermal noise?*

The critical property we need here is *stiffness*. Stiffness is a measure of how far something moves when you push on it. If it moves a lot when you push on it a little, it's not very stiff. If it doesn't budge when you push hard, it's very stiff.

Big positional devices, used in today's Scanning Probe Microscopes (SPMs), have been made stiff enough to image individual atoms despite thermal noise. In SPMs, a very sharp tip is brought down to the surface of the sample being scanned. Like a blind man tapping in front of him with his cane, we can tell that the tip is approaching the surface and so can "feel" the outlines of the surface in front of us. Many different types of physical interactions with the surface are used to detect its presence. Some scanning probe microscopes literally push on the surface — and note how hard the surface pushes back. Others connect the surface and probe to a voltage source, and measure the current flow when the probe gets close to the surface. A host of other probe-surface interactions can be measured, and are used to make different types of SPMs. But in all of them, the basic idea is the same: when the sharp tip of the probe approaches the surface a signal is generated — a signal which lets us map out the surface being probed.

The SPM can not only map a surface, in many cases the probe-surface interaction changes the surface as well. This has already been used experimentally to spell out molecular words, and the obvious opportunities to modify the surface in a controlled way are being investigated both experimentally and theoretically.

A few big SPMs making a few molecular structures won't let us make much — certainly not tons of precisely structured shatterproof diamond. We'll need vast numbers of very small positional devices operating in parallel. Unfortunately, as we make our positional devices smaller and smaller, they will be more and more subject to thermal noise. To make something that's *both* small *and* stiff is more challenging. It helps to get the stiffest material you can find. Diamond, as usual, is stiffer than almost anything else and is an excellent material from which to make a very small, very stiff positional device. *Theoretical analysis gives firm support to the idea that positional devices in the 100 nanometer size range able to position their tips to within a small fraction of an atomic diameter in the face of thermal noise at room temperature should be feasible.* Trillions of such devices would occupy little more than a few cubic millimeters (a speck slightly larger than a pinhead).

Stewart Platforms

While Drexler's proposal for a small robotic arm is easy to understand and should be adequate to the task, more recent work has focused on the Stewart platform. This positional device has the great advantage that it is *stiffer than a robotic arm* of similar size. Conceptually, the Stewart platform is based on the observation that a polyhedron, all of whose faces are triangular, will be rigid. If some of the edges of the polyhedron can be adjusted in length, then the position of one face can be moved with respect to the position of another face. If we want a full six degrees of freedom (X, Y, Z, roll, pitch and yaw) then we must be able to independently adjust the lengths of six different edges of the polyhedron. If we further want one triangular face of the polyhedron to remain of fixed size and hold a "tool," and a second face of the polyhedron to act as the "base" whose size and position is fixed, then we find that the simplest polyhedron that will suit our purpose is the octahedron.

In the Stewart platform, one triangular face of the octahedron is designated the "platform," while the opposing triangular face is designated the "base." The six edges that connect the base to the platform can then be adjusted in length to control the position of the platform with respect to the base. Mechanically, this adjustment is often done using six hydraulic pistons. A picture of the Stewart platform (from *Nanosystems*) is shown above.

The advantage of the Stewart platform can now be seen: because the six adjustable-length edges are either in pure compression or pure tension and are never subjected to any bending force, this positional device is stiffer than a long robotic arm which can bend and flex. The Stewart platform is also conceptually simpler than a robotic arm, having fewer different types of parts; for this reason, we can reasonably expect that making one will be simpler than making a robotic arm.

The utility of diamond: why it's a dream material

From airplanes and space ships to cars and chairs, making products lighter and stronger is almost always an improvement. Sometimes, as *in the case of space flight, this is a quantum leap all by itself.* At other times it's just a convenience: moving heavy furniture is a chore most of us would avoid if we could.

The strength and lightness of materials depends on the number and strength of the bonds that hold their atoms together, and the lightness of those atoms. Light atoms that form many strong bonds are at the heart of strong, light stiff materials. Boron, carbon and nitrogen are lighter and form more and stronger bonds than other atoms. The carbon-carbon bond in particular is very strong, and carbon atoms can form four bonds to four neighboring atoms. In diamond, this makes a very dense network of very strong bonds, which creates a very strong, light and stiff material. Besides light weight and great strength, diamond has a host of materials properties that make it an excellent choice for almost any application. (See the table for a list of diamond's material properties).

Another material with remarkable properties is graphite: carbon atoms arranged in a hexagonal lattice with each carbon atom having three bonds to its three neighbors. While graphite has fewer bonds per atom, the strength of these bonds is greater and so graphitic materials also have remarkable properties.

Diamond is also a wonderful material for making transistors and computer gates, though it takes a bit of explanation to understand why.

Computer gates should switch as quickly as possible: that's what makes computers so fast. To do this, the gates must be made of transistors in which the electrons move as fast as possible over the shortest possible distances. But as electrons move through a material, they generate heat: think of how hot the tungsten wire in a light bulb can get! So the faster we make our computer the faster the electrons move through the material and the hotter it gets. When it gets too hot the computer stops working.

Diamond excels in its electronic properties. Fundamentally, it lets us move charge around much faster before things stop working. There are several reasons for this. First, diamond transistors can operate at much higher temperatures because diamond has a larger "bandgap" than other materials (particularly silicon). Electrons in semiconductors (such as diamond or silicon) are either in the "conduction band" or the "valence band." An electron in the conduction band can move freely and transport charge. Moving an electron from the valence band to the conduction band requires adding a specific amount of energy to the electron. In silicon, this is 1.12 electron volts. In diamond, it is 5.47 electron volts. As the temperature increases, more and more electrons can jump into the conduction band from the valence band because of thermal noise. When too many electrons do this the semiconductor becomes conductive everywhere and the transistors short out and stop working. Because diamond has a wider bandgap, it shorts out at a proportionally higher temperature than silicon.

Diamond also has greater thermal conductivity, which lets us move heat out of a diamond transistor more quickly to prevent it from getting too hot.

To move an electron faster we want to "pull" on it harder. The stronger the electric field, the harder we're pulling the electrons. But as the electric field gets too strong, it tears electrons out of the valence band causing the transistors to short out. This "breakdown" field varies from material to material, and (you guessed it) diamond leads the pack.

Finally, electrons (and holes) move with different speeds through different materials, even when the electric field is the same. Again, electrons and holes in diamond move faster than in silicon.

Because diamond transistors can be hotter, are more easily cooled, can tolerate higher voltages before breaking down, and electrons move more easily in them; they make better transistors than other materials. Diamond would be ideal for electronic devices if only we could manufacture it inexpensively and with precisely the desired structure. In summary: from cars to airplanes to spaceships, from furniture to buildings, from pocket calculators to supercomputers, materials similar to diamond are just *better* than other materials, often a *lot* better. Just as the stone age, the bronze age and the steel age were named after the materials that we could make, this new age we are entering might be called *the diamond age*.

Positional Control and the Synthesis of Diamond

We can in principle make a great many very useful structures from diamond, including very powerful computers and molecular robotic devices able to position molecular components to within a fraction of an atomic diameter. But how can we synthesize diamond? One answer to this question comes from looking at how we grow diamond today using *Chemical Vapor Deposition (CVD)*. In a process somewhat reminiscent of spray painting, we build up layer after layer of diamond on a surface by holding that surface in a cloud of very reactive molecules and atoms — like H; CH_3, C_2H, etc. When these reactive molecules bump into the surface they change it, either by adding, removing, or re-arranging atoms. By carefully controlling the pressure, temperature, and the exact composition of the gas, we can create conditions that favor the growth of diamond on the surface. While these are the right chemical reactions to make diamond, randomly bombarding the growing surface with reactive molecules doesn't offer the finest control over the growth process. It would be like trying to build a wristwatch using a sandblaster. We want the chemical reactions to take place at precisely the places on the surface that we specify — not wherever random gusts of turbulent gas might dictate.

A second problem is that the hydrogenated diamond surface is chemically inert: it's difficult to add carbon (or anything else) to it. We could overcome this problem by removing a hydrogen atom from the surface, leaving behind a very reactive dangling bond. This is a critical first step during CVD diamond growth but it happens at random, anywhere on the growing diamond surface, when the right kind of reactive gas molecule happens to strike the surface in the right way. We want to do something similar but more controlled: remove a *specific* hydrogen atom from a *specific* spot on the diamond surface. To do this, we'll need a "hydrogen abstraction tool".

What might such a tool look like? If we base it on our understanding of diamond CVD, it should be a highly reactive radical with a high affinity for hydrogen. At the same time it must be possible to position it, so it must have a stable region which can serve as a "handle." The tool would be held by (for example) the molecular robotic arm discussed earlier, and would be positioned directly over the hydrogen we wish to abstract.

A simple strategy for finding such radicals is to look through the table of bond strengths in the *Handbook of Chemistry and Physics* and pick the molecule with the strongest bond to hydrogen. The radical created by removing the hydrogen should then have a very high affinity for hydrogen, as desired. Unfortunately, this strategy first produces fluorine. While atomic fluorine does indeed have a very high affinity for hydrogen, there is no obvious way to attach a "handle" to it. Interest therefore turns to the molecule with the second strongest bond to hydrogen: acetylene. Here, we are in luck. Not only does the acetylene radical have a high affinity for hydrogen, it also has a chemically stable region which can be modified into a "handle".

Drexler realized this sometime in the early 1980's. The author subsequently and independently followed the same line of logic and reached the same conclusions. An initial check of the validity of this idea was provided by semi-empirical calculations at Xerox PARC (easily done on a workstation using freely available programs). Discussions with *Bill Goddard and members of his molecular modeling group at Caltech* resulted in *more accurate calculations using high level ab initio quantum chemistry methods* on a more powerful computer. (These methods have the interesting property that the calculations will converge on the correct answer if enough computing power is used). *Drexler published this observation in Nanosystems* in 1992. A *group at NRL interested in diamond growth* found the proposal intriguing and modeled the abstraction of a hydrogen at room temperature using a qualitatively different computational model of molecular behavior (which permitted them to model the behavior of several hundreds of atoms in the vicinty of the abstraction). As the results of three different computational approaches have all produced the same qualitative answer, and as the results are also in accord with our intuition about what *should* happen, it's safe to say that this hydrogen abstraction tool will, when finally implemented, be able to abstract hydrogen.

The hydrogen abstraction tool illustrates some important ideas common to other proposals for molecular tools for the synthesis of diamond. First, the tool has a reactive end which is brought into contact with the molecular workpiece and an inert end which is held by a molecular positional device. Second, the point of application of the tool is controlled by the molecular positional device. Third, the environment around the tool is itself inert (typical proposals involve the use of either vacuum or a noble gas). The inert environment prevents the tool (which is quite reactive) from reacting with anything undesired.

While the hydrogen abstraction tool creates a very reactive spot on the surface we also need a tool that will deposit one or more carbon atoms on the growing surface. One proposal is the dimer deposition tool. (a "dimer" is just two of something stuck together. In this case, two carbon atoms are stuck together by a triple bond, hence the two carbon atoms form a "dimer"). A dimer deposition tool is simply a dimer with two weak bonds to a supporting structure. A proposal by Drexler is shown to the left. A second proposed dimer deposition tool, which has particularly weak bonds to the dimer, is illustrated at the right. [Subsequent note: further investigation of the proposal at right suggests it might not be stable except at very low temperatures. A variety of alternative dimer deposition tools are possible, so even if this particular structure should prove unsastisfactory for room temperature operation, other tools able to carry out the same function should be feasible.]

To illustrate how we might combine the use of the hydrogen abstraction tool with the dimer deposition tool, we proceed in three steps. First, we start with the hydrogenated diamond (111) surface. (The notation (111) refers to a particular surface of the diamond crystal. When diamond shatters, it shatters along certain planes and not along others.

As these different planes have surfaces with different properties, it's important to specify which plane you're talking about). Second, we use the hydrogen abstraction tool to remove two adjacent hydrogen atoms. Third, we use the dimer deposition tool to deposit two carbon atoms on the surface — the two ends of the carbon dimer are originally single-bonded to the rest of the tool. The two dangling bonds on the surface are very reactive, and react with the ends of the carbon dimer. When this happens, the bonds holding the dimer to the dimer deposition tool break, and the bonds in the deposition tool rearrange to eliminate what would otherwise be two radicals. Result: the carbon dimer is transferred from the tool to the surface. The following illustration shows the dimer deposition tool adding a dimer to a small cluster which represents the diamond (111) surface. (Because accurate ab initio calculations are computationally less expensive for a small cluster of atoms, the following illustrations show only the atoms directly involved in the reaction).

This sequence of steps starts with a flat diamond surface and adds two carbon atoms to it. It could be repeated at other sites. A computational investigation of the final step was reported by *Stephen Walch at the recent workshop on computational nanotechnology* sponsored by the NAS program at NASA. Because the energy released during the reaction is much larger than thermal noise (e.g., the reaction is quite exothermic), the reverse reaction is very unlikely to occur. The barrier to the reaction is also small. Taken together, these mean the reaction can take place easily and irreversibly. The dimer will "snap" onto the surface and stay there. (Calculations at the 6-31G* MP2 level show all positive vibrational frequencies for this tool. Further investigation of its stability is in progress). A third proposed tool is the carbene insertion tool. "It is no exaggeration to claim a major role for carbenes in the modern chemists's attitude that he can very probably make anything he wants." (W. J. Baron). A positionally controlled carbene should be correspondingly more useful, as it could be made to react at any position on a growing molecular workpiece. Carbenes are very reactive and can insert readily into double and triple bonds. A plausible reaction would be to insert a carbene into the dimer described above.

A fourth proposal is for a hydrogen deposition tool. Where the hydrogen abstraction tool is intended to make an inert structure reactive by creating a dangling bond, the hydrogen deposition tool would do exactly the opposite: make a reactive structure inert by terminating dangling bonds. This tool could be used during the synthetic process to stabilize a structure that might otherwise undergo a spontaneous and undesired rearrangement or reconstruction, or it could be used at the end of the synthetic process to stabilize the finished product.

There are many possible candidates for such a tool: any molecular structure that has a weak bond to a hydrogen. One particularly attractive candidate is the use of tin. The hydrogen-tin bond is quite weak, and so a tin-based hydrogen deposition tool should be quite effective.

These four tools let us (a) make an inert surface reactive by removing a hydrogen from one or more specific sites, (b) add one or two carbon atoms to a surface at selected sites and (c) make a reactive surface inert by adding hydrogen to it — which lets us prevent undesired surface rearrangements.

The Stiff Hydrocarbons as a Model System for Study

These four molecular tools — by themselves — should be sufficient to make a remarkably wide range of stiff hydrocarbons. This suggests that we define a simplified model of nanotechnology that we might call "hydrocarbon-based nanotechnology." This simplified version would only be able to make things from hydrogen and carbon — a much less ambitious goal than making things from the approximately 100 elements found in the full periodic table. But in exchange for confining ourselves to this more limited class of structures, we make it much easier to analyze the structures that can be fabricated and the synthetic reactions needed to make them. This narrower proposal can be more readily and more thoroughly investigated than full nanotechnology.

At the same time, hydrocarbon-based nanotechnology retains several of the key features of general nanotechnology. Diamond and shatterproof variants of diamond still fall within its purview; and it can still make all the parts needed for the operation of basically mechanical devices, including: struts, bearings, gears, robotic arms, and the like.

It's often useful to define a somewhat broader class than the stiff hydrocarbons, but a class which is still much more restrictive than the class of "most structures consistent with physical law." We'd like to be able to build molecular structures that are similar to diamond but which incorporate some other elements. We might want to add impurities (as in diamond electronic devices) or terminate a bearing surface with fluorine. Perhaps we're bending the diamond structure, and want to add some silicon internally to relieve the strain, (the silicon-carbon bond is longer than the carbon-carbon bond, so adding silicon to a region will "puff it up"). Or we find that adding some nitrogen to the internal surface of a bearing will relieve some strain (the nitogen-carbon bond is slightly shorter than the carbon-carbon bond). The term "diamondoid" is used to cover these (and related) possiblities. Diamondoid materials are made primarily of the elements hydrogen; first row elements like carbon, nitrogen, oxygen or fluorine; and second row elements like silicon, phosphorous, sulfur or chlorine; all held together by many strong covalent bonds. While including only about a tenth of the elements in the periodic table, the diamondoid materials include many structures of striking utility.

Self replication: making things inexpensively

Positional control combined with appropriate molecular tools should let us build a truly staggering range of molecular structures — but a few molecular devices built at great expense would hardly seem to qualify as a revolution in manufacturing. How can we keep the costs down?

Potatoes are a miracle of biology with tens of thousands of genes and proteins and intricate molecular machinery; yet we think nothing of eating this miracle, mashed with a little butter. Potatoes, along with many other agricultural products, cost less than a dollar a pound. The key reason: provide them with a little dirt, water and sunlight and a potato can make more potatoes.

If we could make a general purpose programmable manufacturing device which was able to make copies of itself (the *author does* work at *Xerox*, after all....), then the manufacturing costs for both the devices and anything they made could be kept quite low — likely no more than the costs for growing potatoes.

Drexler called such devices "assemblers."

The first serious analysis of self replicating systems was by *von Neumann* in the 1940's. *He carried out a detailed analysis of one such system in a theoretical cellular automata model* (The best known example of a cellular automata model is *Conway's Game of Life*, which is played out on a giant checkerboard and uses a simple set of rules to decide what happens at each square of the checkerboard at each time step). In von Neumann's cellular automata model he used a universal computer for control and a "universal constructor" to build more automata. The "universal constructor" was a robotic arm that, under computer control, could move in two dimensions and alter the state of the cell at the tip of its arm. By sweeping systematically back and forth, the arm could "build" any structure that the computer instructed it to. In his three-dimensional "kinematic" model, von Neumann retained the idea of a positional device (now able to position in three dimensions rather than two) and a computer to control it.

A *NASA study in 1980* extended the general conclusions of von Neumann and concluded that a fully automated lunar mining and manufacturing operation, able to extend itself using its own mining and manufacturing capabilities, would be feasible and could be built given a multi-billion dollar budget and a few decades of work. (Note that this kind of budget and time frame was in keeping with other proposed aerospace projects of the time).

The architecture for Drexler's assembler is a specialization of the more general architecture proposed by von Neumann. As before, there is a computer and constructor, but now the computer has shrunk to a "molecular computer" while the constructor combines two features: a robotic positional device (such as the robotic arm discussed earlier) and a well defined set of chemical operations that take place at the tip of the positional device (such as the hydrogen abstraction reaction and the other reactions involved in the synthesis of diamond).

The complexity of a *self replicating system* need not be excessive. In this context the complexity is just the size, in bytes, of a "recipe" that fully describes how to make the system. The complexity of an assembler needn't be beyond the complexity that can be dealt with by today's engineering capabilities. As shown in the following table, there are several self replicating systems whose complexity is well within current capabilities. Drexler estimated the complexity of his original proposal for an assembler at about 10,000,000 bytes. *Further work should reduce this.*

Complexity of self replicating systems (bytes)	
Von Neumann's universal constructor	about 60,000
Internet worm	60,000
Mycoplasma genitalia	145,018
E. Coli	1,000,000
Drexler's assembler	12,000,000
Human	800,000,000
NASA Lunar Manufacturing Facility	over 10,000,000,000

What is the "complexity" of a living system? We'll take this to mean the number of bytes in the DNA "blueprints." As each base pair in DNA can be one of four possibilities, it encodes two bits. One byte (8 bits) can be encoded in four base pairs. This means we count the number of base pairs in the DNA and divide by four to get the number of bytes in the "blueprints." For Mycoplasma genitalia, which has 580,070 base pairs, this results in 145,017.5 bytes. For humans, with roughly 3.2 billion base pairs, this results in 800 million bytes. (We're using the haploid base pair count: each cell in your body has DNA from your mother and DNA from your father — but much of this is similar. We're counting only the DNA from one parent). The complexity and sophistication of most living systems goes well beyond anything we might need just to achieve low cost manufacturing. Much of the complexity of the human genome is unrelated to issues of self-replication: people do more than just self-replicate!

The complexity of the internet worm is just an estimate of the number of bytes in its C program. Because the environment in which it operates is highly structured and provides relatively easy access to complex and sophisticated software, it could be argued that its complexity might be much less than the complexity of a system that operated in a "simple" environment. We'll leave it in our table anyway, as it's an interesting data point. This argument is less applicable to von Neumann's universal constructor which operates in a simple environment: just a large two-dimensional checkerboard with a finite number of states at each square. Its complexity is an estimate of the number of bytes needed to describe the constructor. The complexity of the NASA Lunar Manufacturing Facility was estimated in the NASA study.

What will we be able to make?

Today, most airplanes are made from metal despite the fact that diamond has a strength-to-weight ratio over 50 times that of aerospace aluminum. Diamond is expensive, we can't make it in the shapes we want, and it shatters. Nanotechnology will let us inexpensively make shatterproof diamond (with a structure that might resemble diamond fibers) in exactly the shapes we want. This would let us make a Boeing 747 whose unloaded weight was 50 times lighter but just as strong.

Today, travel in space is *very* expensive and reserved for an elite few. *Nanotechnology will dramatically reduce the costs and increase the capabilities of space ships and space flight.* The strength-to-weight ratio and the cost of components are absolutely critical to the performance and economy of space ships: with nanotechnology, both of these parameters will be improved by one to two orders of magnitude. Improvements in these two parameters alone (ignoring other advantages provided by nanotechnology) should improve the overall cost/performance ratio by over three orders of magnitude. This has led the *National Space Society* (NSS) to adopt *a position paper supporting nanotechnology*; Dan Goldin, *NASA*'s chief administrator, to support nanotechnology; and *NAS at NASA Ames Research Center to start a project (now with perhaps half a dozen researchers) to examine molecular manufacturing systems and molecular machines using computational models.*

Beyond inexpensively providing remarkably light and strong materials for space ships, nanotechnology will also provide extremely powerful computers with which to guide both those ships and a wide range of other activities in space.

Today, computer chips are made using lithography — literally "stone writing." It's our finest manufacturing technology, but there seem to be fundamental limits in how much further we can improve it. In lithography, we draw fine lines on a silicon wafer by using methods borrowed from photography. A light-sensitive film — called a "resist" — is spread over the silicon wafer. The resist is exposed to a complex pattern of light and dark, like a negative in a camera, and developed. The exposed resist is then washed away, and a particular chemical is sprayed over the surface. Where the resist has been washed away, the chemical reaches the surface of the silicon and can diffuse a short distance into it. Where the resist has not been washed away, the spray is blocked. Finally, the remaining resist is washed away, along with any chemicals that it prevented from reaching the silicon surface. The result: a fine pattern of some desired chemical is laid out on the silicon surface. By repeating this process, an intricate set of interlocking patterns can be made that defines the complex logic elements of a modern computer chip.

If the computer hardware revolution is to continue at its current pace, in a decade or so we'll have to move beyond lithography to some new post lithographic manufacturing technology. Making patterns on a resist and spraying chemicals around simply can't arrange atoms with the ultimate precision that should be feasible. More precise methods will be needed. Ultimately, each logic element will be made from just a few atoms. Designs for computer gates with less than 1,000 atoms have already been proposed — but each atom in such a small device has to be in exactly the right place. To economically build and interconnect trillions upon trillions of such small and precise devices in a complex three dimensional pattern we'll need a manufacturing technology well beyond today's lithography: we'll need nanotechnology. With it, we should be able to build mass storage devices that can store more than a hundred billion billion bytes in a volume the size of a sugar cube; RAM that can store a mere billion billion bytes in such a volume; and massively parallel computers of the same size that can deliver a billion billion instructions per second.

Today, "smart" weapons are fairly big — we have the "smart bomb" but not the "smart bullet." In the future, even weapons as small as a single bullet could pack more computer power than the largest supercomputer in existence today, allowing them to perform realtime image analysis of their surroundings and communicate with weapons tracking systems to acquire and navigate to targets with greater precision and control. We'll also be able to build weapons both inexpensively *and much more rapidly*, at the same time taking full advantage of the remarkable materials properties of diamond. Rapid and inexpensive manufacture of great quantities of stronger more precise weapons guided by massively increased computational power will alter the way we fight wars. Changes of this magnitude could destabilize existing power structures in unpredictable ways.

Military applications of nanotechnology raise a number of concerns that prudence suggests we begin to investigate before, rather than after, we develop this new technology. While molecular manufacturing will not arrive for many years, its obvious military potential will increasingly attract the interest of strategic planners. For example, in a talk titled *Nanotechnology and global security* at the *Fourth Foresight Conference on Molecular Nanotechnology*, *Admiral David E. Jeremiah, USN (Ret)*, former Vice Chairman of the Joint Chiefs of Staff, said "Military applications of molecular manufacturing have even greater potential than nuclear weapons to radically change the balance of power." As it seems implausible that military applications of this technology will never be developed and deployed, it would seem safer to encourage the relatively early interest of those organizations less prone to the abuse of power and more likely to curb its abuse by others.

To make power today we dig coal and oil from the ground, we dam rivers, and we burn nuclear fuel in nuclear power plants. Yet the sunshine all around us could provide orders of magnitude more power than we use — and do so more cleanly and less expensively — if only we could make low cost solar cells and batteries. We already know how to make solar cells that are efficient enough for this application: it just costs too much to make them. Nanotechnology will cut costs both of the solar cells and the equipment needed to deploy them, making solar power economical. In this application we need not make new or technically superior solar cells: making inexpensively what we already know how to make expensively would move solar power into the mainstream.

Today, our surgical tools are large and crude at the molecular scale — yet the cellular and molecular machinery in our tissue is small and precise. There is a fundamental mismatch between the capabilities of our tools and what's needed to cure the injuries in our tissue. Today's scalpels are, as seen by a cell, large crude scythes that are more suited to cut and tear than to heal and mend. Ripping through tissue, they leave dead and maimed cells in their wake. The only reason that modern surgery works is the remarkable ability of cells to regroup, bury their dead, and heal over the wound.

It is not modern medicine that does the healing, but the cells themselves: we are but onlookers. If we had surgical tools that were molecular both in their size and precision, we could develop a medical technology that for the first time would let us directly heal the injuries at the molecular and cellular level that are the root causes of disease and ill health. With the precision of drugs combined with the intelligent guidance of the surgeon's scalpel, *we can expect a quantum leap in our medical capabilities*.

Nanotechnology should let us make almost every manufactured product faster, lighter, stronger, smarter, safer and cleaner. We can already see many of the possibilities as these few examples illustrate. New products that solve new problems in new ways are more difficult to foresee, yet their impact is likely to be even greater. Could Edison have foreseen the computer, or Newton the communications satellite?

Summary of the core technology

We can now see the fundamental shape of a molecular manufacturing technology. Self replicating assemblers, operating under computer control, let us inexpensively build more assemblers. The assemblers can be reprogrammed to build other products. The assemblers use programmable positional control to position molecular tools and molecular components, permitting the inexpensive fabrication of most structures consistent with physical law. Diamondoid materials in particular become inexpensive and commonplace, and their remarkable properties usher in what has been called the Diamond Age.

Developmental pathways

In the preceding discussion of nanotechnology we gave scant attention to possible developmental pathways that would let us migrate from our current technology to what we might call "mature" systems able to inexpensively manufacture most diamondoid structures. The proposals advanced so far have been driven largely by the desired goal: a system able to inexpensively synthesize most diamondoid structures. The proposal that emerged was a general purpose programmable manufacturing system which uses positionally controlled highly reactive tools in vacuum and is able to self replicate.

No constraint was imposed requiring that the proposed system be easy to fabricate given our current technology — and indeed, this problem appears non trivial. But we must eventually build such systems if they are to be of any use — how can we do this? To give the reader some feeling for the magnitude of the effort involved, we can compare the requirements for a mature system with present capabilities.

A mature system must be able to build an assembler with hundreds of millions or billions of atoms with no atom out of place. If we are to do this, and if each unit operation typically handles one or a few atoms (a hydrogen abstraction, a carbene insertion, etc), then either the error rate per unit operation must be low (less than one in a billion) or we must adopt error detection and correction methods.

A general rule of thumb is that making something right the first time is easier than making it wrong and fixing it. Given the choice between a manufacturing process that has a low enough error rate that most finished products will simply work correctly the first time versus a process which requires elaborate error detection and correction to make any working products at all, the former is greatly to be preferred. Because analysis of the fundamental causes of errors during the manufacture of diamondoid products in general and assemblers in particular supports the idea that error rates substantially lower than one in a billion should be achievable, and hence that an assembler can make another assembler with a high probability of success while using no error detection or correction, existing design proposals adopt the simpler and higher performance approach. Had the opposite conclusion been reached, i.e., that error rates could *not* be driven below one in one billion, then design proposals would have adopted a smaller module size (consistent with the error rate) and incorporated appropriate mechanisms for module testing, the rejection of bad modules, and the assembly of working modules into larger systems. This line of logic results in proposals for assemblers in which error rates are very low. These low error rates, however, are predicated on the prior existence of diamondoid assemblers. Unfortunately, we do not have assemblers today. Worse, current technology would be hard pressed to achieve such error rates. Chemists view a synthesis that provides 99% yield as very good. The synthesis of proteins from amino acids by ribosomes has an error rate of perhaps one in 10,000. Replication of DNA, by using extensive error detection and correction, along with built in redundancy (DNA has two complementary strands), achieves an error rate of roughly one base in a billion (which varies depending on particular circumstances).

The size of structure that can be built by today's SPMs without error is smaller than has been achieved by chemists. The first such work *arranged 35 xenon atoms on a nickel surface at 4 kelvins in vacuum to spell out "IBM."* Later work in Japan spelled out "NANO SPACE" at room temperature in vacuum by removing individual sulfur atoms from the surface. More recently, *6 molecules were arranged in a hexagonal pattern at room temperature in vacuum*. As each molecule had 173 atoms, this modular approach was able to position over 1,000 atoms. While these are great successes we have not yet seen a long sentence spelled out in this fashion, let alone a paragraph or a book. Further, SPMs can "see" the structure that's being built. Current successes use this ability to provide constant feedback to the human operator, who can detect and correct errors.

In short, today's SPMs can build structures that are only a small fraction of the size of proposed assemblers and have error rates high enough that they must use relatively sophisticated error detection and correction methods.

The situation is made more difficult by the additional requirement that unit operations be fast. If an assembler is to manufacture a copy of itself in about a day, and if this takes one hundred million to a billion operations, then each unit operation must take place in a fraction of a millisecond. Ribcsomes take tens of milliseconds to add a single amino acid to a growing protein. Today's SPMs can take hours to arrange a few atoms or molecules.

At the same time, the use of vacuum prevents us from using self assembly or any of the other solution-based techniques developed by chemistry. Abandoning these powerful tools does not seem like a good first step.

While speed, reliability and operation in vacuum do not pose fundamental problems, the conclusion is obvious: directly building a diamondoid assembler using existing technology is a daunting task. Perhaps, rather than attempting to solve all the problems in a single giant leap, we should instead approach them in a more incremental fashion. To this end, we need to define one or more intermediate systems. Almost by definition, an intermediate system is easier to build from current technology, but is still able to make systems that are more powerful than itself. A series of intermediate systems, each able to build the next system in the chain, would let us adopt more incremental methods.

One approach is to eliminate the requirement that the assembler be made from diamondoid structures. We wanted to make diamondoid structures because of their remarkable strength, stiffness, electrical properties, etc. But an intermediate system need only be able to make a more advanced system, and perhaps products that are impressive in comparison with today's products. It doesn't have to be diamondoid itself.

This suggests what might be called "building block based nanotechnology." Rather than building diamond, we'll build some other material from molecular building blocks that are relatively large: tens, hundreds or even thousands of atoms in size. Larger building blocks reduce the number of assembly steps, so fewer unit operations are needed and they need not be as reliable. Soluble building blocks, with "linkage" groups that are selective (the building blocks only stick to other building blocks, not to the solvent or low concentrations of contaminants) eliminate the need for vacuum. We certainly have many choices: any of the wide range of molecules that chemists have synthesized or could reasonably synthesize which has the desired properties. *Krummenacker* concluded that each molecular building block should have at least three sites where it can link to other building blocks (two sites leads to the familiar polymers so ubiquitous in biological systems: proteins, DNA, RNA, etc. Addition of a third site makes the design of three dimensional stiff structures much easier). While such building blocks could be linked to each other using any one of a variety of reactions, a particularly attractive possibility is the Diels-Alder reaction. This reaction, well known to chemists, works in most solvents (or even in vacuum). It involves a reaction between a diene and a dieneophile. The reaction is specific: the diene and the dieneophile react with each other but seldom with other groups (see illustration above, courtesy of Krummenacker). Because it works in vacuum and doesn't produce any small molecules (which would destroy the vacuum), it could be used both in intermediate systems in solution, and later in systems that use vacuum to make diamondoid materials. Solution based systems could use positional control to assemble the building blocks, but can also use the methods of self assembly. In particular, the self assembly of a positional device should be feasible. If you ask someone familiar with the field if it would be feasible to self assemble a robotic arm, the usual response is some variation of "no." If, instead, you ask whether it would be feasible to self assemble an octahedron, the usual response is to point out that *Nadrian Seeman* of New York University has already self assembled a *truncated octahedron from DNA* and plans to do the same for a regular octahedron (both for reasons unrelated to positional control). (The computer generated graphic of the truncated octahedron at right is courtesy of Nadrian Seeman). For this work he received *The 1995 Feynman Prize in nanotechnology*.

As we have seen, an octahedron is the basic structure required for the Stewart platform. The truncated octahedron, because it has sides that aren't triangular, is floppier than a regular octahedron. Even the regular octahedron is likely to be too floppy for this application because of the limited stiffness of DNA. Further, the length of the edges cannot be changed (a requirement if we wish to control the position of the platform with respect to the base). However, self assembling a Stewart platform able to assemble molecular building blocks seems much less difficult than directly building a diamondoid assembler able to synthesize diamond in vacuum. Several ways to deal with the problem of inadequate stiffness and adjustable length edges are possible. One way would be to attach selectively sticky strands of DNA to a stiff molecular structure. This would let us use DNA (which has been intensively studied and whose selective stickiness is well known) to guide the process of self assembly, while the stiff molecular structure would would let us deal with concerns about inadequate stiffness. To change the length of an edge, we would need to use molecular structures that change shape in response to light, pressure, temperature, chemicals, or some other external signal. Many such molecular structures are known. Self assembling a Stewart platform whose edges are stiff enough to make a useful positional device, and the length of whose edges can be changed by a suitable signaling mechanism, is no longer a challenge that seems beyond our reach.

A second approach for positioning molecular building blocks in solution is to use an Atomic Force Microscope (AFM). This type of SPM relies on "touch" to create an image. By pushing on the structure being scanned, and feeling how hard it pushes back, the AFM can build up an image of stiff structures (it doesn't work very well when the applied force is strong enough to deform the structure). Because the AFM touches the surface of the structure being probed, it can also change that surface. To do this, it's very useful if the precise molecular structure of the tip can be controlled, so that the precise nature of the tip-surface interaction can be well defined.

This line of reasoning leads to the "molecular manipulator: " an AFM with a reactive fragment of an *antibody* bound to its tip. By changing the *antibody*, we can change what type of molecule will stick to the AFM tip. If the antibody both sticks selectively to a particular molecular building block, and also is bound to the AFM tip, we can now position the molecular building block by positioning the AFM tip. We can use any one of a wide range of possible molecular building blocks (*antibodies* can today be created that will bind to most small molecules). Again, this no longer seems beyond our reach.

Another developmental pathway is defined by the desire of the semiconductor industry to make ever smaller transistors despite the fact that optical lithography — the current workhorse manufacturing method for making computer chips — will reach a limit in a few more years. Transistors are made today by "drawing" very fine lines on silicon — but the optical methods in use today are limited by the wavelength of light: a few hundred nanometers. SPMs could let us make much smaller circuits by drawing much finer lines. Already demonstrated for making the finest and most critical lines in a transistor, if this approach can be made low cost and reliable it would let the entire semiconductor industry be retooled to use these finer lines to make more and smaller transistors. This, by itself, is not molecular nanotechnology; but once we learn to draw the smallest and finest lines of all, lines where every atom is in the right place (and there will be strong economic incentives to move in this direction), we could use the same techniques to make molecular machines. At first, these molecular machines would be very expensive and useful only for applications where their remarkable precision would justify the cost — but they should eventually lead to machines sophisticated enough to make simple assemblers.

Whether through self-assembly, by improvements in SPMs, some hybrid approach, or perhaps by some other path; we are moving from an era of expensive and imprecise products to an era of inexpensive products of molecular precision. We are going to replace most of the manufacturing base of the world with a fundamentally new and *much* better manufacturing technology.

The critics

Arguments that nanotechnology is infeasible are relatively rare and uniformly poor in technical quality. The best known are the arguments of David Jones, a *Nature* columnist and chemist. Quoted extensively in a *Scientific American* article on nanotechnology (see *http: / / www.foresight.org / SciAmDebate / SciAmOverview.html* for commentary) he advanced arguments like the following: "Single atoms ... are amazingly mobile and reactive. They will combine instantly with ambient air, water, each other, the fluid supporting the assemblers, or the assemblers themselves." However, as the proposals involving reactive molecular tools specify that the environment should be inert (i.e., vacuum) there is no "ambient air" to react with. As the molecular tools are positionally controlled, they will not react with each other or the assembler itself for the same reason that a hot soldering iron does not react with the skin of the person doing the soldering. Jones' criticisms raise greater questions about his understanding of the field than about the feasibility of nanotechnology. Other arguments against the feasibility of nanotechnology have likewise had obvious flaws, reminiscent of the 1920 New York Times editorial which said that rockets to the moon were impossible because there was no air to push against in space.

As should be clear from this article there are major technical challenges that must be overcome if we are to develop molecular manufacturing systems able to synthesize diamondoid structures. The serious questions, however, are about which development pathways should be pursued and how long it will take.

How long?

The single most frequently asked question about nanotechnology is: *how long?*. How long before it will let us make molecular computers? How long before inexpensive solar cells let us use clean solar power instead of oil, coal, and nuclear fuel? How long before we can explore space at a reasonable cost?

The scientifically correct answer is: *I don't know.*

Having said that, it is worth pointing out that the trends in the development of computer hardware have been remarkably steady for the last 50 years. Such parameters as

- the number of atoms required to store one bit
- the size of a transistor
- the energy dissipated by a single logic operation

- the resolution of the finest machining technology
- the cost of a computer gate

have all declined with great regularity, even as the underlying technology has changed dramatically. *From relays to vacuum tubes to transistors to integrated circuits to Very Large Scale Integrated circuits (VLSI) we have seen steady declines in the size and cost of logic elements and steady increases in their performance.*

Extrapolation of these trends suggests we will have to develop molecular manufacturing in the 2010 to 2020 time frame if we are to keep the computer hardware revolution on schedule.

Of course, extrapolating past trends is a philosophically debatable method of technology forecasting. While no fundamental law of nature prevents us from developing nanotechnology on this schedule (or even faster), there is equally no law that says this schedule will not slip.

Much worse, though, is that such trends imply that there is some ordained schedule — that nanotechnology will appear regardless of what we do or don't do. Nothing could be further from the truth. How long it takes to develop this technology depends very much on what we do. If we pursue it systematically, it will happen sooner. If we ignore it, or simply hope that someone will stumble over it, it will take much longer. And by using theoretical, computational and experimental approaches together, we can reach the goal more quickly and reliably than by using any single approach alone.

While some advances are made through serendipitous accidents or a flash of insight, others require more work. It seems unlikely that a scientist would forget to turn off the bunsen burner in his lab one afternoon and return to find he'd accidentally made a Space Shuttle.

Like the first human landing on the moon, the Manhatten project or the development of the modern computer the development of molecular manufacturing will require the coordinated efforts of many people for many years. How long will it take? A lot depends on when we start.

Engines of Healing: The Coming Era of Nanotechnology

We will use *molecular technology* to bring health because the human body is made of molecules. The ill, the old, and the injured all suffer from mis-arranged patterns of *atoms*, whether mis-arranged by invading viruses, passing time, or swerving cars. Devices able to rearrange atoms will be able to set them right. *Nanotechnology* will bring a fundamental breakthrough in medicine.

Physicians now rely chiefly on surgery and drugs to treat illness. Surgeons have advanced from stitching wounds and amputating limbs to repairing hearts and reattaching limbs. Using microscopes and fine tools, they join delicate blood vessels and nerves. Yet even the best micro-surgeon cannot cut and stitch finer tissue structures. Modern scalpels and sutures are simply too coarse for repairing *capillaries*, *cells*, and molecules. Consider "delicate" surgery from a cell's perspective: a huge blade sweeps down, chopping blindly past and through the molecular machinery of a crowd of cells, slaughtering thousands. Later, a great obelisk plunges through the divided crowd, dragging a cable as wide as a freight train behind it to rope the crowd together again. From a cell's perspective, even the most delicate surgery, performed with exquisite knives and great skill, is still a butcher job. Only the ability of cells to abandon their dead, regroup, and multiply makes healing possible.

Yet as many paralyzed accident victims know too well, not all tissues heal. Drug therapy, unlike surgery, deals with the finest structures in cells. Drug molecules are simple molecular devices. Many affect specific molecules in cells. Morphine molecules, for example, bind to certain receptor molecules in brain cells, affecting the neural impulses that signal pain. Insulin, beta blockers, and other drugs fit other receptors. But drug molecules work without direction. Once dumped into the body, they tumble and bump around in solution haphazardly until they bump a target *molecule*, fit, and stick, affecting its function.

Surgeons can see problems and plan actions, but they wield crude tools; drug molecules affect tissues at the molecular level, but they are too simple to sense, plan, and act. But molecular machines directed by nanocomputers will offer physicians another choice. They will combine sensors, programs, and molecular tools to form systems able to examine and repair the ultimate components of individual cells. They will bring surgical control to the molecular domain.

These advanced molecular devices will be years in arriving, but researchers motivated by medical needs are already studying molecular machines and molecular engineering. The best drugs affect specific molecular machines in specific ways. Penicillin, for example, kills certain *bacteria* by jamming the nanomachinery they use to build their cell walls, yet it has little effect on human cells.

Biochemists study molecular machines both to learn how to build them and to learn how to wreck them. Around the world (and especially the Third World) a disgusting variety of viruses, bacteria, protozoa, fungi, and worms parasitize human flesh. Like penicillin, safe, effective drugs for these diseases would jam the parasite's molecular machinery while leaving human molecular machinery unharmed. Dr. Seymour Cohen, professor of pharmacological *science* at SUNY (Stony Brook, New York), *argues that biochemists should systematically study the molecular machinery of these parasites*. Once biochemists have determined the shape and function of a vital protein machine, they then could often design a molecule shaped to jam it and ruin it. Such drugs could free humanity from such ancient horrors as schistosomiasis and leprosy, and from new ones such as AIDS.

Drug companies are already redesigning molecules based on knowledge of how they work. Researchers at Upjohn Company have *designed and made modified molecules of vasopressin*, a hormone that consists of a short chain of *amino acids*. Vasopressin increases the work done by the heart and decreases the rate at which the kidneys produce urine; this increases blood pressure. The researchers designed modified vasopressin molecules that affected receptor molecules in the kidney more than those in the heart, giving them more specific and controllable medical effects. More recently, they designed a modified vasopressin molecule that binds to the kidney's receptor molecules without direct effect, thus blocking and *inhibiting* the action of natural vasopressin.

Medical needs will push this work forward, encouraging researchers to take further steps toward protein design and molecular engineering. Medical, military, and economic pressures all push us in the same direction. Even before the *assembler* breakthrough, molecular technology will bring impressive advances in medicine; trends in biotechnology guarantee it. Still, these advances will generally be piecemeal and hard to predict, each exploiting some detail of biochemistry. Later, when we apply assemblers and technical AI systems to medicine, we will gain broader abilities that are easier to foresee.

To understand these abilities, consider cells and their self-repair mechanisms. In the cells of your body, natural radiation and noxious chemicals split molecules, producing reactive molecular fragments. These can misbond to other molecules in a process called *cross-linking*. As bullets and blobs of glue would damage a machine, so radiation and reactive fragments damage cells, both breaking molecular machines and gumming them up.

If your cells could not repair themselves, damage would rapidly kill them or make them run amok by damaging their control systems. But *evolution* has favored organisms with machinery able to do something about this problem. The self-replicating factory system sketched in this encyclopaedia repaired itself by replacing damaged parts; cells do the same. So long as a cell's *DNA* remains intact, it can make error-free tapes that direct ribosomes to assemble new protein machines.

Unfortunately for us, DNA itself becomes damaged, resulting in mutations. Repair *enzymes* compensate somewhat by detecting and repairing certain kinds of damage to DNA. These repairs help cells survive, but existing repair mechanisms are too simple to correct all problems, either in DNA or elsewhere. Errors mount, contributing to the aging and death of cells—and of people.

Life, Mind, and Machines

Does it make sense to describe cells as "machinery," whether self-repairing or not? Since we are made of cells, this might seem to reduce human beings to "mere machines," conflicting with a holistic understanding of life.

But *a dictionary definition of holism* is "the theory that reality is made up of organic or unified wholes that are greater than the simple sum of their parts." This certainly applies to people: one simpler sum of our parts would resemble hamburger, lacking both mind and life. The human body includes some ten thousand billion *billion* protein parts, and no machine so complex deserves the label—mere." Any brief description of so complex a system cannot avoid being grossly incomplete, yet at the cellular level a description in terms of machinery makes sense. Molecules have simple moving parts, and many act like familiar types of machinery. Cells considered as a whole may seem less mechanical, yet biologists find it useful to describe them in terms of molecular machinery.

Biochemists have unraveled what were once the central mysteries of life, and have begun to fill in the details. They have traced how molecular machines break food molecules into their building blocks and then reassemble these parts to build and renew tissue. Many details of the structure of human cells remain unknown (single cells have billions of large molecules of thousands of different kinds), but biochemists have mapped every part of some viruses. Biochemical laboratories often sport a large wall chart showing how the chief molecular building blocks flow through bacteria. Biochemists understand much of the process of life in detail, and what they don't understand seems to operate on the same principles. The mystery of heredity has become the industry of genetic engineering. Even embryonic development and memory are being explained in terms of changes in biochemistry and cell structure.

In recent decades, the very quality of our remaining ignorance has changed. Once, biologists looked at the process of life and asked, "How can this be?" But today they understand the general principles of life, and when they study a specific living process they commonly ask, "Of the many ways this could be, which has nature chosen?" In many instances their studies have narrowed the competing explanations to a field of one. Certain biological processes—the coordination of cells to form growing embryos, learning brains, and reacting immune systems—still present a real challenge to the imagination. Yet this is not because of some deep mystery about how their parts work, but because of the immense complexity of how their many parts interact to form a whole.

Cells obey the same natural laws that describe the rest of the world. Protein machines in the right molecular environment will work whether they remain in a functioning cell or whether the rest of the cell was ground up and washed away days before. Molecular machines know nothing of "life" and "death."

Biologists—when they bother—sometimes define life as the ability to grow, replicate, and respond to stimuli. But by this standard, a mindless system of replicating factories might qualify as life, while a conscious *artificial intelligence* modeled on the human brain might not. Are viruses alive, or are they "merely" fancy molecular machines? No experiment can tell, because nature draws no line between living and nonliving.

Biologists who work with viruses instead ask about viability: "Will this *virus* function, if given a chance?" The labels of "life" and "death" in medicine depend on medical capabilities: physicians ask, "Will this patient function, if we do our best?" Physicians once declared patients dead when the heart stopped; they now declare patients dead when they despair of restoring brain activity. Advances in cardiac medicine changed the definition once; advances in brain medicine will change it again.

Just as some people feel uncomfortable with the idea of machines thinking, so some feel uncomfortable with the idea that machines underlie our own thinking. The word "machine" again seems to conjure up the wrong image, a picture of gross, clanking metal, rather than signals flickering through a shifting weave of neural fibers, through a living tapestry more intricate than the mind it embodies can fully comprehend. The brain's really machinelike machines are of molecular size, smaller than the finest fibers.

A whole need not resemble its parts. A solid lump scarcely resembles a dancing fountain, yet a collection of solid, lumpy molecules forms fluid water. In a similar way, billions of molecular machines make up neural fibers and synapses, thousands of fibers and synapses make up a neural cell, billions of neural cells make up the brain, and the brain itself embodies the fluidity of thought.

To say that the mind is "just molecular machines" is like saying that the Mona Lisa is "just dabs of paint." Such statements confuse the parts with the whole, and confuse matter with the pattern it embodies. We are no less human for being made of molecules.

Protein Design as a Pathway to Molecular Manufacturing

Feynman's 1959 talk entitled "There's Plenty of Room at the Bottom" discussed microtechnology as a frontier to be pushed back, like the frontiers of high pressure, low temperature, or high vacuum. He suggested that ordinary machines could build smaller machines that could build still smaller machines, working step by step down toward the molecular level; he also suggested using particle beams to define two-dimensional patterns. Present microtechnology (exemplified by integrated circuits) has realized some of the potential outlined by Feynman by following the same basic approach: working down from the macroscopic level to the microscopic.

Present microtechnology handles statistical populations of atoms. As the devices shrink, the atomic graininess of matter creates irregularities and imperfections, so long as atoms are handled in bulk, rather than individually. Indeed, such miniaturization of bulk processes seems unable to reach the ultimate level of microtechnology — the structuring of matter to complex atomic specifications. In this paper, I will outline a path to this goal, a general molecular engineering technology. The existence of this path will be shown to have implications for the present.

Although the capabilities described may not prove necessary to the achievement of any particular objective, they will prove sufficient for the achievement of an extraordinary range of objectives in which the structuring and analysis of matter are concerned. The claim that devices can be built to complex atomic specifications should not, however, be construed to deny the inevitability of a finite error rate arising from thermodynamic effects (and radiation damage). Such errors can be minimized through the use of free energy in error-correcting procedures (including rejection of faulty components before device assembly); the effects of errors can be minimized through fault-tolerant design, as in macroscopic engineering. The emphasis on devices that have general capabilities should be taken in the spirit of early work on the theoretical capabilities of computers, which did not attempt to predict such practical embodiments as specialized or distributed computation systems. The present argument, however, will proceed from step to step by close analogies between the proposed steps and past developments in nature and technology, rather than by mathematical proof. We commonly accept the feasibility of new devices without formal proof, where analogies to existing systems are close enough: consider the feasibility of making a clock from zirconium. The detailed design of many specific devices to render them describable by dynamical equations would be a task of another order (consider designing a clock from scratch) and appears unnecessary to the establishment of the feasibility of certain general capabilities.

Protein design

Biochemical systems exhibit a "microtechnology" quite different from ours: they are not built down from the macroscopic level but up from the atomic. Biochemical microtechnology provides a beachhead at the molecular level from which to develop new molecular systems by providing a variety of "tools" and "devices" to use and to copy. Building with these tools, themselves made to atomic specifications, we can begin on the far side of the barrier facing conventional microtechnology. What can be built with these tools? Gene synthesis and recombinant DNA technology can direct the ribosomal machinery of bacteria to produce novel proteins, which can serve as components of larger molecular structures. One might think assembly of such components into complex systems would require a preexisting technology able to handle molecules and assemble them; fortunately, biochemistry demonstrates that intermolecular attraction between complementary surfaces can assemble complex structures from solution. For example, the complex machinery of the ribosome self-assembles from more than 50 different protein molecules and can do so in vitro.

At present, the design of protein systems as complex as a ribosome seems an awesome task. Indeed, chemists cannot yet predict the three-dimensional conformation of a natural protein from its amino acid sequence, an ability that might seem requisite to the design of new proteins. Two considerations suggest that this obstacle is surmountable: first, the continuing improvement in protein science and, second, the difference between natural science and design engineering.

Regarding the first, computer simulation of protein molecules in solution (5) shows promise. As computer technology and chemical knowledge improve, simulations will increase in accuracy, speed, and size. Improvement promises new insight into protein behavior and may permit the designer to modify (simulated) molecules quickly and to observe their behavior directly.

Regarding the second consideration, natural scientists seek a more general understanding than design engineers require. Science seeks the ability to predict the conformations of all natural polypeptides. In attempting this, protein chemists can search for a minimum-energy chain conformation (in hope that the protein assumes not a local but a global minimum-energy conformation) (6) or can attempt to follow the chain-folding mechanism to find the final conformation (7). Prediction will be easier if the natural conformation has outstanding stability or if its folding mechanism proceeds in a sequence of strongly preferred steps. Unfortunately, natural selection accepts polypeptides that have natural conformations of low stability (in energetic terms) so long as they exhibit long lifetimes on the cellular time scale (or renature readily). Similarly, natural selection accepts any folding process so long as the chain reaches its natural conformation with essentially 100% yield. Moreover, random mutations are unlikely to enhance the stability of a particular conformation (or the predictability of its folding mechanism). Thus, natural proteins tend to accumulate disruptive changes until they reach the threshold of poor stability or reduced yield of the natural conformation; only then does natural selection come into play. Thus, it is little wonder that chemists cannot yet predict the conformations of natural proteins; they are not designed to fold predictably. Engineers (in contrast to scientists) need not seek to understand all proteins but only enough to produce useful systems in a reasonable number of attempts. An engineer designing a protein that has 1000 amino acids may choose among some 10^{1300} different amino acid sequences. It might be that only one in 10^9 (or even 10^{700}) randomly selected sequences would yield a predictable conformation, yet this tiny fraction represents a vast number of proteins. Through use of strategically placed charged groups, polar groups, disulfide bonds, hydrogen bonds, and hydrophobic groups, the engineer should be able to design proteins that not only fold predictably to a stable structure (sometimes) but that serve a planned function as well. Even a low success rate will lead to an accumulation of successful designs. Thus, the difficulties encountered in predicting the conformations of natural proteins do not seem insurmountable obstacles to protein engineering. Computer modeling and chemical understanding of biclogical targets have already found use in pharmaceutical design (8), and an artificial 34-residue polypeptide designed to interact with RNA has been synthesized and found active (9). It has been proposed to give microcircuitry special sensitivities by adsorbing engineered proteins onto selected surfaces (10). The promise of enzyme design in chemical engineering is evident. As protein science has great promise and difficulties in understanding natural proteins need not block engineering, the substantial payoffs for improved capabilities should lead to development of protein design technology. It would be foolish to underestimate the time and effort that will be required to develop basic design capabilities and then a broad family of working molecular devices; still, the path seems clear to achieving the capabilities exhibited by existing biochemical systems, by copying their features if need be.

Molecular Machinery

A comparison f biochemical to macroscopic components will show the possibilities of the former by analogy to the latter. With structural members, moving parts, bearings, and motive power, versatile mechanical systems can be built. Molecular assemblages of atoms can act as solid objects, occupying space and holding a definite shape. Thus, they can act as structural members and moving parts. Sigma bonds that have low steric hindrance can serve as rotary bearings able to support ~ 10^{-9} N. A line of sigma bonds can serve as a hinge. Conformation-changing proteins (such as myosin) can serve as sources of motive power for linear motion; the reversible motor of the bacterial flagellum can serve as a source of motive power for rotary motion. The existence of this range of components in nature indicates that power-driven mechanical systems can be constructed on a molecular scale. By analogy with macroscopic devices, feasible molecular machines presumably include manipulators able to wield a variety of tools. Thermal vibrations in typical structures are a modest fraction of interatomic distances; thus, such tools can be positioned with atomic precision. As present microtechnology can lay down conductors on a molecular scale (10 nm) and molecular devices can respond to electric potentials (through conformation changes, etc.), such devices can be controlled by human operators or macroscopic machines. Further, by analogy with biological sensors, molecular scale instruments can evidently produce macroscopic signals, indicating the feasibility of feedback control in molecular manipulations.

Together, these arguments indicate the feasibility of devices able to move molecular objects, position them with atomic precision, apply forces to them to effect a change, and inspect them to verify that the change has indeed been accomplished. It would be foolish to minimize the time and effort that will be required to develop the needed components and assemble them into such complex and versatile systems. Still, given the components, the path seems clear.

Ordinary chemical synthesis relies on thermal agitation to bring reactant molecules in solution together in the correct orientation and with sufficient energy to cause the desired reaction. Enzyme-like molecular machines can hold reactants in the best relative positions as bonds are strained or polarized. Like some enzymes, they can do work on reactant molecules to drive reactions not otherwise thermodynamically favored.

These are clearly techniques of great power, yet the synthetic capabilities of systems based on polypeptide chains might seem limited by amino acid properties. However, enzymes show that other molecular structures bound to the polypeptide (such as metal ions and complex ring structures) (11) can extend protein capabilities.

The range of such tools is large and greater than found in nature. Thus, the synthetic capabilities of enzymes set only a lower bound on the capabilities of engineered protein systems. Indeed, as tool-wielding protein systems can control the chemical environment of a reaction site completely, they should be able, at a minimum, to duplicate the full range of moderate-temperature synthetic steps achieved by organic chemists. Further, where chemists must resort to complex strategies to make or break specific bonds in large molecules, molecular machines can select individual bonds on the basis of position alone. Conventional organic chemistry can synthesize not only one-, two-, and three-dimensional covalent structures but also exotic strained and fused rings. With the addition of controlled site-specific synthetic reactions, a broad range of large complex structures can doubtless be built.

Still, the synthetic abilities of protein machines will be limited by their need for a moderate temperature aqueous environment (although applied forces can sometimes replace or exceed thermal agitation as a source of activation energy and reaction sites and reactive groups can be protected from the surrounding water, as in some enzymatic active sites). These limits may be sidestepped by using the broad synthetic capabilities outlined above to build a second generation of molecular machinery whose components would not be coiled hydrated polypeptide chains but compact structures having three-dimensional covalent bonding. There is no reason why such machines cannot be designed to operate at reduced pressure or extreme temperatures; synthesis can then involve highly reactive or even free radical intermediates, as well as the use of mechanical arms wielding molecular tools to strain and polarize existing bonds while new molecular groups are positioned and forced into place. This may be done at high or low temperature as desired. The class of structures that can be synthesized by such methods is clearly very large, and one may speculate that it includes most structures that might be of technological interest.

Firmness of the argument

The development path described above should lead to advanced molecular machinery capable of general synthesis operations. As the results of this path can be shown to have consequences for the present, it is of interest to discuss the degree of confidence that should be placed in its feasibility.

It might be argued that complex protein or nonprotein machines are impossible or useless, on the grounds that, if they were possible and useful, organisms would be using them. A similar argument would, however, conclude that bone is a better structural material than graphite composite, that neurons can transmit signals faster than wires, and that technology based on the wheel is impossible or useless. Nature has been constrained less by what is physically possible than by what could be evolved in small steps. Thus, the absence of a proposed kind of molecular machinery in

To deny the feasibility of advanced molecular machinery, one must apparently maintain either (i) that design of proteins will remain infeasible indefinitely, or (ii) that complex machines cannot be made of proteins, or (iii) that protein machines cannot build second-generation machines.

In light of the expected improvements in computation, the simplified task of design engineers (compared with scientists), the possibilities offered by sheer trial-and-error modification of natural proteins, and the progress already made in protein design, the first seems difficult to maintain. Further, even if protein design were to prove intractible (because of difficulties in predicting conformations), this would in no way preclude developing an alternative polymer system with predictable coiling and using it as a basis for further development. In light of the presence of the needed components for mechanical devices in the cell, the second scems difficult to maintain. Indeed, the cytoskeleton provides a fair counterexample. In light of the results of synthetic organic chemistry and the ability of molecular machines to make reactions site specific, it seems difficult to maintain that nonprotein machine components cannot be built and assembled.

Each of the development steps outlined above seems closely analogous to past steps taken by nature or by technology. Each of these steps can be accomplished in many ways. To argue their infeasibility would seem to require some general principle precluding success, and it is difficult to see what such a principle might be like. Thus, the claim that advanced molecular technology can be developed seems well founded.

Although the existence of molecular machinery in cells indicates the feasibility of some sort of artificial molecular machinery, errors in assembly might limit the synthesis of structures of great complexity. In the cell, molecular machinery uses DNA to direct the assembly of DNA and other molecules. In some eukaryotic cells, DNA directs DNA synthesis with an error rate of ~ 10^{-11} per nucleotide added (12). As engineers commonly design systems to function reliably with many more failed components than 1 in 10^{11}, such an error rate seems no barrier to the construction of quite complex devices.

The possibility of low error rates is not surprising. For synthesis systems permitting error detection and correction (such as DNA synthesis), the net error rate in assembly can be reduced to roughly the product of the raw error rate in assembly and the rate at which errors are falsely identified as correct. As no uncertainty principle prohibits accurate discrimination between objects of different kinds (such as correctly and incorrectly assembled molecular structures), no limits to the detection and correction of errors are apparent.

Applications to computation

Molecular technology has obvious application to the storage and processing of information. A crude approach would involve literal "molecular machinery" patterned on the Babbage machine. In a more subtle approach, bits could be represented by protons, bound electrons, reactive groups, or conformation changes and transferred by movement of protons or of well-localized electrons (13), excitons, or phonons. The range of plausible device speeds is suggested by the 10^{-6} -sec turnover time for a fast enzyme, by the 10^{-13} -sec scale of collisional interactions (11), and by the 10^{-16} sec taken for an electron to cross an interatomic distance at a typical Fermi velocity.

It seems highly likely that a cubic cell 0.1 micrometers on a side (containing some 10^8 optimally arranged atoms) can hold a bit or perform a logic operation and, at the same time, transmit bits through itself to provide communication from cell to cell in a lattice. If so, then computers can be built with at least 10^{15} active elements per cubic centimeter. In a well-designed computer (with elements closer to their true technological limit and not laid out in regular cubical cells), this volume estimate should prove quite conservative. Elements so small will be sensitive to radiation damage; to be reliable, systems will require a large measure of redundancy.

Concern might be raised about the cost of such intricately patterned matter, either because of labor or energy requirements. It seems clear, however, that molecular-scale production systems can be completely automated (what use is there for hands?). Thus, labor costs of production (including production of additional production equipment) can approach zero. The energy needed to produce molecularly engineered material will generally be greater than the energy needed to produce ordinary materials of similar bulk composition, but analogy suggests that the energy cost need not be vastly greater than for the production of biological materials. In many cases (e.g., advanced computers or any of a number of applications not discussed here), the unique value of the products would make such energy costs unimportant, even if energy costs remained high.

Some biological applications

Molecular devices can interact directly with the ultimate molecular components of the cell and thus serve as probes of unique value in studying processes within the cell. Further, molecular devices can characterize a frozen cell in essentially arbitrary detail by removal and characterization of successive layers of material (atomically thin layers, if desired). Although the amount of data involved is large (a typical cell contains billions of protein molecules), the physical bulk of a device able to store and manipulate this amount of data will be quite small.

The change of temperature and water distribution during freezing modifies cell structures in several ways, primarily by physical displacement of structures by ice crystals and denaturation of proteins by concentration of solutes in the residual liquid (14). With frozen tissue, knowledge of normal structures (membrane geometries, natural protein structures) and analysis of frozen structures (position of ice crystals, position of denatured proteins) should permit quite accurate reconstruction of the nature of the tissue before freezing.

Such procedures would have special utility in analyzing the structure of tissue in the brain. Unlike, say, muscle or liver tissue, the function of brain tissue depends on the detailed three-dimensional structure of intertwined cells and their interfaces. The freezing process is far too slow to stop such dynamic processes as action potentials and synaptic transmission; short-term memory, however, is suspected to involve chemical modification of the neurons, and long-term memory is believed to involve the growth and modification of neuronal structures, particularly synapses. At the modest freezing rates possible in substantial pieces of tissue, ice crystals may be expected to nucleate and grow in the intercellular fluid, displacing the cell membranes as they do so. Electron micrographs, however, show that synapses (like many intercellular junctions) involve complementary structures on both sides of the intercellular gap, which should provide information enough to reconstruct the pre-freezing configurations of the cells almost regardless of ice crystal locations.

The ability to reconstruct the prefreezing structure of tissue, when combined with the general synthetic capabilities outlined above, will make feasible the physical restoration of tissue damaged by ordinary freezing through characterization, reconstruction, and restoration of successive segments of frozen material. Although restored to a frozen condition, such tissue would lack the characteristic damage caused by the freezing process. As many tissues can survive the gross insult of ordinary freezing, it seems likely that most could survive freezing followed by repair. The remaining mode of damage would seem to be denaturation of proteins sensitive to cold alone during the thawing process. Should cell components of some species prove sensitive to short periods of cold, they could presumably be modified to resemble those of hardier species without changing either cell function or DNA.

Implications for the present

The existence of a path to an advanced molecular technology has implications for the present. As with all technologies, long-range promise should tend to increase interest in undertaking the early steps, even beyond the interest springing from more immediate benefits. The longer the expected wait, however, the less the interest.

On the other hand, molecular engineering of materials and devices can extend the capabilities of technology many fold in many areas. The implications of the feasibility of molecular technology are important to present day speculations concerning the probable behavior (and likelihood of existence) of extraterrestrial technological civilizations. Similarly, those concerned with the long-range future of humanity must concern themselves with the opportunities and dangers arising from this technology. Finally, the eventual development of the ability to repair freezing damage (and to circumvent cold damage during thawing) has consequences for the preservation of biological materials today, provided a sufficiently long-range perspective is taken.

Development of the ability to design protein molecules will, by analogy between features of natural macromolecules and components of existing machines, make possible the construction of molecular machines. These machines can build second-generation machines able to perform extremely general synthesis of three-dimensional molecular structures, thus permitting construction of devices and materials to complex atomic specifications. This capability has implications for technology in general and in particular for computation and characterization, manipulation, and repair of biological materials.

REFERENCES

Beck, J.S., J.C. Vartuli, W.J. Roth, M.E. Leonowicz, C.T. Kresge, K.D. Schmitt, C.T.-W. Chu, D.H. Olsen, E.W. Shepard, S.B. McCullen, J.B. Higgins, and J.L. Schlenker. 1992. *J. Am. Chem. Soc.* 114:10834.

Becker, M.F., J.R. Brock, H. Cai, N. Chaudhary, D. Henneke, L. Hilsz, J.W. Keto, J. Lee, W.T. Nichols, and H.D. Glicksman. 1997. Nanoparticles generated by laser ablation. In *Proc. of the Joint NSF-NIST Conf. on Nanoparticles.*

Berger, R., C. Gerber, H.P. Lang, and J.K. Gimzewski. 1996. Micromechanics: a toolbox for femtoscale science: "Towards a laboratory on a tip." *Microelectronic Engineering*, 35:373-9. (International Conference on Micro- and Nanofabrication, Glasgow, U.K., 22- 25 Sept. 1996).

Berkowitz, A.E., J.R. Mitchell, M.J. Carey, A.P. Young, S. Zhang, F.E. Spada, F.T. Parker, A. Hutten, and G. Thomas. 1992. *Phys. Rev. Lett.* 68:3745.

Berndt, C.C., J. Karthikeyan, T. Chraska, and A.H. King. 1997. Plasma spray synthesis of nanozirconia powder. In *Proc. of the Joint NSF-NIST Conf. on Nanoparticles.*

Bimberg, D., N. Kirstaedter, N.N. Ledentsov, Zh.I. Alferov, P.S. Kopev, and V.M. Ustinov. 1998. InGaAs-GaAs quantum-dot lasers. *IEEE J. of Selected Topics in Quant. Electronics* 3:196-205.

Braun, P.V., P. Osenar, and S.I. Stupp. 1996. *Nature*. 368: 2. Brinker, C.J. 1996. *Curr. Opin. Solid State Mater. Sci.* 1:798. Chen, C-Y., S.L. Burkett, H.-X. Li, and M.E. Davis. 1993. *Microporous Mater.* 2:27.

Brotzman, R. 1998. Nanoparticle Dispersions. In *R&D status and trends*, ed. Siegel et al. Bu, X., P. Feng, and G.D. Stucky. 1998. Large-cage zeolite structures with multidimensional 12-ring channels. *Science* 278:2080-2085.

Brotzman, R. 1998. Nanoparticle dispersions. In *R&D status and trends*, ed. Siegel et al. Brus, L.E. 1996. Theoretical metastability of semiconductor crystallites in high pressure phases with applications to beta tin structures of silicon. *J. Am. Chem. Soc.* 118:4834-38.

Brus, L. 1996. Semiconductor colloids: Individual nanocrystals, opals and porous silicon. *Current Opinion in Colloid & Interface Science* 1:197-201. *Chemical Engineering News.* 1997. Particulate matter health studies to be reanalyzed (August 18):33.

Chance, B., Mueller, P., DeVault, D. & Powers, L. (1980) *Phys. Today* 33 (10), 32-38.

Chang, K. "Smaller Computer Chips Built Using DNA as Template." New York Times, November 21, 2003.

Chantrell. 1994. *J. Appl. Phys.* 76:6811. Erb, U., G. Palumbo, R. Zugic, and K.T. Aust. 1996. In *Processing and properties of nanocrystalline materials*, ed. C. Suryanarayana, J. Singh, and

F.H. Froes. Warrendale, PA: TMS. Gertsman, V.Y., M. Hoffman, H. Gleiter and R. Birringer. 1994. *Acta Metall. Mater*. 42:3539-3544.

Chopra, N.G., R.J. Luyken, K. Cherrey, H. Crespi, M.L. Cohen, S.G. Louie, and A. Zettl. 1995. *Science* 269:966.

Clasen, R. 1990. *Int. Journal of Glass and Science Technology* 63: 291. Crandall, B.C. and J. Lewis, (eds.). 1992. *Nanotechnology: Research and perspectives*. Cambridge: MIT Press.

Czekai, D., et al. 1994. *Use of smaller milling media to prepare nanoparticulate dispersions*. U.S. Patent application. Docket 69802(02) Filed 2/25/94.

Drexler, E., Phoenix, C.: 2004, 'Self-Replication in Nanotechnology: Feasible, Potentially Safe, and Unnecessary' (unpublished paper, in preparation).

Drexler, E.: 1986, *Engines of Creation: The Coming Era of Nanotechnology*, Anchor Books, New York (expanded edition with a new afterword, 1990).

Drexler, K.E.: 1992, *Nanosystems. Molecular machinery, manufacturing and computation*, John Wiley & Sons, New York.

Drexler, K.E.: 2001, 'Machine-Phase nanotechnology', *Scientific American*, (Sept.), 66-67.

Drexler, K.E.: 2003a, 'Open Letter', *Chemical & Engineering News*, 81, no. 48, 37-42.

Drexler, K.E.: 2003b, 'An open letter to Richard Smalley', April 16.

Dupuy, J.P. "Two Temporalities, Two Rationalities: A New Look at Newcomb's Paradox." in Bourgine, P.&Walliser, B. (eds.), Economics and Cognitive Science. New York: Pergamon. 1992, 191-220.

Fennema, O. R. (1973) in *Low-Temperature Preservation of Foods and Living Matter*, eds. Fennema, O. R., Powrie, W. D. & Marth, E. H. (Dekker, New York), pp. 436-475.

Feyerabend, P.: 1981, *Realism, Rationalism and Scientific Method* (Philosophical Papers, volume 1), Cambridge University Press, Cambridge.

Feynman R. "There's Plenty of Room At the Bottom." Talk given on at the annual meeting of the American Physical Society at the California Institute of Technology, 1959.

Feynman, R.: 1960, 'There's Plenty of Room at the Bottom', *Engineering and Science*, 23, 22-36.

Fleming, D. "The Economics of Taking Care: An Evaluation of the Precautionary Principle." in Freestone, D.&Hey, E. (eds.), The Precautionary Principle and International Law. La Haye: Kluwer Law International, 1996.

Foerster, H.v.: 1962, 'Bio-Logic', in: E.E. Bernard & M.R. Kare (eds.), *Biological prototypes and synthetic systems*, Plenum Press, New York, vol. 1., pp. 1-12

Friedlander, S.K. 1993. Controlled synthesis of nanosized particles by aerosol processes. *Aerosol Sci. Technol*. 19:527.

Friedlander, S.K. 1998. Synthesis of nanoparticles and their agglomerates: Aerosol reactors. In *R&D status and trends*, ed. Siegel et al.

Friedlander, S.K., H.D. Jang, and K.H. Ryu. 1998. *Appl. Phy. Lett*. 72(2):173.

Goddard, W.A. 1998. Nanoscale theory and simulation. In *R&D status and trends*, ed. R. Siegel et al.

Godsell, D.: 2003, *Living Machinery. Bionanotechnology: Lessons from Nature*, Wiley-Liss, New-York.

Guarnieri, F., M. Fliss, and C. Bancroft. 1996. Making DNA Add. *Science* 273:220-223. Hanes, J., J.L. Cleland, and R. Langer. 1997.

Held, R., T. Heinzel, A.P. Studerus, K. Ensslin, and M. Holland. 1997. Semiconductor quantum point contact fabricated by lithography with an atomic force microscope. *Appl. Phys. Lett.* 71:2689-91.

Hellemans, A. 1998. X-rays find new ways to shine. *Science* 277:1214-15.

Hoch, H.C., L.W. Jelinski, and H.G. Craighead, eds. 1996. *Nanofabrication and biosystems*. New York: Cambridge University Press.

Howland, H.: 1962, Structural, hydraulic and economic aspects of leaf venation and shape', in: E.E. Bernard & M.R. Kare (eds.), *Biological prototypes and synthetic systems*, Plenum Press, New York, vol. 1, pp. 183-192 .

Hoyningen-Huene, P.: 1993, *Reconstructing Scientific Revolutions: Thomas S. Kuhn's Philosophy of Science* (trans. by A. Levin), University of Chicago Press, Chicago.

Huang, J.Y., Y.K. Wu, and H.Q. Ye. 1996. *Acta Mater.* 44:1211. Inoue, A. 1997. Private communication. Inturi, R.B., and Z. Szklavska-Smialowska. 1992. *Corrosion* 48:398. Karch, J., R. Birringer, and H. Gleiter. 1987. *Nature* 330:556.

Jena, P., S.N. Khanna, and B.K. Rao. 1996. In *Science and technology of atomically engineered materials*, ed. P. Jena. River Edge, NJ: World Scientific.

Jonas, H. The Imperative of Responsibility. In Search of an Ethics for the Technological Age. Chicago: University of Chicago Press, 1985.

Karak, N., and S. Maiti. 1997. Dendritic polymers: A class of novel material. *J. Polym. Mater.*14:105.

Karger, J., and D.M. Ruthven. 1992. *Diffusion in zeolites*. New York: J. Wiley. Krätschmer, W., L.D. Lamb, K. Fostiropoulos, and D.R. Huffman. 1990. *Nature* 347:354.

Karplus, M. & Weaver, D. L. (1976) *Nature (London)* 260, 404-406.

Kirschvink, J.L., A. Koyayashi-Kirschvink, and B.J. Woodford. 1992. Magnetite biomineralization in the human brain. *Proc. Nat'l. Acad. Sci.* USA 89:7683-7687.

Kishida, M., T. Fujita, K. Umakoshi, J. Ishiyama, H. Nagata, and K. Wakabayashi. 1995. *Chem. Commun.* 763.

Klein, J.D., et al. 1993. *Chem. Mater.* 5:902. Koch, C.C. 1989. Materials synthesis by mechanical alloying. *Annual Review of Mater. Sci.* 19:121-143.

Kourilsky, Ph.&Viney, G. Le Principe de précaution. Report to the Prime Minister, Paris, Éditions Odile Jacob, 2000.

Krejchi, M.T., E.D.T. Atkins, A.J. Waddon, M.J. Fournier, T.L. Mason, and D.A. Tirrell. 1994. Chemical sequence control of beta-sheet assembly in macromolecular crystals of periodic proteins. *Science* 265:1427-1432.

Krejchi, M.T., S.J. Cooper, Y. Deguchi, E.D.T. Atkins, M.J. Fournier, T.L. Mason, and D.A. Tirrell. 1997. Crystal structures of chain-folded antiparallel beta-sheet assemblies from sequence-designed periodic polypeptides. *Macromolecules* 30:5012-5024.

Kurzweil, R.: 1998, *The Age of Spiritual Machines, How We Will Live, Work and Think in the New Age of Intelligent Machines*, Phoenix, New York.

Kyprianidou-Leodidou, T., W. Caseri, and V. Suter. 1994. *J. Phys. Chem.* 98:8992. Kung, H.H., and E.I. Ko, 1996. *Chem. Eng. J.* 64:203.

Leonard, D., M. Krishnamurthy, C.M. Reaves, S.P. Denbaars, and, P.M. Petroff. 1993. Direct formation of quantum-sized dots from uniform coherent islands of InGaAs on GaAs surfaces. *Appl. Phys. Lett.* 63:3203-5.

Liang, G., Z. Li, and E. Wang. 1996. *J. Mater. Sci.* Makino, A., A. Inoue, T. Hatanai, and T. Bitoh. 1997. *Materials Science Forum* 235-238: 723.

Mao, C., W. Sun, and N.C. Seeman. 1997. Construction of Borromean rings from DNA. *Nature* 386:137-138.

Martin, T.P., N. Malinowski, U. Zimmerman, U. Naher, and H. Schaber. 1993. *J. Chem. Phys.* 99:4210.

Martin, T.P., U. Naher, H. Schaber, U. Zimmerman. 1993. *Phys. Rev. Lett.* 70:3079. Prigogine, I., and S. Rice. 1988. *Advances in chemical physics*, Vol. 70, Parts 1 & 2. New York: J. Wiley.

McCammon, J. A., Gelin, B. R. & Karplus, M. (1977) *Nature (London)* 267, 585-590.

McConnell, H.M. 1996. Light-addressable potentiometric sensor: Applications to drug discovery. In *Nanofabrication and biosystems*, ed. Hoch et al.

McCulloch, W.S.: 1962, 'The imitation of one form of life by another – Biomimesis', in: E.E. Bernard & M.R. Kare (eds.), *Biological prototypes and synthetic systems*, Plenum Press, New York, vol. 1., p. 393-97.

Mehl, R.F., and R.W. Cahn. 1983. Historical development. In *Physical metallurgy*. North Holland. Milligan, W.W., S.A. Hackney, M. Ke, and E.C. Aifantis. 1993. *Nanostructured Materials* 2:267.

Minsky, M.: 1995, 'Virtual Molecular Reality', in: Krummenacker, M. & Lewis, J. (eds.), *Prospects in Nanotechology. Proceedings of the 1st general conference on nanotechnology: developments, applications, and opportunities, November 11-14, 1992, Palo-Alto*, John Wiley & Sons, New York, pp. 187-205.

Mishra, R.S., and A.K. Mukherjee. 1997. Oral presentation at TMS meeting, Indianapolis, Indiana, 16-18 September 1997, to be published in proceedings of Symp. "Mechanical Behavior of Bulk Nano-Materials."

Mishra, R.S., R.Z. Valiev, and A.K. Mukherjee. 1997. *Nanostructured Materials* 9:473. Morris, D.G., and M.A. Morris. 1991. *Acta Metall. Mater.* 39:1763-1779.

Moore, J.C., H.M. Jin, O. Kuchner, and F.H. Arnold. 1997. Strategies for the in vitro evolution of protein function: Enzyme evolution by random recombination of improved sequences. *J. Mol. Biol.* 272:336-347.

Morris, D.G., and M.A. Morris. 1997. *Materials Science Forum* 235-238:861. Nagpal, P., and I. Baker. 1990. *Scripta Metall. Mater.* 24:2381.

Nieman, G.W., J.R. Weertman, and R.W. Siegel. 1991a. *Mater. Res. Soc. Symp. Proc.* 206:581-586.

Nomura, M. & Held, W. (1974) in *Ribosomes,* eds. Nomura, M., Tissiers, A. & Lengyel, P. (Cold Spring Harbor Laboratory, Cold Spring Harbor, NY), pp. 193-203.

Nordmann, A. (rapp.) Converging Technologies—Shaping the Future of European Societies, European Commission report, 2004.

Parker, J.C., et al. 1995. U.S. Patent 5,460,701. POST (Parliamentary Office of Science and Technology). 1995.

Phoenix, C.; Drexler, E.: 2004, 'Safe Exponential Manufacturing', *Nanotechnology*, 15, 869-72.

Rademann, K., B. Kaiser, U. Even, F. Hensel. 1987. *Phys. Rev. Lett.* 59:2319. Rao, C.N.R., B.C. Satishkumar, and A. Govindaraj. 1997. *Chem. Commun.* 1581.

Ramanan, V.R. 1998. Nanocrystalline soft magnetic alloys for applications in electrical and electronic devices. In *R&D Status and Trends*, ed. Siegel et al.

Rao, M.B., and S. Sircar. 1993. *Gas Separation and Purification* 7:279. Reetz, M.T., et al. 1995. *Science* 267:367.

Rofagha, R., R. Langer, A.M. El-Sherik, U. Erb, G. Palumbo, and K.T. Aust. 1991. *Scr. Metall. Mater.* 25:2867.

Roher, H. 1993. *Jpn. J. Appl. Phys.* 32:1335. Rohlfing, E.A., D.M. Cox, and A. Kaldor. 1984. *J. Chem. Phys.* 81:3846.

Romanov, A.E., V.I. Vladimirov. 1992. In *Dislocations in solids*, ed. F.R.N. Nabarro, Vol. 9. Amsterdam: North-Holland. Salishekev, G.A., O.R. Valiakhmetov, V.A. Valitov, and S.K. Mukhtarov. 1994. *Materials Science Forum*. 170-172:121.

Ruthven, D.M., S. Farooq, K.S. Knaebel. 1994. *Pressure swing adsorption*. New York: VCH Publishers. Sayari, A. 1996. *Chem. Mater.* 8:1840.

S. P. Ho and W. F. DeGrado, "Design of a 4-Helix bundle protein: Synthesis of peptides which self-associate into a helical protein." *J. Am. Chem. Soc.*, 109: 6751-6758 (1987).

Sankey, H.: 1994, *The Incommensurability Thesis*, Avebury, Aldershot.

Sarikaya, M.; Aksay, I. (eds.), 1995, *Biomimetics. Design and Processing of Materials*, AIP Press, Woodbury, New York.

Scanlan, R.M., W.A. Fietz, and E.F. Koch. 1975. *J. Appl. Phys.* 46:2244.

Schiefsky, M.J.: forthcoming, 'Art and Nature in Ancient Mechanics', in: W.R. Newman & B. Bensaude-Vincent (eds.), *The Artificial and the Natural: An Ancient Debate and its Modern Descendants*, MIT Press, Cambridge, MA.

Schultz, L., J. Wecker, and E. Hellstern. 1987. *J. Appl. Phys.* 61:3583.

Schultz, L., K. Schnitzke, and J. Wecker. 1989. *J. Magn. Mater.* 80:115.

Schwarz, R.B. 1998. Storage of hydrogen powders with nanosized crystalline domains. In *R&D Status and Trends*, ed. Siegel et al.

Shen, T.D., C.C. Koch, T.Y. Tsui, and G.M. Pharr. 1995. *J. Mater. Res.* 10: 2892.

Shull, R.D., R.D. McMichael, and J.J. Ritter. 1993. *Nanostructured Mater.* 2:205.

Siegel et al. Ogunnaike, B and W. Ray. 1994. *Process dynamics, modeling and control*. Oxford University Press, pp 5-21; 1033-48.

Siegel R.W. and G.E. Fougere. 1994. In *Nanophase materials*, ed. G.C. Hadjipanayis and R.W. Siegel. Netherlands: Kluwer Acad. Publ.

Siegel, H.: 1980, 'Objectivity, Rationality, Incommensurability, and More', *British Journal for the Philosophy of Science*, 31, 359-384.

Siegel, R.W., E. Hu, and M.C. Roco, eds. 1998. *R&D status and trends in nanoparticles, nanostructured materials, and nanodevices in the United States*. Baltimore: Loyola College, International Technology Research Institute.

2

Nanotechnology and Nano-velcro: An Overview

Study of Nanotechnology

Students interested in nanotechnology often ask what they should study. This web page provides a partial answer to that question. Foresight has a briefing on the subject by Eric Drexler.

Nanosystems

The standard text in the field is *Nanosystems: molecular machinery, manufacturing and computation* by K. Eric Drexler. Buy a copy and study it.

Molecular Mechanics

Any manufacturing technology must move atoms from where they are to where we want them to be. How atoms move and the forces that act upon them during their motion are therefore critical areas of study in nanotechnology. This field is called molecular mechanics. A very brief discussion of molecular mechanics and its significance for nanotechnology is available on the web in *Computational Nanotechnology*. It contains references to further reading. A classic introduction to molecular mechanics is *Molecular Mechanics*, by Ulrich Burkert and Norman L. Allinger, ACS Monograph 177, American Chemical Society, 1982. While out of print, it can often be found in campus libraries.

Nanosystems introduces the basic concepts of molecular mechanics in this study. A great advantage of Drexler's treatment is the adoption of consistent SI units. A slow and careful reading of this chapter is worth the effort. Many other introductions to molecular mechanics are available. Software packages implementing specific approaches to molecular mechanics are available, and can be very useful in learning the concepts.

Positional Control, Stiffness and Elasticity

A central idea in nanotechnology is that of positional control. This can be provided by fairly standard robotic devices. A major difference between conventional robotic devices and the molecular robotic devices used to position molecular components is the need to consider thermal noise. At the molecular scale, components wiggle and jiggle because of Brownian motion. To control this, the component must be held stiffly, i.e., there must be a restoring force which acts to return the component to an equilibrium position if it is displaced. (The existence of a restoring force serves as a good abstract definition of "positional control.") The restoring force is usually assumed to be linear in the displacement: restoring force = k_s times displacement. The constant k_s is a measure of the stiffness of the system. The greater the stiffness, the greater the restoring force and the smaller the deviations of the system from its equilibrium position. The basic equation relating stiffness and positional uncertainty is:

$$\text{sigma}^2 = kT / k_s$$

This equation should be memorized and basic applications understood. To use it, it is essential to determine the stiffness k_s. The stiffness of a structure can be determined from its geometry and material properties. The basic concepts are introduced by Feynman in chapters 38 and 39 of *The Feynman Lectures on Physics*, by R.P. Feynman, R.B. Leighton, and M. Sands, M. Addison-Wesley (1964). Buying the Feynman Lectures is recommended.

Application of these equations to some robotic devices (including the Stewart platform which, because of its great stiffness, is very attractive for molecular robotic applications) is illustrated in *A New Family of Six Degree of Freedom Positional Devices*.

Self Replication

A second central idea in nanotechnology is that of self replication. The student should read the web page introducing self replication and select some of the references therein for further reading. The recursion theorem is basic to self replicating systems. This theorem should be understood. Exercise: write a program which prints out an exact copy of itself. Buy a copy of *Kinematic Self-Replicating Machines*, it is an excellent survey of the literature.

There are, of course, a great many other subjects relevant to the development of nanotechnology. It seemed more useful to provide a short and focused list of critical subject areas that could be mastered with a reasonable effort than a longer and more unwieldy list which included everything of importance. The student can rest assured that there is no shortage of material to study relevant to this new field of research.

Nanotechnology and Worlds Stickiest Material

Scientists in Britain and USA are taking inspiration from the gecko to try and develop the ultimate in adhesive tape. The gecko has evolved with toe surfaces covered in the perfect adhesive that allows them to stick in the most difficult of circumstances—even upside down on wet glass. Geckos have micro and nano-scale hairs on their toes. Each main hair is a thousand times thinner than a human hair. On the end of each hair are more hairs—each a thousand times smaller again. Molecular attraction then occurs by way of Van der Waals forces that allow the gecko to cling on. Synthetic versions of these hairs have now been developed. Applications include incorporating the new adhesive material into lightweight rescue robots that can readily climb walls in search and rescue missions. NASA has plans to use the material on space flights as early as next year. The theory is that by being able to have astronauts stick to the exterior of a space station or the space shuttle, then spacewalks would become far less dangerous.

Nano-velcro

Researchers in the US are working on a Nano-velcro that could have a holding power tens of times greater than epoxy adhesives. The researchers at Michigan State University are working on the reusable material made from carbon nanotubes. Unlike conventional velcro that uses loops and hooks, nano-velcro only has hooks. It is believed that the material could become a strong, self-repairing, heat resistant micro-fastening system to hold together components in ultra-small robots and electrical devices.

Assuming that various elementary molecular parts fabrication and assembly operations can be performed for certain limited classes of structures, perhaps as described, then what structures would be interesting and useful to build? This Section briefly describes a small set of possibly useful molecular mechanical components and then concludes with a discussion of the concept of the molecular assembler.

In order to lay a foundation for molecular manufacturing, it is necessary to create and to analyze possible designs for nanoscale mechanical components that could, in principle, be manufactured. Because these components could not yet be built in 1998, such designs could not be subjected to rigorous experimental testing and validation. Designers were forced instead to rely upon ab initio structural analysis and computer studies including molecular dynamics simulations. Noted Drexler: "Our ability to model molecular machines (systems and devices) of specific kinds, designed in part for ease of modeling, has far outrun our ability to make them. Design calculations and computational experiments enable the theoretical studies of these devices, independent of the technologies needed to implement them."

In nanoscale design, building materials do not change continuously as they are cut and shaped, but rather must be treated as being formed from discrete atoms. A nanoscale component is a supermolecule, not a finely divided solid. Any stray atoms or molecules within such a structure may act as dirt that can clog and disable the device, and the scaling of vibrations, electrical forces, thermal expansion, magnetic interaction and surface tension with size lead to dramatically different phenomena as system size shrinks from the macroscale to the nanoscale.

Molecular bearings are perhaps the most convenient class of components to design because their structure and operation is fairly straightforward. One of the simplest examples is Drexler's overlap-repulsion bearing design, shown with end views and exploded views using both ball-and-stick and space-filling representations. This bearing has exactly 206 atoms including carbon, silicon, oxygen and hydrogen, and is comprised of a small shaft that rotates within a ring sleeve measuring 2.2 nm in diameter. The atoms of the shaft are arranged in a 6-fold symmetry, while the ring has 14-fold symmetry, a combination that provides low energy barriers to shaft rotation. An exploded view of a 2808-atom strained-shell sleeve bearing designed by Drexler and Merkle using molecular mechanics force fields to ensure that bond lengths, bond angles, van der Waals distances, and strain energies are reasonable. This 4.8-nm diameter bearing features an interlocking-groove interface which derives from a modified diamond surface. Ridges on the shaft interlock with ridges on the sleeve, making a very stiff structure. Attempts to bob the shaft up or down, or rock it from side to side, or displace it in any direction (except axial rotation, wherein displacement is-extremely smooth) encounter a very strong resistance. Whether these bearings would have to be assembled in unitary fashion, or instead could be assembled by inserting one part into the other without damaging either part, had not been extensively studied or modeled by 1998.

Molecular gears are another convenient component system for molecular manufacturing design-ahead. For example, Drexler and Merkle designed a 3557-atom planetary gear, shown in side, end, and exploded views. The entire assembly has twelve moving parts and is 4.3 nm in diameter and 4.4 nm in length, with a molecular weight of 51,009.844 daltons and a molecular volume of 33.458 nm^3. An animation of the computer simulation shows the central shaft rotating rapidly and the peripheral output shaft rotating slowly. The small planetary gears rotate around the central shaft, and they are surrounded by a ring gear that holds the planets in place and ensures that all of the components move in the proper fashion. The ring gear is a strained silicon shell with sulfur atom termination; the sun gear is a structure related to an oxygen-terminated diamond surface; the planet gears resemble multiple hexasterane structures with oxygen rather than CH_2 bridges between the parallel rings; and the planet carrier is adapted from a Lomer dislocation array created by R. Merkle and L. Balasubramaniam, and linked to the planet gears using C-C bonded bearings. View (c) retains the elastic deformations that are hidden in (a) — the gears are bowed. In the macroscale world, planetary gears are used in automobiles and other machines where it is necessary to transform the speeds of rotating shafts.

W. Goddard and colleagues at CalTech performed a rotational impulse dynamics study of this "first-generation" planetary gear. At the normal operational rotation rates for which this component was designed (e.g., <1 GHz for <10 m/sec interfacial velocities), the gear worked as intended and did not overheat. Started from room temperature, the gear took a few cycles to engage, then rotated thermally stably at ~400 K. However, when the gear was driven to ~100 GHz, significant instabilities appeared although the device still did not self-destruct. One run at ~80 GHz showed excess kinetic energy causing gear temperature to oscillate up to 450 K above baseline. One animation of the simulation shows that the ring gear wiggles violently because it is rather thin. In an actual nanorobot incorporating numerous mechanical components, the ring gear would be part of a larger wall that would hold it solidly in place and would eliminate these convulsive motions which, in any case, are seen in the simulation only at unrealistically high operating frequencies.

Drexler and Merkle later proposed a "second-generation" planetary gear design with 4235 atoms, a molecular weight of 72,491.947 daltons and a molecular volume of 47.586 nm^3. This new version was indeed more stable but still had too much slip at the highest frequencies. Commenting on the ongoing design effort, Goddard suggested that an optimal configuration could have the functionality of a planetary gear but might have an appearance completely different from the macroscopic system, and offered an example: "Because a gear tooth in the xy plane cannot be atomically smooth in the z-direction, we may develop a Vee design so that the Vee shape of the gear tooth in the z-direction nestles within a Vee notch in the race to retain stability in the z-direction as the teeth contact in the xy plane. This design would make no sense for a macroscopic gear system since the gear could never be placed inside the race. However, for a molecular system one could imagine that the gear is constructed and that the race is constructed all except for a last joining unit. The parts could be assembled and then the final connections on the face made to complete the design" analogous to the ZARBI system of Rebek.

Another class of nanodevice that has been designed is a gas-powered molecular motor or pump. The pump and chamber wall segment contain 6165 atoms with a molecular weight of 88,190.813 daltons and a molecular volume of 63.984 nm^3. The device can serve either as a pump for neon gas atoms or (if run backwards) as a motor that can convert neon gas pressure into rotary power. The helical rotor has a grooved cylindrical bearing surface at each end, supporting a screw-threaded cylindrical segment in the middle. In operation, rotation of the shaft moves a helical groove past longitudinal grooves inside the pump housing. There is room enough for small gas molecules only where facing grooves cross, and these crossing points move from one side to the other as the shaft turns, moving the neon atoms along. Goddard reported that preliminary molecular dynamics simulations of the device showed that it could indeed function as a pump, although "structural deformations of the rotor can cause instabilities at low and high rotational frequencies. The forced translations show that at very low perpendicular forces due to pump action, the total energy rises significantly and again the structure deforms." Merkle acknowledged that the pump moves neon atoms at an energy cost of 185 Kcal/mole-Angstrom (12,900 zJ/atom-nm), which is not very energy-efficient. Further refinement of this crude design is clearly warranted.

Conveyor systems would also be useful. Employing a primitive molecular CAD software package called Crystal Sketchpad, G. Leach created a design for a 5-nanopart, ~2500-atom conveyor belt system comprised of two rollers, two axles, and a belt consisting of a strained thin-walled diamond sheet. The design has not been subjected to further computational analysis, either to minimize rotational energy barriers or to optimize rotational dynamics or operational stability.

Drexler and Merkle have also produced a preliminary design for a 2596-atom fine-motion controller. A general-purpose molecular assembler arm must be able to move its "hand" by many atomic diameters, position it with fractional-atomic-diameter accuracy, and then execute finely-controlled motions, perhaps to transfer one or a few atoms in a guided chemical reaction. Human arms use large muscles and joints for large motions and more finely-controlled finger motions for precision; the device presented here can execute precise finger-like motions over several atomic diameters with associated 90-degree rotations. The core of the device consists of a shaft linking two hexagonal endplates, sandwiching a stack of eight rings, making a modified Stewart platform. In a complete system, each ring would be rotated by a lever driven by a cam mechanism. Each ring supports a strut linked to a central platform (here shown raised, displaced, and twisted). Rotating a ring moves a strut; moving a strut moves the platform; positioning all eight rings (over)determines a platform position in x, y, z, roll, pitch, and yaw. (If the struts were rigid, six would do the job; here, two struts have been added to increase stiffness.) Notes Drexler: "The chief design problem is to enable an adequate range of motion without mechanical interference or unacceptable bond strains, and within the size constraints set by available modeling tools and patience."

Almost all current design research in molecular nanotechnology is restricted to computer simulation, which allows the design and testing of large structures or complete nanomachines, and the compilation of growing libraries of molecular designs. The work is relatively inexpensive and does not require the support of a large team. Of course, calculations of many-body systems are notoriously difficult, with many computer packages making a number of simplifying assumptions — e.g., nuclei as point masses, electrons treated as a continuous charge distribution, and 3-D potential energy functions derived semi-empirically from experimental data and treated as a classical field despite their true quantum mechanical character (for ease of computation). Notwithstanding these shortcuts, Tuzun and colleagues claim that classical simulations may overestimate the rate of energy transfer between vibrational modes even at low energies, in which case "current designs for various nanocomponents [would] actually perform better and be more stable than recent molecular dynamics simulations suggest."

Goddard notes that future nanosystem simulations may require 1-100 million atoms to be considered explicitly, demanding major improvements in molecular dynamics methodologies. By 1998, new algorithms for parallel processing of massive molecular dynamics simulations were being developed, producing methods and optimized parallelized computer programs efficient for high capacity molecular dynamics simulations of 10,000-1,000,000 atoms for finite molecular structures.

Ultimately, Computer-Aided (Molecular) Design (CAD) systems will be needed to efficiently design and analyze molecular components and their higher-order assemblies. CAD systems are commonplace in macroscale engineering and architecture, as are Computer-Aided Manufacturing (CAM) systems in macroscale manufacturing. In the molecular realm, Computer Aided Synthesis Design (CASD) has been avidly pursued by computational chemists since the 1980s, and by 1998 many university and commercial molecular modeling software packages were already well-known, including Alchemy (Tripos Inc.; http://www.tripos.com/), Cerius2 (Molecular Simulations Inc.; http://www.msi.com/), Chem3D (CambridgeSoft; http://www.camsoft.com/), Conformer (Princeton Simulations; http://www.conformer.com/), Gaussian94 (Gaussian Inc.; http://www.gaussian.com/), Hyperchem (Hypercube Inc.; http://www.hyper.com/), Molecular Operating Environment (Chemical Computing Group Inc.; http://www.chemcomp.com/), MOPAC97 (Fujitsu Ltd.; http://www.winmopac.com/), RealMol/CAVE/NAMD (Fraunhofer Institute for Computer Graphics; www.igd.fhg.de/www/igda4/research/chemie), Sculpt (Interactive Simulations Inc.; www.intsim.com/products/index.htm), and Spartan (Wavefunction Inc.; http://www.wavefun.com/).

In 1998, the capacity of such systems to design and manipulate large nanoscale mechanical components was extremely limited. A few very primitive molecular nanotechnology design packages had been attempted, including Crystal Sketchpad, DiamondCAD (www.zyvex.com/diamond.html), Molecular Assembly Sequence Software (www.carol.com/mass.shtml), Molecular Modelling Toolkit (starship.python.net/crew/hinsen/mmtk.html), and NanoCAD (world.std.com/~wware/ncad.html). The possible design of molecules for specific purposes using genetic software techniques had also been investigated.

Drexler has proposed the molecular assembler, a device resembling an industrial robot which would be capable of holding and positioning reactive moieties in order to control the precise location at which chemical reactions take place. This general approach would allow the construction of large atomically precise objects by a sequence of precisely controlled chemical reactions. Much like the ribosome in biology, an assembler would build various classes of useful molecular structures following a sequence of instructions. During this process, the assembler would provide three-dimensional positional and full orientational control over the molecular component (analogous to the individual amino acid in the ribosome model) that is being added to a growing complex molecular structure (analogous to the growing polypeptide in the ribosomal model). In one approach, a molecular assembler may be capable of forming any one of several different kinds of chemical bonds (e.g., by changing tool tips), not just a single kind such as the peptide bond that the ribosome makes. In bonding atoms or molecules to one another, the assembler would provide any needed energy (especially if the reaction happens not to be energetically favored) through physical force, thus performing mechanosynthesis (as opposed to the traditional means of chemical synthesis in solution). In another approach, a molecular assembler might be capable only of noncovalent assembly operations, wherein nanoparts are fabricated by other means and then presented to the assembler, which assembles the nanoparts into working nanomachines.

The first simple molecular assembler will almost certainly be a macroscale device, perhaps a modified SPM system as was being pursued by Zyvex in 1998. Multiple SPM heads could be equipped with a small number of nanoscale tool tips. In one scenario, nanoparts fabricated using bulk chemistry techniques would be inspected and selected by the SPM, then assembled one by one into working nanomachines (e.g., the desired useful nanoscale products). Such assembly operations will be very slow, because the placement of each new component may require simultaneous rotations and translations of large macroscale SPM components. Assembly time scales roughly linearly with assembler size because smaller assembler components moving at a given velocity need to travel less distance to accomplish a given physical operation, hence consuming less time and energy per physical operation. An important early developmental goal thus will be to design and fabricate nanoscale molecular assemblers.

At its most basic level, the simplest possible nanoscale molecular assembler may be comprised of one or more nanoscale manipulators. For example, the diamondoid telescoping manipulator arm described in has about 4 million atoms excluding the base and control and power structures; doubling the size to account for support structures gives a molecular weight per arm of ~100 megadaltons and a total molecular volume of ~140,000 nm^3. Designed for high strength and stiffness, this robot arm should be able to hold a molecular fragment stiffly enough to make it react with a chosen end of a carbon-carbon double bond with an error rate of only 10^{-15}. The robot arm could guide chemical reactions with high reliability at room temperature in vacuo, with little need for sensing the positions of the molecules with which it is working. Alternatively, with appropriate tool tips, such an arm could grasp and manipulate individual prefabricated nanoparts in solution phase. Assembly of nanoparts without fabrication may require only a very small tool set.

The manipulator arm must be driven by a detailed sequence of control signals, just as the ribosome needs mRNA to guide its actions. However, such detailed control signals can be provided by external acoustic, electrical, or chemical signals that are received by the robot arm via an onboard sensor or power transducer, using a simple "broadcast architecture", a technique which can also be used to import power. Such transducers may be extremely small, on the order of $(\sim 10\ nm)^3$ each. Interestingly, the biological cell may be regarded as an example of a broadcast architecture: the nucleus, located external to the cytoplasm, broadcasts mRNA chemical signals to millions of spatially diverse cytoplasmic ribosomes, thereby remotely controlling the construction of cellular proteins.

Thus a two-armed mechanical molecular assembler that receives power and instructions from some external agency may have a total molecular volume of ~300,000 nm^3 (a cube ~67 nm on an edge), containing ~16 million atoms with a molecular weight of ~200 megadaltons — very roughly the mass and scale of a medium-size virus particle, such as an adenovirus. For comparison, the average enzyme (a biochemical crimping tool) weighs about ~0.1 megadalton, while a typical ribosome (a primitive "protein assembler") weighs ~4.2 megadaltons. A simple mechanical assembler in the 10-100 megadalton range cannot be ruled out. However, such a device might have a very small set of tool tips and an extremely limited manufacturing repertoire.

Multiple nanoscale assemblers each capable of independent simultaneous actuation may require control signals more conveniently provided by an onboard nanocomputer. This programmable nanocomputer must be able to accept stored instructions which are sequentially executed to direct the manipulator arm to place the correct moiety or nanopart in the desired position and orientation, thus giving precise control over the timing and locations of chemical reactions or assembly operations. The mechanical nanocomputer analyzed by Drexler requires ~16 nm^3 per logic gate and ~40 nm^3 per data register. If such components could be used to construct the equivalent of the most primitive 4-bit Intel 4004 microprocessor, then the nanocomputer could process ~10^5 bits/sec (~25,000 ops/sec) at a ~1 KHz clock speed within a mechanism volume of ~36,000 nm^3, again neglecting power supply, I/O linkages, and the like. An additional ~160,000 nm^3 of rod logic registers adds ~1 kilobyte of onboard RAM memory or ~40 kilobytes of internal tape memory (conservatively assuming a tape storage density comparable to linear DNA). (A memory tape containing all bits needed for complete self-description may be considerably longer, even allowing for substantial data compression.) Doubling the total volume to account for support structure and other overhead gives a minimum mechanical nanocomputer molecular volume of ~400,000 nm^3, roughly 70 million atoms with a molecular weight of ~800 megadaltons. Thus the smallest nanocomputer-driven nanoscale molecular assembler with two manipulator arms may have a total molecular volume of ~700,000 nm^3 (a cube ~88 nm on an edge), containing ~86 million atoms with a molecular weight of ~1 gigadalton.

By 1998, only a small amount of research targeted at actual assembler design had begun. Following Drexler's original discussions, during 1991-1998 R. Merkle authored or coauthored a continuing series of papers discussing various operational aspects and specific components of assembler design, including mechanosynthetic positional control, general design considerations for assemblers, the broadcast architecture, convergent assembly, binding sites, positioning devices, mechanosynthetic path sets, designs for a neon pump and a fine motion controller, and possible assembler casings. J.S. Hall has considered high-level designs for a nanoscale parts-fabrication and parts-assembly nanorobot. Zyvex, founded in 1996 by James von Ehr, has set itself the task of building the first programmable nanoassembler in a 5-10 year time frame (e.g., by ~2006). W. Goddard and colleagues have proposed a series of molecular dynamics simulations of simple assemblers, although by 1998 these studies evidently had not yet begun. According to Goddard's original proposal:

"Ultimately we need a programmable synthetic system to make a real device. Even though we may not have tools for all the chemical steps and may not have designs for all the pumps, engines, and transmissions needed, we propose to study the dynamics of simplified prototype assemblers. In these studies we anticipate having (1) a reservoir or supplies area for providing the various building units (atoms and fragments) required, (2) a work area in which we construct the nanomachine device (initially we will consider assembling the structure on top of a diamond surface), and (3) a molecular scale nanohand which will extract the atoms from (1) and carry them to (2). We will then use extensions of our massive molecular dynamics program to operate the system: moving the tip from reservoir to work area, moving it to contact the appropriate surface site, moving it to regenerate the active tip, and then moving it back to add new atoms and molecules. This will include proper temperature effects, molecular vibrations, energy release upon the various chemical steps, etc. The ground rules here are that a realistic force field be used and that all pieces be treated at the atomic level (but some might be semi-rigid). This will use the force field developed for nanosynthesis. The purpose of these simulations is to examine issues of vibration caused by chemical forces as the tool picks up and delivers atoms to the growing surface. Also we want to consider the effect of energy release in the chemical steps on the thermal fluctuations in these systems (which may cause displacements and vibrations)."

Building mega-atom or giga-atom nanoproducts one at a time would be incredibly expensive and time consuming. For example, imagine that we wish to construct a simple medical nanorobot such as the 1-micron spherical respirocyte described in which consists of ~18 billion atoms (dry structure). A factory employing a coordinated team of 100 macroscale SPM assemblers, each able to place ~1 atom/sec-SPM on a convergently-assembled workpiece achieves a pitiful manufacturing throughput rate of ~2 respirocytes per decade. Nanoscale assemblers with appendages ~10^6 times smaller than macroscale SPM assemblers might plausibly achieve net assembly rates of ~10^6 atoms/sec-nanoassembler, but even a production line employing ~300 such nanoassemblers (which the abovementioned 100-SPM factory team could build in ~1 year, assuming ~10^7 atoms/nanoassembler) can only manufacture ~1 respirocyte nanorobot per minute. At that rate, it would take ~2 million years to build the first ~1 cm^3 therapeutic dosage containing ~10^{12} respirocyte nanorobots.

The necessary solution to this mass-production bottleneck is to employ any of several massively parallel approaches to manufacturing, including such techniques as self-assembly, convergent assembly, or, most usefully, self-replication. The basic advantage of self-replication is readily illustrated. Consider a single ~10^8 atom "seed" nanoassembler (having an onboard nanocomputer) that has been painstakingly built in ~10^6 sec using the abovementioned 100-SPM macroscale assembler team, working at ~1 atom/sec-SPM. The seed nanoassembler is first programmed to build a copy of itself, which it accomplishes (working at ~10^6 atoms/sec-nanoassembler) in ~100 sec. These two nanoassemblers then each build a copy of themselves in another ~100 sec; now there are four nanoassemblers. After ~48 generations, requiring a total of ~80 minutes to complete, there are ~3 x 10^{14} nanoassemblers. These nanoassemblers are then reprogrammed for the manufacture of 18-billion-atom respirocytes and are fed the appropriate, presumably different, feedstock. This vastly expanded nanoassembler manufacturing system can now produce ~10^{12} respirocytes (~1 cm^3 therapeutic dose) per minute.

The design of machines able to make copies of themselves was first described by von Neumann. Many variations on this theme have been reviewed by Freitas and Gilbreath and Sipper, and design for self-replication in the context of nanoscale assemblers has been considered by Drexler, Merkle, and Hall. As Drexler notes: "It may seem somehow paradoxical that a machine can contain all the instructions needed to make a copy of itself, including those selfsame complex instructions, but this is easily resolved. In the simplest approach, the machine reads the instructions twice: first as commands to be obeyed, and then as data to be copied. Adding more data does not increase the complexity of the data-copying process, hence the set of instructions can be made as complex as is necessary to specify the rest of the system. By the same token, the instructions transmitted in a replication cycle can specify the construction of an indefinitely large number of other artifacts."

The estimated information content of self-replicating systems — the length of the instruction tape — can be surprisingly small. Von Neumann's original analysis concluded that perhaps twelve different kinds of units of unknown complexity could be required as building materials, and Haldane inferred that as many as ~10^5 individual parts might be needed to make a replicator. This inference was refuted just three years later with the arrival of the first in a stream of very simple but ingenious designs for mechanical self-replicating machines that were assembled and operated in the late 1950s. In the first example, a pair of joined ratchet-like blocks, when placed in a "sea" of left and right blocks and then physically agitated, replicated itself from this well-ordered input substrate, making more block-pairs until all the single blocks were used up, a process similar to chemical autocatalysis. More complex congeries of blocks, including clever four-block, eight-block, and 12-block replicators that could replicate themselves in a sea of blocks were presented by Penrose. Jacobson demonstrated a 3-unit replicator consisting of toy train engines circulating on HO model railroad tracks, and Morowitz designed a simple two-unit device with one unit comprised of about a dozen components including switches, batteries and electromagnets, that could assemble copies of itself from parts floating on the water surface of a bathtub. In 1998, Lohn and colleagues gave two designs for simple self-replicating systems that could be constructed out of wood, batteries and electromagnets, with explicit analogies to nanoscale fabrication. Several important conclusions may be drawn from these examples:

First, replication is fundamentally so simple a task that machines capable of displaying this behavior predate most of the modern electronic computer era. Second, assembly is an inherently simpler operation than fabrication. A complex part may embody hundreds or thousands of prior fabrication and assembly operations, yet may be installed within an assemblage in a single step. Hence nanopart assembly may be an easier candidate for early implementation in first-generation self-replicating molecular manufacturing systems than is molecular fabrication. Third, and most important, the simplest nanoreplicator may require the ability to assemble, but not to (atomically) fabricate, in order to replicate itself. A nanomachine capable of assembly alone can replicate itself only from a very limited set of well-ordered input materials of relatively high complexity, but such a machine can be extremely simple both in structure and function. Certainly a nanomachine capable of both assembly and fabrication can replicate itself from a more diverse and disordered set of more elementary input materials but only at the expense of far greater internal structural and functional complexity. At least two distinct models of replication have been identified in replicating systems design. The first model may be called the "unit replication" or organismic model, in which the replicator is an independent unit which employs the surrounding substrate to directly produce an identical copy of itself; both the original and the copy remain fertile and may replicate again, thus exponentiating their numbers. The second model may be called the "unit growth" or factory model, in which a population of specialist devices, each one individually incapable of self-replication, can collectively fabricate and assemble all necessary components comprising all specialist devices within the system, hence the factory is capable of expanding its size (or of manufacturing duplicate factory systems) indefinitely in an appropriate environment. The factory model is sometimes called "bootstrapping," which is any production of more productive capacity, from less. T. McKendree [personal communication, 1999] likens the situation to an ant colony: "There necessarily is the line of reproducing queens, making a sufficient number of ants feasible; [however,] most of the queen's products are 'workers' that perform useful functions, but themselves are incapable of reproduction."

It cannot be emphasized too strongly that mechanical medical nanodevices will not self-replicate inside the human body, nor will they have any need for self-replication themselves. Machines that perform medical tasks are fundamentally different from machines that manufacture other machines. As R. Merkle explains: "While self-replicating systems are the key to low cost [manufacture], there is no need (and little desire) to have such systems function in the outside world. Instead, in an artificial and controlled environment they can manufacture simpler and more rugged systems that can then be transferred to their final destination. Medical devices designed to operate in the human body don't have to self-replicate: we can manufacture them in a controlled environment and then inject them into the patient as needed. The resulting medical device will be simpler, smaller, more efficient and more precisely designed for the task at hand than a device designed to perform the same function and self-replicate. This conclusion should hold generally: optimize [product] device design for the desired function, manufacture the [product] device in an environment optimized for manufacturing, then transport the [product] device from the manufacturing environment to the environment for which it was designed. A single device able to do everything would be harder to design and less efficient."

In an effort to stimulate scientific and engineering interest in constructing the first nanoassembler, in November 1995 the Foresight Institute created the $250,000 Feynman Grand Prize, with funding contributed by Zyvex founder James von Ehr and St. Louis venture capitalist Marc Arnold. The prize will be awarded to the individual or group that first achieves both of two significant nanotechnology breakthroughs — first, the design and construction of a functional nanometer-scale robotic arm, and second, the design and construction of a functional 8-bit adder computing nanodevice. According to Grand Prize rules, the robotic nanomanipulator must fit entirely inside a 100-nm cube, carry out actions directed by input signals of specified types, be able to move to a directed sequence of positions anywhere within a 50-nm cube, complete all directed actions with a positioning accuracy of 0.1 nanometer or better, and perform at least 1,000 accurate, nanometer-scale positioning motions per second for at least 60 consecutive seconds. The adder must fit entirely within a 50-nm cube, and must be capable of adding accurately any pair of 8-bit binary numbers, discarding overflow, accepting input signals of specified types, and producing its output as a pattern of raised nanometer-scale bumps on anatomically precise and level surface. (J.S. Hall notes that a conventional 8-bit adder may be constructed using a total of 94 AND, OR, and NOT gates; using XOR gates, the total may be reduced to 37 gates.) Both devices may accept inputs from acoustic, electrical, optical, diffusive chemical, or mechanical means, although any mechanical driving mechanism used for input must be limited to a single linkage that either slides or rotates on a single axis. To demonstrate the capacity for mass production, at least 32 copies of each device must be provided for analysis and destructive testing by judges.

Once nanoassemblers are available, the design of far more complex nanomachines each containing tens or hundreds of billions of precisely arranged atoms will require new molecular CAD tools and techniques, including automated hierarchical design decomposition of objects (that must be built up from an array of nanoparts) and a shape description language. Nanosystem design compilers, conceptually similar to silicon compilers that generate a complex pattern of transistors and conductors from an abstract specification of the properties of a digital circuit, will also be required for efficient nanorobot design. Given a fully specified design that meets all constraints necessary to physically permit assembly, all operations of the assembly process must be specified using an assembly process compiler. Explains Drexler: "The design of structures and assembly procedures by hierarchical decomposition directly generates a tree of assembly steps. If this tree has been chosen to generate parts of the appropriate sizes and numbers, then it can be mapped onto a manufacturing layout. The vast number of manipulator motions to be specified at the finest, earliest, and most-dispersed levels of the assembly process can be planned in parallel by identical software systems running almost independently on separate processors. The result of this parallel assembly process compilation is a set of instructions that, when executed by the transportation system and manipulator controllers of a manufacturing mechanism, will result in the assembly of an object corresponding to the initial design." Scale-up to larger systems, including the control, coordination, and programming of large aggregates of cooperating nanorobots.

Since molecular manufacturing systems can be used to make more molecular manufacturing systems, it is believed that the capital costs of production can be quite low. Drexler claims that an analysis of inputs, outputs, and productivity suggests the total cost of production can be in the range familiar in agriculture and in the production of industrial chemicals — on the order of tens of cents per pound. Merkle has made similar assertions, for example: "Eventually (after amortization of possibly quite high development costs), the price of assemblers (and of the objects they build) should be no higher than the price of other complex structures made by self-replicating systems. Potatoes — which have a staggering design complexity involving tens of thousands of different genes and different proteins directed by many megabits of genetic information — cost well under a dollar per pound." Such analogies from conventional agriculture or industrial chemistry are not strictly applicable to the medical products and pharmaceutical areas, where structural precision, product repeatability, and intensive quality control are of supreme importance, and where most products are subject to rigorous government regulation — unlike potatoes, which vary widely in size, shape, and molecular constitution, and generally do not require stringent product liability insurance, multiphase clinical trials, or FDA approval.

Still, it is likely that molecular manufacturing will eventually allow reasonably inexpensive production of large batches of designed medical nanorobots. As a very crude analogy, even in 1995 a typical mail-order bioscience vendor could manufacture 15 micromoles (~9 x 10^{18} product items) of customized oligonucleotides for $18/base up to at least 110 bases with a convenient 48-hour turnaround on orders; other online vendors offer similar or slightly higher prices up to 200-mer sequences with additional charges for necessary purifications. If a billion-base structure could be induced to self-assemble from 10 million different sets of 100-mer custom-built oligonucleotides (total ~20 billion atoms), forming a desired ~20-billion-atom nanomachine, then, optimistically assuming a ~100% yield, the cost would be ~$2 x 10^9 per nanomachine or ~$2000 per ~1 cm^3 of product volume which would include 10^{12} finished nanomachines. In 1998, many specialty biotechnology-related drug treatments were comparably expensive. For example, treatments for multiple sclerosis using two closely related ~22.5 kilodalton interferons were administered in dosages of ~1 milligram/week at a treatment cost of

~\$10,000/yr, which is equivalent to a treatment cost of ~\$2 x 10^9 (2 "nanodollars") per giga-atom. The famous HeLa (Henrietta Lacks) cell line is an example of a once tiny population of unique cancerous cells that were purposely replicated in vitro (~24-hour replication time) over many decades and are now widely available worldwide at very low cost (e.g., \$216 per 1-cm^3 ampule of the ATCC CCL-2 HeLa cell line). Swallowable therapeutic pills containing bacteria (i.e., natural biological nanomachines) are already widely available over the counter for gastrointestinal refloration, as for example Salivarex which "contains a minimum of ~2.9 billion beneficial bacteria per capsule", and Alkadophilus which "contains 1.5 billion organisms per capsule", both at a 1998 price of ~\$0.2 x 10^9 per bacterium.

Another useful analogic approach to cost estimation is suggested by the plunging price of computer power during the 20th century. The cost, in constant 1998 U.S. dollars, of a computer processing capability equivalent to ~1 TeraFLOP (10^{12} floating-point operations per second, ~10^{14} bits/sec assuming 64-bit words). The plotted data is from a compilation of 67 historical computers by Moravec through 1989, with an additional 27 data points for other computers spanning 1973-1999 added by the author. The chart shows that the cost of 1 TFLOP has fallen from \$2 x 10^{20} in 1908 (the mechanical Hollerith Tabulator) to just \$6 x 10^6 in 1998 (the SGI/Cray T3E-1200E), a 90-year price decline of almost 14 orders of magnitude — on average, a halving every ~2 years, a century-long example of Moore's Law. This trend may give us some confidence that even if the first individual 1 micron^3 nanorobot costs ~\$100,000 to assemble, eventually the unit price may fall at least 14 orders of magnitude to below ~\$$10^9$ per nanorobot or <\$1000 per ~1 cm^3 dosage (10^{12} nanorobots). Recycling or remanufacturing each nanomachine ~1000 times before final discard would then imply a net ~\$1/cm^3 treatment cost. Of course, in 1998 all such estimates were mere guesswork, but the ultimate expectation of relatively low-cost nanomedical treatments was certainly rational, if not provable.

This Chapter has demonstrated that there are many different pathways leading toward molecular nanotechnology, each one of which provides incremental benefits to motivate further travel down the paths. The likelihood of all the paths being blocked is low, even if we have little confidence that we can predict which approach will succeed first. Progress along the many pathways will provide precursor products, including some early medical applications, but the ultimate achievement of molecular manufacturing will finally make nanomedicine feasible.

As Drexler concluded Nanosystems: "Each step along [these pathways] will present great practical challenges, but each step will also bring valuable new capabilities. The long-term rewards, measured in terms of scientific and technological capabilities, appear large." This author agrees. For the remainder of the present work, we shall assume that a molecular manufacturing technology will someday exist that can economically manufacture macroscopic batches of microscopic machines comprised of atomically-precise nanoscale components.

REFERENCES

Asada, M., Y. Mayamoto, and Y. Suematsu. 1986. *IEEE Journ. Quantum Electronics* QE- 22(9): 1915-1921.

Attard, G.S., et al. 1997. *Science* 278:838. Averback, R.S., J. Bernholc, and D.L. Nelson. 1991. *MRS Symposium Proceedings* Vol. 206.

Axelbaum, R.L. 1997. Developments in sodium/halide flame technology for the synthesis of unagglomerated non-oxide nanoparticles. In *Proc. of the Joint NSF-NIST Conference on Nanoparticles: Synthesis, Processing into Functional Nanostructures and Characterization* (May 12-13, Arlington, VA).

B. W. Matthews, H. Nicholson, and W. J. Becktel, "Enhanced protein thermostability from site-directed mutations that decrease the entropy of unfolding." *Proc. Nat. Acad. Sci., 84:* 6663-6667 (1987), and included references.

Baibich, M..N., J.M. Broto, A. Fert, F. Nguyen Van Dau, F. Petroff, P. Etienne, G. Creuzet, A. Friederich, and J. Chazelas. 1998. *Phys. Rev. Lett.* 61:2472.

Baird, D.: 2004, *Thing Knowledge: A Philosophy of Scientific Instruments*, University of California Press, Berkeley.

Baker, R.T.K. Synthesis, properties and applications of graphite nanofibers. In *R&D status and trends*, ed. Siegel et al.

Beck, J.S., J.C. Vartuli, W.J. Roth, M.E. Leonowicz, C.T. Kresge, K.D. Schmitt, C.T.-W. Chu, D.H. Olsen, E.W. Shepard, S.B. McCullen, J.B. Higgins, and J.L. Schlenker. 1992. *J. Am. Chem. Soc.* 114:10834.

Becker, M.F., J.R. Brock, H. Cai, N. Chaudhary, D. Henneke, L. Hilsz, J.W. Keto, J. Lee, W.T. Nichols, and H.D. Glicksman. 1997. Nanoparticles generated by laser ablation. In *Proc. of the Joint NSF-NIST Conf. on Nanoparticles*.

Berger, R., C. Gerber, H.P. Lang, and J.K. Gimzewski. 1996. Micromechanics: a toolbox for femtoscale science: "Towards a laboratory on a tip." *Microelectronic Engineering*, 35:373-9. (International Conference on Micro- and Nanofabrication, Glasgow, U.K., 22- 25 Sept. 1996).

Berkowitz, A.E., J.R. Mitchell, M.J. Carey, A.P. Young, S. Zhang, F.E. Spada, F.T. Parker, A. Hutten, and G. Thomas. 1992. *Phys. Rev. Lett.* 68:3745.

Berndt, C.C., J. Karthikeyan, T. Chraska, and A.H. King. 1997. Plasma spray synthesis of nanozirconia powder. In *Proc. of the Joint NSF-NIST Conf. on Nanoparticles*.

Bimberg, D., N. Kirstaedter, N.N. Ledentsov, Zh.I. Alferov, P.S. Kopev, and V.M. Ustinov. 1998. InGaAs-GaAs quantum-dot lasers. *IEEE J. of Selected Topics in Quant. Electronics* 3:196-205.

Bowes, C.L., A. Malek, and G.A. Ozin. 1996. *Chem. Vap. Deposition* 2:97.

Braun, P.V., P. Osenar, and S.I. Stupp. 1996. *Nature*. 368: 2. Brinker, C.J. 1996. *Curr. Opin. Solid State Mater. Sci.* 1:798. Chen, C-Y., S.L. Burkett, H.-X. Li, and M.E. Davis. 1993. *Microporous Mater.* 2:27.

Brotzman, R. 1998. Nanoparticle Dispersions. In *R&D status and trends*, ed. Siegel et al. Bu, X., P. Feng, and G.D. Stucky. 1998. Large-cage zeolite structures with multidimensional 12-ring channels. *Science* 278:2080-2085.

Brotzman, R. 1998. Nanoparticle dispersions. In *R&D status and trends*, ed. Siegel et al. Brus, L.E. 1996. Theoretical metastability of semiconductor crystallites in high pressure phases with applications to beta tin structures of silicon. *J. Am. Chem. Soc.* 118:4834-38.

Brus, L. 1996. Semiconductor colloids: Individual nanocrystals, opals and porous silicon. *Current Opinion in Colloid & Interface Science* 1:197-201. *Chemical Engineering News*. 1997. Particulate matter health studies to be reanalyzed (August 18):33.

Calcote, H.F., and D.G. Keil. 1997. Combustion synthesis of silicon carbide powder. In *Proc. of the Joint NSF-NIST Conf. on Nanoparticles*.

Canguilhem, G.: 1952, *Machine et organisme' in La connaissance de la vie*, Hachette, Paris [quoted from the fourth edition Vrin, Paris, 1971].

Carsley, J.E., R. Shaik, W.W. Milligan, and E.C. Aifantis. 1997. In *Chemistry and physics of nanostructures and related non-equilibrium materials*. ed. E. Ma, B. Fultz, R. Shull, J. Morral, and P. Nash. Warrendale, PA: TMS.

Chance, B., Mueller, P., DeVault, D. & Powers, L. (1980) *Phys. Today* 33 (10), 32-38.

Chang, K. "Smaller Computer Chips Built Using DNA as Template." New York Times, November 21, 2003.

Chantrell. 1994. *J. Appl. Phys.* 76:6811. Erb, U., G. Palumbo, R. Zugic, and K.T. Aust. 1996. In *Processing and properties of nanocrystalline materials*, ed. C. Suryanarayana, J. Singh, and F.H. Froes. Warrendale, PA: TMS. Gertsman, V.Y., M. Hoffman, H. Gleiter and R. Birringer. 1994. *Acta Metall. Mater.* 42:3539-3544.

Chopra, N.G., R.J. Luyken, K. Cherrey, H. Crespi, M.L. Cohen, S.G. Louie, and A. Zettl. 1995. *Science* 269:966.

Clasen, R. 1990. *Int. Journal of Glass and Science Technology* 63: 291. Crandall, B.C. and J. Lewis, (eds.). 1992. *Nanotechnology: Research and perspectives*. Cambridge: MIT Press.

Czekai, D., et al. 1994. *Use of smaller milling media to prepare nanoparticulate dispersions*. U.S. Patent application. Docket 69802(02) Filed 2/25/94.

Dai, H., E.W. Wong, Y.Z. Lu, S. Fan, and C.M. Leiber. 1995. *Nature* 375:769. deHeer, W.A., W.S. Bacsa, A. Chatelain, T. Gerfin, R. Humphrey-Baker, L. Forro, and D. Ugarte. 1995. *Science* 268:845.

Dresselhaus, M.S., G. Dresselhaus, and P. Eklund. 1996. *Science of fullerenes and carbon nanotubes*. San Diego: Academic Press. Gleiter, H. 1989. *Prog. Mater. Sci.* 33:223.

Drexler K.E.: 1995, 'Introduction to nanotechnology', in: Krummenacker, M. & Lewis, J. (eds.), *Prospects in Nanotechology. Proceedings of the 1st general conference on nanotechnology: developments, applications, and opportunities, November 11-14, 1992, Palo-Alto*, John Wiley & Sons, New York, pp. 1-20.

Drexler, E., Phoenix, C.: 2004, 'Self-Replication in Nanotechnology: Feasible, Potentially Safe, and Unnecessary' (unpublished paper, in preparation).

Drexler, E.: 1986, *Engines of Creation: The Coming Era of Nanotechnology*, Anchor Books, New York (expanded edition with a new afterword, 1990).

Drexler, E.: 1992, *Nanosystems: Molecular Machinery, Manufacturing, and Computation*, John Wiley & Sons, New York.

Drexler, K.E.: 1981, 'Molecular engineering: An approach to the development of general capabilities for molecular manipulation', *Proceedings of the National Academy of Sciences*, 78, no. 9, chemistry section, 5275-78.

Drexler, K.E.: 1992, *Nanosystems. Molecular machinery, manufacturing and computation*, John Wiley & Sons, New York.

Drexler, K.E.: 2001, 'Machine-Phase nanotechnology', *Scientific American*, (Sept.), 66-67.

Drexler, K.E.: 2003a, 'Open Letter', *Chemical & Engineering News*, 81, no. 48, 37-42.

Drexler, K.E.: 2003b, 'An open letter to Richard Smalley', April 16, [published on KurzweilAI.net].

Dupuy, J.P. "Common Knowledge, Common Sense." Theory and Decision 27 (1989): 37-62.

Dupuy, J.P. "Two Temporalities, Two Rationalities: A New Look at Newcomb's Paradox." in Bourgine, P.&Walliser, B. (eds.), Economics and Cognitive Science. New York: Pergamon. 1992, 191-220.

Dupuy, J.P.: 2000, *The Mechanization of the Mind*, Princeton University Press, Princeton N.J.

Ellsberg, D. "Risk, Ambiguity and the Savage Axioms." Quarterly Journal of Economics 75 (1961): 643-669.

Entingh, D., Dunn, A., Glassman, E., Wilson, J. E., Hogan, E. & Damstra, T. (1975) in *Handbook of Psychobiology*, eds. Gazzinga, M. S. & Blakemore, C. (Academic, New York), pp. 201-238.

Estermann, M., L.B. McCusker, C. Baerlocher, A. Merrouche, and H. Kessler. 1991. *Nature* 331:698.

Esumi, A., A. Suzuki, N. Aihara, K. Uswi, and K. Torigoe. 1998. *Langmuir*, 14:3157.

Evans, D.F.; Wennerstrom, H.: 1999, *The Colloidal Domain: Where Physics, Chemistry, Biology and Technology Meet*, John Wiley & Sons, New York.

Faraday, M.: 1857, 'Experimental relations of gold (and other metals) to light', *Philosophical Transactions of the Royal Society London*, 147, 145-181.

Fennema, O. R. (1973) in *Low-Temperature Preservation of Foods and Living Matter*, eds. Fennema, O. R., Powrie, W. D. & Marth, E. H. (Dekker, New York), pp. 436-475.

Feyerabend, P.: 1981, *Realism, Rationalism and Scientific Method* (Philosophical Papers, volume 1), Cambridge University Press, Cambridge.

Feynman R. "There's Plenty of Room At the Bottom." Talk given on at the annual meeting of the American Physical Society at the California Institute of Technology, 1959.

Feynman, R.: 1960, 'There's Plenty of Room at the Bottom', *Engineering and Science*, 23, 22-36.

Fleming, D. "The Economics of Taking Care: An Evaluation of the Precautionary Principle." in Freestone, D.&Hey, E. (eds.), The Precautionary Principle and International Law. La Haye: Kluwer Law International, 1996.

Foerster, H.v.: 1962, 'Bio-Logic', in: E.E. Bernard & M.R. Kare (eds.), *Biological prototypes and synthetic systems*, Plenum Press, New York, vol. 1., pp. 1-12

Friedlander, S.K. 1993. Controlled synthesis of nanosized particles by aerosol processes. *Aerosol Sci. Technol*. 19:527.

Friedlander, S.K. 1998. Synthesis of nanoparticles and their agglomerates: Aerosol reactors. In *R&D status and trends*, ed. Siegel et al.

Friedlander, S.K., H.D. Jang, and K.H. Ryu. 1998. *Appl. Phy. Lett*. 72(2):173.

Gleiter, H. 1989. *Prog. Mater. Sci*. 33:223. Goddard, W.A. 1998. Nanoscale theory and simulation. In *R&D status and trends*, ed. Siegel et al.

Gleiter, H. 1990. *Progress in Materials Science* 33:4. Guinier, A. 1938. *Nature* 142:569; Preston, G.D. 1938. *Nature* 142:570.

Goddard, W.A. 1998. Nanoscale theory and simulation. In *R&D status and trends*, ed. R. Siegel et al.

Godsell, D.: 2003, *Living Machinery. Bionanotechnology: Lessons from Nature*, Wiley-Liss, New-York.

Guarnieri, F., M. Fliss, and C. Bancroft. 1996. Making DNA Add. *Science* 273:220-223. Hanes, J., J.L. Cleland, and R. Langer. 1997.

Günther, B., A. Baalmann, and H. Weiss. 1990. *Mater. Res. Soc. Symp. Proc.* 195:611-615.

Gutte, B., Dannigen, M. & Wittschieber, E. (1979) *Nature (London)* 281, 650-655.

Hadjipanayis, C.G. 1998. Nanostructured magnetic materials. In *R&D Status and Trends*, ed. Siegel et al.

Han, W., S. Fam, O. Li, and Y. Hu. 1998. Synthesis of gallium nitride nanorods through a carbon nanotube-confined reaction. *Science* 277:1287-1289.

Hansson, S.O. "The False Promises of Risk Analysis." Ratio 6 (1993): 16-26.

Held, R., T. Heinzel, A.P. Studerus, K. Ensslin, and M. Holland. 1997. Semiconductor quantum point contact fabricated by lithography with an atomic force microscope. *Appl. Phys. Lett.* 71:2689-91.

Hellemans, A. 1998. X-rays find new ways to shine. *Science* 277:1214-15.

Hietpas, P.B., S.D. Gilman, R.A. Lee, M.R. Wood, N. Winograd, and A.G. Ewing. 1996. Development of votammetric methods, capillary electrophoresis and tof sims imaging for constituent analysis of single cells. In *Nanofabrication and biosystems*, ed.

Higgins, R.J. 1997. An economical process for manufacturing of nano-sized powders based on microemulsion-mediated synthesis. In *Proc. of the Joint NSF-NIST Conf. on Nanoparticles.*

Hiruma, K., M. Yazawa, T. Katsoyama, K. Ogawa, K. Haraguchi, M. Koguchi, and H. Kakibayashi. 1995. *J. Appl. Phys.* 77(2):476.

Hoch et al. Ho, S.V., P.W. Sheridan, and E. Krupetsky. 1996. Supported polymeric liquid membranes for removing organics from aqueous solutions.

Hoch, H.C., L.W. Jelinski, and H.G. Craighead, eds. 1996. *Nanofabrication and biosystems*. New York: Cambridge University Press.

Howland, H.: 1962, Structural, hydraulic and economic aspects of leaf venation and shape', in: E.E. Bernard & M.R. Kare (eds.), *Biological prototypes and synthetic systems*, Plenum Press, New York, vol. 1, pp. 183-192 .

Hoyningen-Huene, P.: 1993, *Reconstructing Scientific Revolutions: Thomas S. Kuhn's Philosophy of Science* (trans. by A. Levin), University of Chicago Press, Chicago.

Huang, J.Y., Y.K. Wu, and H.Q. Ye. 1996. *Acta Mater.* 44:1211. Inoue, A. 1997. Private communication. Inturi, R.B., and Z. Szklavska-Smialowska. 1992. *Corrosion* 48:398. Karch, J., R. Birringer, and H. Gleiter. 1987. *Nature* 330:556.

Hubbell, J.A., and R. Langer. 1995. Tissue engineering. *Chem. Eng. News* (March 13): 42- 54.

Huo, Q., D.I. Margolese, U. Ciesla, P. Feng, T.E. Gier, P. Sieger, R. Leon, P.M. Petroff, F. Schuth, and G.D. Stucky. 1994. *Nature* 368:317.

Imae, Y., and T. Atsumi. 1989. T. Na+-driven bacterial flagellar motors: A mini-review. *J. Bioenergetics and Biomembranes* 21:705-716.

Itakura, K. & Riggs, A. D. (1980) *Science* 209, 1401-1405.

Jaworek, T., D. Deher, G. Wegner, R.H. Wieringa, and A.J. Schouten. 1998. Electromechanical properties of an ultrathin layer of directionally aligned helical polypeptides. *Science* 279:57-60.

Jena, P., S.N. Khanna, and B.K. Rao. 1996. In *Science and technology of atomically engineered materials*, ed. P. Jena. River Edge, NJ: World Scientific.

Jonas, H. The Imperative of Responsibility. In Search of an Ethics for the Technological Age. Chicago: University of Chicago Press, 1985.

Jounet, C., W.K. Maser, P. Bernier, A. Loiseau, M. Lamy de la Chapelle, S. Lefrant, P. Deniard, R. Lee, and J.E. Fischer. 1997. *Nature* 388:756.

Junno, T., S.-B. Carlsson, H. Xu, L. Montelius, and L. Samuelson. 1998. Fabrication of quantum devices by angstrom-level manipulation of nanoparticles with an atomic force microscope. *Appl. Phys. Lett.* 72:548-550.

K. E. Drexler, "Molecular engineering: An approach to the development of general capabilities for molecular manipulation." *Proc. Nat. Acad. Sci.,* 78: 5275-5258 (1981).

K. E. Drexler, *Engines of Creation*, Anchor/Doubleday (New York, 1986).

Karak, N., and S. Maiti. 1997. Dendritic polymers: A class of novel material. *J. Polym. Mater.*14:105.

Karger, J., and D.M. Ruthven. 1992. *Diffusion in zeolites*. New York: J. Wiley. Krätschmer, W., L.D. Lamb, K. Fostiropoulos, and D.R. Huffman. 1990. *Nature* 347:354.

Karplus, M. & Weaver, D. L. (1976) *Nature (London)* 260, 404-406.

Kasuga, T., M. Hiramatsu, A. Hoson, T. Sekino, and K. Niihara. 1998. *Langmuir* 14:3160.

Katari, J.E.B., V.L. Colvin, and A.P. Alivisatos. 1994. *J. Phys. Chem.* 98:4109.

Ke, M., S.A. Hackney, W.W. Milligan, and E.C. Aifantis. 1995. *Nanostructured Mater.* 5:689.

Kear, B., and G. Skandan. 1998. Nanostructural bulk materials: Synthesis, processing, properties and performance. *In R&D status and trends*, ed.

Kear, B.H., R.K. Sadangi, and S.C. Liao. 1997. Synthesis of WC/Co/diamond nanocomposites. In *Proc. of the Joint NSF-NIST Conf. on Nanoparticles.*

Kirschvink, J.L., A. Koyayashi-Kirschvink, and B.J. Woodford. 1992. Magnetite biomineralization in the human brain. *Proc. Nat'l. Acad. Sci.* USA 89:7683-7687.

Kishida, M., T. Fujita, K. Umakoshi, J. Ishiyama, H. Nagata, and K. Wakabayashi. 1995. *Chem. Commun.* 763.

Klein, J.D., et al. 1993. *Chem. Mater.* 5:902. Koch, C.C. 1989. Materials synthesis by mechanical alloying. *Annual Review of Mater. Sci.* 19:121-143.

Knight, F.H. Risk, Uncertainty and Profit. London School of Economics and Political Science, London, 1933.

Koch, C.C. 1998. Bulk behavior. In *R&D status and trends*, ed. Siegel et al. 32 *Evelyn L. Hu and David T. Shaw.*

Kortan, A.R., R. Hull, R.L. Opila, M.G. Bawendi, M.L. Steigerwald, P.J. Carroll, and L.E. Brus. 1990. *J. Am. Chem. Soc.* 112:1327.

Kourilsky, Ph.&Viney, G. Le Principe de précaution. Report to the Prime Minister, Paris, Éditions Odile Jacob, 2000.

Krejchi, M.T., E.D.T. Atkins, A.J. Waddon, M.J. Fournier, T.L. Mason, and D.A. Tirrell. 1994. Chemical sequence control of beta-sheet assembly in macromolecular crystals of periodic proteins. *Science* 265:1427-1432.

Krejchi, M.T., S.J. Cooper, Y. Deguchi, E.D.T. Atkins, M.J. Fournier, T.L. Mason, and D.A. Tirrell. 1997. Crystal structures of chain-folded antiparallel beta-sheet assemblies from sequence-designed periodic polypeptides. *Macromolecules* 30:5012-5024.

Kresge, C.T., M.E. Leonowicz, W.J. Roth, J.C. Vartuli, and J.S. Beck. 1992. *Nature* 359:710. Kupperman, A., S. Nadimi, S. Oliver, G. Ozin, J. Garcés, and M. Olken. 1993. *Nature* 365:239.

Kroes, P. & Meijers, A. "The Dual Nature of Technical Artifacts—Presentation of a New Research Programme." Techné 6, 2 (2002): 4-8.

Kuhn, T.: 1970, *The Structure of Scientific Revolutions* (second edition), University of Chicago Press, Chicago.

Kumar, A., and G.M. Whitesides. 1993. Features of gold having micrometer to centimeter dimensions can be formed through a combination of staming with an elastomeric stamp and an alkanethiol ink followed by chemical etching. *App. Phys. Lett.* 63:2002-2004.

Kurzweil, R.: 1998, *The Age of Spiritual Machines, How We Will Live, Work and Think in the New Age of Intelligent Machines*, Phoenix, New York.

Kyprianidou-Leodidou, T., W. Caseri, and V. Suter. 1994. *J. Phys. Chem.* 98:8992. Kung, H.H., and E.I. Ko, 1996. *Chem. Eng. J.* 64:203.

L. J. Perry and R. Wetzel, "Disulfide bond engineered into T4 lysozyme: stabilization of the protein toward thermal inactivation." *Science,* 226: 555-557 (1984).

L. Regan and W. F. DeGrado, "Characterization of a helical protein designed from first principles." *Science,* 241: 976-978 (1988).

Laudan, L.: 1984, *Science and Values: The Aims of Science and their Role in Scientific Debate*, University of California Press, Berkeley.

Lehn, J.M., 2004: 'Une chimie supramoléculaire foisonnante', *La lettre de l'Académie des sciences*, 10, 12-13.

Leonard, D., M. Krishnamurthy, C.M. Reaves, S.P. Denbaars, and P.M. Petroff. 1993. Direct formation of quantum-sized dots from uniform coherent islands of InGaAs on GaAs surfaces. *Appl. Phys. Lett.* 63:3203-5.

Liang, G., Z. Li, and E. Wang. 1996. *J. Mater. Sci.* Makino, A., A. Inoue, T. Hatanai, and T. Bitoh. 1997. *Materials Science Forum* 235-238: 723.

Lide, D.R., ed. 1993-1994. *CRC Handbook of Chemistry and Physics*, 74th ed. Lin, H.-P., and C.-Y. Mou. 1996. *Science* 273:765.

Lukens, R.: 2004, 'Historic Nanotechnology', *Chemical Heritage,* 22, 29.

Majetich, S.A. and A.C. Canter. 1993. *J. Phys. Chem.* 97:8727.

Mao, C., W. Sun, and N.C. Seeman. 1997. Construction of Borromean rings from DNA. *Nature* 386:137-138.

Martin, T.P., N. Malinowski, U. Zimmerman, U. Naher, and H. Schaber. 1993. *J. Chem. Phys* 99:4210.

Martin, T.P., U. Naher, H. Schaber, U. Zimmerman. 1993. *Phys. Rev. Lett.* 70:3079. Prigogine, I., and S. Rice. 1988. *Advances in chemical physics*, Vol. 70, Parts 1 & 2. New York: J. Wiley.

McCammon, J. A., Gelin, B. R. & Karplus, M. (1977) *Nature (London)* 267, 585-590.

McConnell, H.M. 1996. Light-addressable potentiometric sensor: Applications to drug discovery. In *Nanofabrication and biosystems*, ed. Hoch et al.

McCulloch, W.S.: 1962, 'The imitation of one form of life by another – Biomimesis', in: E.E. Bernard & M.R. Kare (eds.), *Biological prototypes and synthetic systems*, Plenum Press, New York, vol. 1., p. 393-97.

Mehl, R.F., and R.W. Cahn. 1983. Historical development. In *Physical metallurgy*. North Holland. Milligan, W.W., S.A. Hackney, M. Ke, and E.C. Aifantis. 1993. *Nanostructured Materials* 2:267.

Minsky, M.: 1995, 'Virtual Molecular Reality', in: Krummenacker, M. & Lewis, J. (eds.), *Prospects in Nanotechology. Proceedings of the 1st general conference on nanotechnology: developments, applications, and opportunities, November 11-14, 1992, Palo-Alto*, John Wiley & Sons, New York, pp. 187-205.

Mishra, R.S., and A.K. Mukherjee. 1997. Oral presentation at TMS meeting, Indianapolis, Indiana, 16-18 September 1997, to be published in proceedings of Symp. "Mechanical Behavior of Bulk Nano-Materials."

Mishra, R.S., R.Z. Valiev, and A.K. Mukherjee. 1997. *Nanostructured Materials* 9:473. Morris, D.G., and M.A. Morris. 1991. *Acta Metall. Mater.* 39:1763-1779.

Moore, J.C., H.M. Jin, O. Kuchner, and F.H. Arnold. 1997. Strategies for the in vitro evolution of protein function: Enzyme evolution by random recombination of improved sequences. *J. Mol. Biol.* 272:336-347.

Morris, D.G., and M.A. Morris. 1997. *Materials Science Forum* 235-238:861. Nagpal, P., and I. Baker. 1990. *Scripta Metall. Mater.* 24:2381.

Nieman, G.W., J.R. Weertman, and R.W. Siegel. 1991a. *Mater. Res. Soc. Symp. Proc.* 206:581-586.

Nomura, M. & Held, W. (1974) in *Ribosomes,* eds. Nomura, M., Tissiers, A. & Lengyel, P. (Cold Spring Harbor Laboratory, Cold Spring Harbor, NY), pp. 193-203.

Nordmann, A. (rapp.) Converging Technologies—Shaping the Future of European Societies, European Commission report, 2004.

Parker, J.C., et al. 1995. U.S. Patent 5,460,701. POST (Parliamentary Office of Science and Technology). 1995.

Phoenix, C.; Drexler, E.: 2004, 'Safe Exponential Manufacturing', *Nanotechnology*, 15, 869-72.

Rademann, K., B. Kaiser, U. Even, F. Hensel. 1987. *Phys. Rev. Lett.* 59:2319. Rao, C.N.R., B.C. Satishkumar, and A. Govindaraj. 1997. *Chem. Commun.* 1581.

Ramanan, V.R. 1998. Nanocrystalline soft magnetic alloys for applications in electrical and electronic devices. In *R&D Status and Trends*, ed. Siegel et al.

Rao, M.B., and S. Sircar. 1993. *Gas Separation and Purification* 7:279. Reetz, M.T., et al. 1995. *Science* 267:367.

Rechard, R.P. "Historical Relationship Between Performance Assessment for Radioactive Waste Disposal and Other Types of Risk Assessment." Risk Analysis 19, 5 (1999): 763-807.

Rietman, E.A.: 2001, 'Drexler hypothesis of a universal assembler is supported not by theoretical arguments alone but by existence proof in the form of biological life', in: *Molecular Engineering of Nanosystems*, Springer, New York & Berlin.

Rip, A., Misa, Th. J.,&Schot, J. W. (eds.). Managing Technology in Society. The Approach of Constructive Technology Assessment. London: Pinter Publishers, 1995.

Roco, M.&Bainbridge, W. (eds.). Converging Technologies for Improving Human Performances, National Science Foundation report, 2002.

Rofagha, R., R. Langer, A.M. El-Sherik, U. Erb, G. Palumbo, and K.T. Aust. 1991. *Scr. Metall. Mater*. 25:2867.

Roher, H. 1993. *Jpn. J. Appl. Phys*. 32:1335. Rohlfing, E.A., D.M. Cox, and A. Kaldor. 1984. *J. Chem. Phys*. 81:3846.

Romanov, A.E., V.I. Vladimirov. 1992. In *Dislocations in solids*, ed. F.R.N. Nabarro, Vol. 9. Amsterdam: North-Holland. Salishekev, G.A., O.R. Valiakhmetov, V.A. Valitov, and S.K. Mukht ırov. 1994. *Materials Science Forum*. 170-172:121.

Rosenberg, N. "Why Technology Forecasts Often Fail." Futurist July/August 1995, 16- 21.

Ruthven, D.M., S. Farooq, K.S. Knaebel. 1994. *Pressure swing adsorption*. New York: VCH Publishers. Sayari, A. 1996. *Chem. Mater*. 8:1840.

S. P. Ho and W. F. DeGrado, "Design of a 4-Helix bundle protein: Synthesis of peptides which self-associate into a helical protein." *J. Am. Chem. Soc.*, 109: 6751-6758 (1987).

Sanders, P.G., J.A. Eastman, and J.R. Weertman. 1996. In *Processing and properties of nanocrystalline materials*, ed. Suryanarayana et al.

Sanders, P.G., M. Rittner, E. Kiedaisch, J.R. Weertman, H. Kung, and Y.C. Lu. 1997. *Nanostructured Mater*. 9:433.

Sankey, H.: 1994, *The Incommensurability Thesis*, Avebury, Aldershot.

Sarikaya, M.; Aksay, I. (eds.), 1995, *Biomimetics. Design and Processing of Materials*, AIP Press, Woodbury, New York.

Scanlan, R.M., W.A. Fietz, and E.F. Koch. 1975. *J. Appl. Phys*. 46:2244.

Scheraga, H. A. (1978) in *Versatilty of Proteins*, ed. Li, C. H. (Academic, New York), pp. 119-132.

Schiefsky, M.J.: forthcoming, 'Art and Nature in Ancient Mechanics', in: W.R. Newman & B. Bensaude-Vincent (eds.), *The Artificial and the Natural: An Ancient Debate and its Modern Descendants*, MIT Press, Cambridge, MA.

Schultz, L., J. Wecker, and E. Hellstern. 1987. *J. Appl. Phys*. 61:3583.

Schultz, L., K. Schnitzke, and J. Wecker. 1989. *J. Magn. Mater*. 80:115.

Schwarz, R.B. 1998. Storage of hydrogen powders with nanosized crystalline domains. In *R&D Status and Trends*, ed. Siegel et al.

Shen, T.D., C.C. Koch, T.Y. Tsui, and G.M. Pharr. 1995. *J. Mater. Res*. 10:2892.

Shull, R.D. 1998. NIST activities in nanotechnology. In *R&D Status and Trends*, ed.

Shull, R.D., R.D. McMichael, and J.J. Ritter. 1993. *Nanostructured Mater*. 2:205.

Siegel et al. Ogunnaike, B and W. Ray. 1994. *Process dynamics, modeling and control*. Oxford University Press, pp 5-21; 1033-48.

Siegel et al. Siegel, R.W. 1997. *Materials Science Forum* 235-238:851.

Siegel R.W. and G.E. Fougere. 1994. In *Nanophase materials*, ed. G.C. Hadjipanayis and R.W. Siegel. Netherlands: Kluwer Acad. Publ.

Siegel, H.: 1980, 'Objectivity, Rationality, Incommensurability, and More', *British Journal for the Philosophy of Science*, 31, 359-384.

Siegel, R.W., E. Hu, and M.C. Roco, eds. 1998. *R&D status and trends in nanoparticles, nanostructured materials, and nanodevices in the United States*. Baltimore: Loyola College, International Technology Research Institute.

Smalley, R.: 2001, 'Of Chemistry, Love and Nanobots', *Scientific American*, 285, 76-77.

Smalley, R.: 2003a, 'Smalley Responds', *Chemical & Engineering News,* 81, 39-40.

Smalley, R.: 2003b, 'Smalley Concludes', *Chemical & Engineering News,* 81, 41-42.

Smalley, R.E.: 1999, 'Prepared written statement and supplemental material', Rice University, 22 June [http//www.house.gov/science/smalley_062299.htm].

Smalley, R.E.: 2001, 'Of Chemistry, Love and Nanobots', *Scientific American*, (Sept.), 76-77.

Smith, C.S. 1981. *A search for structure*. Cambridge, Mass.: MIT Press.

Snow, E.S., P.M. Campbell, and F.K. Perkins. 1997. Nanofabrication with proximal probes. *Proceedings of the IEEE* 85:601-11.

Staden, H.v.: forthcoming, '*Physis* and *Technê* in Greek Medicine', in: W.R. Newman & B. Bensaude-Vincent (eds.), *The Artificial and the Natural: An Ancient Debate and its Modern Descendants*, MIT Press, Cambridge, MA.

Stupp, S., and P.V. Braun. 1997. Molecular manipulation of microstructures: biomaterials, ceramics, and semiconductors. *Science* 277:1242-1248.

Sunstrom, J.E., IV, W.R. Moser, and B. Marshik-Guerts. 1996. *Chem. Mater*. 8:2061.

Surowiecki, J.: 2004, 'Bring on the Nanobubble', *The New Yorker*, March 15, p. 68.

Thompson, D.: 1992 [1942], *On Growth and Forms*, Cambridge University Press, Cambridge.

3

Global Issues Regarding Nanoparticles, Nanostructured Materials & Polymer Nanotechnology

Issues in Nanoparticles & Nanostructured Materials

During the course of this study, the WTEC panelists were privileged to hear of and to observe a wide variety of current work on nanostructured dispersions and coatings in various laboratories around the world. Descriptions follow of a number of research projects and the preparation issues being addressed in the United States and in some of the foreign laboratories that panelists visited. The nanoparticle preparation techniques that are currently in use, many of which will be discussed in the paragraphs that follow.

Synthesis and Application of Polymer Nanotechnology

Polymer nanoparticles have been produced for decades for use in a variety of high performance materials such as high impact resistant polymers and specialty coatings long before it was fashionable to use the "nano" label. The extraordinarily large surface area on the nanoparticles presents diverse opportunities to place functional groups on the surface. Particles can be created that can expand/contract with changes in pH, or interact with anti-bodies in special ways to provide rapid ex-vivo medical diagnostic tests. Important extensions have been made in combining inorganic materials with polymers and in combining different classes of polymers together in nanoparticle form.

More recent advances in chemistry, processing techniques, and analytical instrumentation allow a whole host of new types of polymer particles to be made. We now have particles that are hollow, multi-lobed, magnetic, functionalized with reactive groups on the surface, conductive, etc. Our ability to devise new process control strategies have led to the ability to control the shape, chemical composition, internal structure, and morphology of the nanoparticles so as to develop new levels of product performance. Advanced analytical techniques allow us to measure structure at ever-decreasing length scales. Computer simulations of the events occurring during particle formation have also benefited us in developing control strategies to produce structured particles

The first part of the course focuses on the wet synthesis methods and chemistries used to produce polymer particles in the 10-1000 nanometer range. Careful attention is paid to the analytical techniques that allow confident measurement of phase-separated structure within these composite particles and the varied applications of such polymer particles. The objectives are to:

- Introduce the diversity of polymer particle sizes, shapes, and chemical composition with structural control at the nano scale
- Describe how particles can be made from either reactive or non-reactive processing techniques, including emulsion polymerization and self assembly

The second part focuses on characterization (analysis and verification) and the wide range of application areas of economic importance. The objectives are to:

- Demonstrate how particles can be characterized by modern instrumental techniques
- Discuss the multitude of applications for polymer nanoparticles
- Provide an interactive learning experience through the use of individual and group exercises and case studies

Polymeric nanoparticles are predominantly prepared presently, by wet synthetic routes. Several industrial processes will be described. Emphasis will be placed on the type of polymers and morphology structures that can be synthesized using each process. Controlled radical polymerization will be explored for their ability to provide structural control of polymer chains.

A brief overview of suspension and dispersion-precipitation polymerization relevant to nanoparticles is included. Solution polymerization methods are provided to allow a more detailed presentation of a non-reactive pathway to nanoparticle formation. The third day is dedicated to polymeric nanoparticles characterization. Of critical importance is the measurement of particle size, bulk and surface chemical composition. Importantly, the internal structure of the particles, can be measured by combinations of microscopy and thermal analysis. All techniques will be presented for their utility toward polymeric nanoparticles characterization.

The fourth day of the course is focused on the applications of polymeric nanoparticles. The most traditional field of application is waterborne paints, adhesives, and coatings.

The details of some formulation and application issues will be presented. Of more recent emergence is the field of redispersible lattices, powder coatings, and pressure sensitive adhesives. A recent boom in the range of application of polymeric nanoparticles is the sector of biotechnology, and more specifically biomedical products. These include the critical delivery of sensitive drugs and medical diagnostics. Last but not least, polymeric nanoparticles have found their ways in EMO devices.

All through the course practical problems will be solved both individually and in groups.

Synthesis

- Polycondensation polymerization
- Radical polymerization
- Living polymerization
- SFRP, ATRP, RAFT for controlled radical polymerization
- Problem solving
- Suspension and dispersion-precipitation polymerization
- Emulsion polymerization
- Semi-batch and batch processes
- Single stage and multistage products.
- Problem solving
- Micro-emulsion and mini-emulsion
- Nanoencapsulation in direct and reverse phases.
- Self-assembly
- Directed assembly
- Patterned substrate effects
- Solvent removal processes
- Problem solving

Characterization

- Size and size distribution.
- CHDF, Capillary Hydro Dynamic Fractionation
- MALS, Multi Angle Light Scattering
- DLS, Dynamic Light Scattering
- Field flow fractionation
- UAC, Analytical Ultra Centrifugation

- SEM, TEM, AFM, microscopy techniques.
- Problem solving
- Bulk chemical composition
- NMR, Nuclear Magnetic Resonance
- Spectroscopic techniques, FTIR, Raman, UV, and fluorescence,
- Problem solving
- Pyrolysis-gas chromatography-mass spectroscopy,
- Mass-spectroscopy
- Surface composition
- Surfactant titration
- XPS, X-ray Photon Spectroscopy
- EDS, Electron Dispersion Spectroscopy
- SERS, Surfaced Enhanced Raman Spectroscopy
- ITC, Isothermal Titration Calorimetry
- Internal structure, TEM, STXM, thermal analysis
- Problem solving

Applications

- Waterborne paints
- Adhesives
- Coatings
- Redispersible latices
- Powder coatings
- Pressure sensitive adhesives
- Hot melts
- Biotechnology
- Biomedical products
- Drug delivery
- Medical diagnostics
- Problem solving
- Electronics
- Magnetics
- Optoelectronics

Polymeric Nanocomposites

Since the predominant feature of nanoparticles lies in their ultra-fine dimension, a large fraction of the filler atoms can reside at the interface and can lead to a strong interfacial interaction (Roy, 1986), but only if the nanoparticles are well dispersed on a nanometer level in the surrounding polymer matrix. As the interfacial structure plays a critical role in determining the composites' properties, nanocomposites coupled with a great number of interfaces could be expected to provide unusual properties, and the shortcomings induced by the heterogeneity of conventional particle filled composites would also be avoided (Rong et al., 2001). Recent progress indicated that polymer-based nanocomposites acquired reduced frictional coefficients and wear rates. For example, Wang et al. (1998) reported that the incorporation of nanosized ZrO_2 into polyetheretherketone caused a considerable improvement in the tribological characteristics. It was found that the dominant wear mechanism changed from melting adhesive transfer wear to slight transfer wear, and finally to abrasive wear with increasing nano-ZrO_2 content. Wang and his co-workers (1998) prepared nanoscale ceramic (TiC, Si_3N_4, B_4C)/polytetrafluoroethylene multilayers by ion beam sputtering deposition at room temperature. Ball-on-disk tribological tests showed that the multilayers with optimized layer thickness arrangement experienced good performance in wear resistance. There was no obvious periodic variation in the friction coefficient. Schadler et al. (1997) produced silica/polyamide nanocomposite coatings using high-speed oxyfuel thermal spray processing; they found that the surface chemistry of the nano-silica affected the final coating properties. Silica particles with a hydrophobic surface resulted in higher scratch resistance than those with a hydrophilic surface. However, very few works have been presented to date concerning the relationship between microstructural details and tribological properties of bulk, polymer-based nanocomposites. This is very surprising, since this is a topic which has both a fundamental and an applied significance for the development of nanocomposites for new technical applications. Therefore, the objective of the present work is focused on the tribological performance of composites with identical species of the reinforcing components, but with different ways of compounding and different particle sizes. In particular, the effects of the dispersion status and the size of nanoparticles on the friction and wear behavior of such nanocomposites are studied, so as to provide knowledge for an optimum material preparation.

Commercially available epoxy (E-51) and tetraethylenepentamine (TEPA), provided by Guangzhou Adhesive Co. and Shanghai Dye Chemical Engineering Co., China, were used as matrix polymer and curing agent, respectively. Its reinforcement was carried out by the use of home-synthesized TiO_2 particles, having average diameters of 10nm and 40mm, respectively. To prepare nanocomposites with different filler arrangements, three approaches were employed (Liu, 1997):

1. (1) Nanosized TiO_2 particles were directly mixed into the epoxy resin, which was put into an oven, preset at 80°C. After half an hour, the mixture was taken out

and stirred mechanically for another half an hour. To improve the homogeneity of nanoparticles dispersion, the nano-TiO_2/epoxy compounds were further treated under ultrasonic environment for 10min. Finally, the curing agent was added, and the solid nanocomposites were obtained after holding the mixture for 2hr at 80°C.

2. (2) At first, nanosized TiO_2 particles were dispersed in chloroform, aided by sonication. This suspension was then mixed with the epoxy resin by mechanical stirring for half an hour and subsequent sonication for 10min, respectively. When the solvent was evaporated through vacuum drying, the mixture of nano-TiO_2 and epoxy was again stirred and ultrasonically vibrated for half an hour and 10min, respectively. Then the curing agent was added to the mixture, and the above curing process was followed.

3. (3) The nano-TiO_2 suspension in chloroform made according to the last approach was mixed with epoxy. The mixture was then stirred mechanically for 2.5hr, and subsequently ultrasonically vibrated for 10min. In the next step, the mixture was heated to 80°C to remove the solvent while being stirred once more for 2hr. Eventually, the TiO_2/epoxy nanocomposite was cured as described in the last part of approach (1).

When the epoxy was filled with coarse TiO_2 particles (diameter: 40ìm), the composites were manufactured in accordance with the third approach mentioned above. Unlubricated sliding wear tests were carried out on a pin-on-ring apparatus under a pressure of 1MPa and a constant velocity of 0.4m/s. The carbon steel ring (0.42~0.45wt% C, 0.17~0.37wt% Si and 0.5~0.80wt% Mn, HRC 50) had a diameter of 40mm and an initial surface roughness of 0.1ìm. Prior to wear testing, all the samples were pre-worn so as to average surface conditions and reduce the running-in period. The actual steady-state test period was set to 2hr. After that, a weight measurement of the specimens was carried out. Microindentation Vickers hardness values of the composites were determined by a HDX-1000 digitized microhardness tester. The reported hardness data are the average of five successful measurements. Dynamic mechanical tests were conducted through a DMTA BMARK-III analyzer at a frequency of 1Hz. The internal friction, expressed by the tangential loss angle, tan ä, of the materials was measured under three-point bending configuration at a heating rate of 2úC/min.

The specific wear rates ($wÿ_s$) and the frictional coefficients (ì) of nanocomposites which were prepared through a direct compounding process. In general, it can be stated that especially the wear rates are very high, which is, however, mainly due to the low stiffness nature at elevated temperature of the epoxy system used for this study. With a rise in the nano-TiO_2 concentration, $wÿ_s$ keeps almost unchanged at the beginning, but increases furthermore when the filler content exceeds 3phr (phr = amount of filler by weight per 100 parts matrix by weight). At the same time, ì exhibits a fluctuating mode of variation. As can be imagined, approach (1) of composite manufacturing, as described in the experimental part, is bound to generate a rather heterogeneous dispersion of the nanoparticles. This is due to two facts, a strong tendency of the particles to agglomerate, and a high viscosity of the epoxy matrix. As a result, an extensive material loss due to disintegration and crumbling of the nanoparticle agglomerates under tribological conditions occurs (Bely et al., 1982). Besides, the unevenly distributed nanoparticles also provide the pin surface with an inhomogeneous modification of the mechanical properties, leading to the irregular dependence of ì on the TiO_2 loading. It is generally believed that solvent blending helps to improve the homogeneity of composites' microstructure. Therefore, TiO_2/epoxy nanocomposites were produced by using a nano-TiO_2/chloroform suspension. The tribological properties of the nanocomposites made in accordance with approach (2). In comparison to the values of $wÿ_s$ were slightly reduced in the TiO_2 content range, between 1 and 7phr. In the case of 3phr of nano-TiO_2 particles, for example, the wear rate of the composite prepared with the aid of chloroform is more than two times lower than both the wear rate of the neat epoxy ($2.67 \times 10^{-2}mm^3/Nm$) and that of the composite prepared through a direct mixing of the components. This should be attributed to the improved dispersion status of the nanoparticles. However, it can also be seen that the profiles of the $wÿ_s$ vs TiO_2 content-curves in both resemble each other in general, implying that the two types of composites have a similar microstructure. When the TiO_2 loading is higher, splitting of the agglomerated nanoparticles likely occurs and increases the wear loss.

When nano-TiO_2 particles are incorporated into the epoxy, the microhardness of the composites is increased. This corresponds to an increase in the wear resistance of the composites, especially if a rather uniform dispersion of the nanoparticles exists. On the other hand, the deviation of Hv of the composites (prepared according to approaches (1) and (2)) is not remarkably higher than that of the neat epoxy matrix. This means that the effect of reinforcement imparted by the additives, which is closely related to the dispersion state of the particles, was probably not fully utilized. In other words, it can be concluded that approach (2) is not effective enough to produce a homogeneous dispersion of the particles, and as a result, the expected, positive influence of the nano-TiO_2 particles on the wear performance of the epoxy resin could not be brought into full play. With respect to the frictional coefficient, its dependence on the TiO_2 content is characterized by a "U"-shape, which manifests a stable friction process different from that shown. It suggests that the addition of nano-TiO_2 is able to provide certain lubricating effects, but the more severe removal of material under higher filler loading might lead to a three-body abrasive wear situation. Hence, ì increases with increasing TiO_2 content above 5phr. The friction and wear properties of the nanocomposites manufactured through the third approach. The distinct step in approach (3), which is different from approach (2), lies in the fact that chloroform was removed while the mixture of epoxy and TiO_2 particles was being stirred continuously. It is believed that such a way of solvent removal results in a more uniform distribution of the particulate fillers, which is substantiated by the greatly increased microhardness and a rather small deviation (±0.4MPa) close to that of neat epoxy measurements (0.6MPa). The wear resistance of the epoxy resin can be increased by factors of four to 16 for the observed range of nano-TiO_2 loading, while the frictional coefficient shows a decreasing tendency when the particle loading exceeds 5phr. This means the tribological behavior is completely different from that reflected.

Clearly indicates that much more chain motion is restricted by the nanoparticles when the composite is produced by approach (3). This means the mobility of the macromolecules at filler/matrix interfaces becomes clearly lower, as characterized by the unusual increase in the á transition temperature, implying increased contacts between the fillers and the matrix. It further demonstrates the relatively uniform microstructure of the nanocomposites tested. It is interesting to find that the decrease in frictional coefficient corresponds to a rise in wear rate when the nano-TiO_2 content is higher than 5phr. This is somewhat like the case of polytetrafluoroethylene, where the extremely low frictional coefficient wear resistance due to the spec wear resistance due to the specific lamellae arrangement. In the case of lower filler content, however, the composites and the unfilled matrix exhibit almost the same frictional coefficient. This can be attributed to the size effect of the additives. That is, when the nano-fillers are well dispersed, they can have the same size as the segments of the surrounding polymer chains so that the polymer filler phase can make a predominant contribution to friction due to its structural integrity. When the content of the nanoparticles is increased to a certain threshold (e.g. 5phr for the current work), at which approach (3) is no longer effective to prevent agglomeration of the particles, the increasing amount of detached nano-TiO_2 particles at the interface between the steel counterpart and the composite pin can significantly reduce the friction. Further investigation of the micromechanisms involved in the present systems are, however, needed.

It is worth noting that approach (3) is not universally applicable. When nano-TiO_2 was replaced by micron-sized TiO_2, the particles were poorly dispersed. The resultant composites have a performance similar to that of the composites prepared in terms of approaches (1) and (2). This implies, on one hand, that TiO_2 nanoparticles can effectively improve the wear resistance of epoxy, while conventional micron-TiO_2 particles cannot. On the other hand, when the nanoparticles are unevenly dispersed in the matrix, the nanocomposites perform like composites filled with micrometer-sized particulate fillers, in which crack-initiation and coalescence occur more easily in the particulate-rich phases.

Friction and wear properties of polymer nanocomposites are a function of the dispersion state of the inorganic type, nanosized reinforcing nanoparticles. A uniform distribution of the nanoparticles helps to enhance the wear resistance. TiO_2 nanoparticles are able to significantly reduce the wear rate and the frictional coefficient of epoxy. Depending on the dispersion state and content of the nanoparticles, these positive effects do not appear simultaneously. That means, an optimum design of the nanocomposites is required. Mixing of epoxy and the nanoparticles with the aid of chloroform can tentatively lead to a relatively homogeneous phase dispersion, and a continuous stirring associated with heating maintains this microstructure until the composite is fully cured. Further efforts should be made to explore more viable techniques to break up the nanoparticle agglomerates and to improve the interfacial adhesion between the particles and the matrix resin. This is quite important for making full use of nanoparticles in tribological applications.

High Surface Area Materials

The trend to smaller and smaller structures, that is, miniaturization, is well known in the manufacturing and microelectronics industries, as evidenced by the rapid increase in computing power through reduction on chips of the area and volume needed per transistor (Roher 1993). In the materials area this same trend towards miniaturization also is occurring, but for different reasons. Smallness in itself is not the goal. Instead, it is the realization, or now possibly even the expectation, that new properties intrinsic to novel structures will enable breakthroughs in a multitude of technologically important areas (Siegel 1991; Gleiter 1989). Of particular interest to materials scientists is the fact that nanostructures have higher surface areas than do conventional materials. The impact of nanostructure on the properties of high surface area materials is an area of increasing importance to understanding, creating, and improving materials for diverse applications. High surface areas can be attained either by fabricating small particles or clusters where the surface-to-volume ratio of each particle is high, or by creating materials where the void surface area (pores) is high compared to the amount of bulk support material. Materials such as highly dispersed supported metal catalysts and gas phase clusters fall into the former category, and microporous (nanometer-pored) materials such as zeolites, high surface area inorganic oxides, porous carbons, and amorphous silicas fall into the latter category. There are many areas of current academic and industrial activity where the use of the nanostructure approach to high surface area materials may have significant impact:

- microporous materials for energy storage and separations technologies, including nanostructured materials for highly selective adsorption/separation processes such as H2O, H2S, or CO2 removal; high capacity, low volume gas storage of H2 and CH4 for fuel cell applications and high selectivity; high permeance gas separations such as O2 enrichment; and H2 separation and recovery
- thermal barrier materials for use in high temperature engines
- understanding certain atmospheric reactions
- incorporation into construction industry materials for improved strength or for fault diagnostics
- battery or capacitor elements for new or improved operation
- biochemical and pharmaceutical separations
- product-specific catalysts for almost every petrochemical process In catalysis

the key goal is to promote reactions that have high selectivity with high yield.

It is anticipated that this goal will be more closely approached through tailoring a catalyst particle via nanoparticle synthesis and assembly so that it performs only specific chemical conversions, performs these at high yield, and does so with greater energy efficiency. In the electronics area one may anticipate manufacture of single electron devices on a grand scale. Manufacture of materials with greatly improved properties in one or more areas such as strength, toughness, or ductility may become commonplace. In separations science new materials with well defined pore sizes and high surface areas are already being fabricated and tested in the laboratory for potential use in energy storage and separations technologies.

In addition, many laboratories around the world are actively pursuing the potential to create novel thermal barrier materials, highly selective sensors, and novel construction materials whose bonding and strength depend upon the surface area and morphology of the nanoscale constituents. Many are also engaged in developing molecular replication technologies for rapid scaleup and manufacturing. The nanoscale revolution in high surface area materials comes about for several reasons. First, since the late 1970s the scientific community has experienced enormous progress in the synthesis, characterization, and basic theoretical and experimental understanding of materials with nanoscale dimensions, i.e., small particles and clusters and their very high surface-tovolume ratios.

Second, the properties of such materials have opened a third dimension to the periodic table, that is, the number of atoms, N (for a recent example see Rosen 1998). N now becomes a critical parameter by which the properties for "small" systems are defined. As a simple example, for metals we have known for decades that the atomic ionization potential (IP) is typically about twice the value of the bulk work function (Lide 1993).

It is only relatively recently that experiments have shown that the IP (and electron affinity) for clusters containing a specific number N of (metal) atoms varies dramatically and non-monotonically with N for clusters containing less than 100-200 atoms. Other properties such as chemical reactivity, magnetic moment, polarizability, and geometric structure, where they have been investigated, are also found to exhibit a strong dependence on N. Expectations for new materials with properties different from the atom or the bulk material have been realized (e.g., see Jena 1996 and reports therein). The opportunity is now open to precisely tailor new materials through atom-by-atom control of the composition (controlling the types as well as the numbers of atoms) in order to generate the clusters or particles of precision design for use in their own right or as building blocks of larger-scale materials or devices—that is, nanotechnology and fabrication at its ultimate. Such precision engineering or tailoring of materials is the goal of much of the effort driving nanoscale technology. Scientists and engineers typically have approached the synthesis and fabrication of high surface area nanostructures from one of two directions:

1. The "bottom up" approach in which the nanostructures are built up from individual atoms or molecules. This is the basis of most "cluster science" as well as crystal materials synthesis, usually via chemical means. Both high surface area particles and micro- and mesoporous crystalline materials with high void volume (pore volume) are included in this "bottom up" approach.

2. The "top down" approach in which nanostructures are generated from breaking up bulk materials.

This is the basis for techniques such as mechanical milling, lithography, precision engineering, and similar techniques that are commonly used to fabricate nanoscale materials, which in turn can be used directly or as building blocks for macroscopic structures. A fundamental driving force towards efforts to exploit the nanoscale or nanostructure is based upon two concepts or realizations:

(1) that the macroscopic bulk behavior with which we are most familiar is significantly different from quantum behavior, and

(2) that materials with some aspect of quantum behavior can now be synthesized and studied in the laboratory. Obviously, quantum behavior becomes increasingly important as the controlling parameter gets smaller and smaller. There are numerous examples of quantum behavior showing up in high surface materials: the fact introduced above that clusters are found to exhibit novel (compared to the bulk) electronic, magnetic, chemical, and structural properties; the fact that the diffusivity of molecules through molecular sieving materials cannot be predicted or modeled by hard sphere molecule properties or fixed wall apertures; and the fact that catalysts with one, two, or three spatial dimensions in the nanometer size range exhibit unique (compared to the bulk) catalytic or chemical activity.

Clusters and Nanocrystalline Materials

Clusters are groups of atoms or molecules that display properties different from both the smaller atoms or molecules and the larger bulk materials. Many techniques have been developed to produce clusters, beams of clusters, and clusters in a bottle for use in many different applications including thin film manufacture for advanced electronic or optical devices, production of nanoporous structures as thermal barrier coatings, and fabrication of thin membranes of nanoporous materials for filtration and separation. An apparatus developed at the University of Göteborg to measure cluster reactivity and sticking probability as a function of the number of metal atoms in the cluster. The unique properties of nanoparticles make them of interest. For example, nanocrystalline materials composed of crystallites in the 1-10 nm size range possess very high surface to volume ratios because of the fine grain size. These materials are characterized by a very high number of low coordination number atoms at edge and corner sites which can provide a large number of catalytically active sites. Such materials exhibit chemical and physical properties characteristic of neither the isolated atoms nor of the bulk material.

One of the key issues in applying such materials to industrial problems involves discovery of techniques to stabilize these small crystallites in the shape and size desired. This is an area of active fundamental research, and if successful on industrially interesting scales, is expected to lead to materials with novel properties, specific to the size or number of atoms in the crystallite. Schematic drawing of the experimental setup used in Göteborg for studies of chemical reactivity and/or sticking probability of various molecules with the clusters. The production of clusters is via laser vaporization of metal substrates and detection via photoionization time-of-flight mass spectrometry (A. Rosen, University of Göteborg, Sweden). A typical objective of nanoscale catalyst research is to produce a material with exceedingly high selectivity at high yield in the reaction product or product state, i.e., chemicals by design, with the option of altering the product or product state simply by changing the surface functionality, elemental composition, or number of atoms in the catalyst particle. For instance, new catalysts with increasing specificity are now being fabricated in which the stoichiometry may be altered due to size restrictions or in which only one or two spatial dimensions are of nanometer size. Five recent examples where nanocrystalline metallic and ceramic materials have been successfully investigated for catalytic applications are discussed below (Trudeau and Ying 1996).

Catalytic Properties of Nanostructured Gold Catalysts

In the study of transition metal catalytic reactions the group at Osaka National Research Institute has discovered that nanoscale gold particles display novel catalytic properties (Haruta 1997). Highly selective low temperature catalytic activity is observed to switch on for gold particles smaller than about 3-5 nanometers in diameter. Accompanying this turn on in catalytic activity is the discovery that these nanoscale gold particles (crystals) also have icosahedral structure and not the bulk fcc structure, again a nanoscale phenomena not available with bulk samples. In fabricating these novel catalytic materials several issues appear to be crucial. For instance, the Osaka group has shown that

(a) the preparation method plays a crucial role for generating materials with high catalytic activity and selectivity; that

(b) the catalytic activity, selectivity and temperature of operation is critically dependent on the choice of catalyst support, and that

(c) water (moisture) even in parts per million (ppm) levels dramatically alters the catalytic properties.

The effect of moisture on the conversion profiles for CO oxidation for nanoscale gold catalysts supported on cobalt oxide. Examples of novel catalytic behavior of nanoscale gold particles include the following:

- CO oxidation at temperatures as low as -70°C.
- very high selectivity in partial oxidation reactions, such as 100% selectivity at 50°C for oxidation of propylene to propylene oxide as well as near room temperature reduction of nitric oxide with H_2 using alumina-supported gold nanoparticles. Effect of moisture on conversion profiles for CO oxidation over Co_3O_4 and Au/Co_3O_4.

The fundamental work on gold catalysts has led to "odor eaters" for the bathroom, based on nanoscale gold catalysts supported on a-Fe_2O_3, a recent commercialization from Osaka National Research Institute in Japan.

For catalysts based on the layered compound MoS_2, maximum hydrodesulfurization (HDS) activity is obtained only on well-crystallized nanosized materials, while HDS selectivity is determined by the number of layers or "stack height" of the nanocrystalline MoS_2. In the hydrodesulfurization reaction, cyclohexylbenzene occurs only on the MoS_2 "rim sites," or those around the edges of the stack, whereas the pathway to biphenyl requires both "rim" and "edge" sites. Thus, the reaction selectivity can be controlled by controlling the aspect ratio of the nanoparticles of MoS_2. Such control of one- and two-dimensional nanostructures for selective chemical advantage is an exciting new area of research. Of course, a major industrial challenge will be to fabricate such nanocrystals in a commercializable form (Chianelli 1998).

The cerium oxide (CeO_{2-x}) materials have been found to possess a significant concentration of Ce^{3+} and oxygen vacancies, even after high temperature (500°C) calcination. Such nanoclusters give rise to a substantial reduction in the temperature of selective SO_2 reduction by CO and exhibit excellent poisoning resistance against H_2O and CO_2 in the feed stream compared to that for conventional high surface area cerium oxide.

Electrochemical reduction of metal salts is yet another option available to control the size of nanoscale catalyst particles.

A combination of scanning tunneling microscopy (STM) and high-resolution transmission electron microscopy (TEM) has allowed surfactant molecules to be visualized on nanostructured palladium clusters.

Materials with higher hydrogen storage per unit volume and weight are considered by many to be an enabling technology for vehicular fuel cell applications. Scientists at Los Alamos National Laboratories have developed an approach that enables materials such as Mg to be used for hydrogen storage (Schwartz 1998). Magnesium is of interest because it can store about 7.7 wt % hydrogen, but its adsorption/desorption kinetics are slow, i.e., the rate of charge (hydrogen dissociation and hydride formation) is much slower than in metal hydrides. At Los Alamos, high surface area mixtures of nanoscale Mg and Mg2Ni particles are produced by mechanical means, ball milling. The addition of Mg2Ni catalyzes the H2 dissociation such that the rate of hydrogen adsorption increases to a rate comparable to that of LaNi5. Once a hydride is formed, the hydrogen "spillover" leads to magnesium hydride formation. The hydrogen adsorption/desorption characteristics of the mixture of Mg with 23 atomic % Ni. As can be seen, a low pressure plateau at about 1500 torr is obtained for this particular sample. Experiments show that the pressure plateau can be tailored through such alloying. Studies with other nanoscale materials, including other catalysts such as FeTi and LaNi5, are presently ongoing to further improve both the capacity and the rate of hydrogen storage.

A Case Study I: United States of America

Significant work on coatings of dispersions is underway in the United States, as described in the proceedings of the 1997 WTEC workshop report, *R&D status and Trends in Nanoparticles, Nanostructured Materials, and Nanodevices in the United States* (Siegel et al. 1998). Some of the highlights of that volume are summarized below. Similar work is ongoing in other countries.

Controlling the pore structure and synthesizing the building blocks are two technical challenges facing future work in this area. However, Aksay has shown that he can grow silicate films onto a wide variety of substrates. The layer structure of the bound surfactant molecules is key. Atomic force microscopy can reveal some details on the morphology of these films. The nanostructural patterns obtained in processing ceramics that contain organic/inorganic composites allow self-assembly to take place at lower temperatures. Another example of thin films or coating is the work of Gell (1998, 124-130), focused on improving both the physical and the mechanical properties of materials. Nanostructured coating can lead to high diffusivity, improved toughness and strength, and better thermal expansion coefficients, with lower density, elastic modulus, and thermal conductivity. In comparisons of nanostructures and conventional materials, use of nanostructured WC/cobalt composites have resulted in as much as a two-fold increase in abrasion resistance and hardness. Deposition is often accomplished by utilizing a sputtering chamber where the nanomaterial is coated on substrates. This can lead to such benefits as resistance to oxidation and cracks in addition to resistance to wear and erosion. An area that offers exciting possibilities in the area of dispersions is the sol-gel process described by P. Wiltzius (1998, 119-121). Here, a concentrated dispersion of colloids is chemically converted into a gel body. Drying followed by sintering produces a ceramic or glass product. This process can create nanoparticles, fibers, film, plates, or tubes. All processing is at low temperatures. Lucent Technologies has developed a silica casting process that is reproducible for making tubes of pure silica of one meter in length for use in manufacturing optical fibers. Technical challenges include obtaining pure starting materials, removing refractory particles that lead to breakage in the fiber drawing process, and achieving very tight dimensional tolerances. Colloidal dispersions of this type play a critical role in chemical mechanical polishing. To obtain good yield and high quality, it is necessary to achieve very tight process control. R. Brotzman at Nanophase Technologies Corporation describes the gas phase condensation process for synthesizing inorganic and metallic powders (Brotzman 1998, 122-123). Such a process was invented by R. Siegel and his colleagues at Rensselaer Polytechnic Institute in New York (Siegel et al. 1994). The process involves production of physical vapor from elemental or reacted material followed by sudden condensation and reaction of the vapor into small nanometer particles.

The condensation process is rapid and involves dilution to prevent the formation of hard agglomerates and coalescence. Nanophase Technologies has developed a production system where forced convection flow controls the particle/gas stream and enhances particle transport from the particle growth region so as to generate more metal vapor (Parker et al. 1995). The convection flow helps in forming oxides and nitrides from metal crystallites. This process produces commercial quantities of nanosized inorganic powders that have a spherical shape with narrow size distributions. Nanophase Technologies Corporation has developed a coating process that encapsulates nanoparticles with a surface treatment that ensures individual integrity of the particles in subsequent coating steps. Applications span their use in low dielectric media all the way to water and, if needed, steric stabilizers. Nanophase also has directed efforts in the electronics and industrial catalyst areas. Friedlander (1998, 83-8) describes in good detail aerosol reaction engineering where attention is paid to design of the process and the importance of material properties and process conditions. Key process parameters include time, temperature, and volume concentration. Most commercially produced particles have polydispersity. Important full-scale processes are flame reactors for preparing pigments and powdered materials for optical fibers. Pyrolysis reactors have long been used to prepare carbon blacks. Kear and Skandan (1998, 102-4) have discussed two divergent approaches toward fabrication of bulk materials. The first is a powder processing route involving preparations by physical or chemical means followed by condensation and sintering. Materials for cutting tools such as tungsten carbide/cobalt powders were prepared in a controllable way to about 50 nm. Liquid phase sintering completes the formation of the solid dispersion phase. However, a gap in fabrication technology remains that of controlling grain growth during liquid phase sintering. A second approach to fabrication of bulk materials is spray forming. This procedure avoids contamination and coarsening of the dispersion and its particles when the process has a controlled atmosphere of inert gas at low pressure. In this approach there is need to establish process/product codesign where we understand cause-effect relationships between processing parameters and properties of nanophase spray/deposited materials.

The Case Studies II: Europe and Japan

Due to time constraints, the WTEC panel was able to visit only a few labs in Europe and Japan that deal with issues associated with dispersions and coatings. Cited below are a visit to the Institute for New Materials at Saarbrücken in Germany and a visit to Japan's Industrial Research Institute of Nagoya, both of which are exploring direct applications of nanostructured dispersions and coatings. Drs. Rudiger Nass and Rolf Clasen at the Institute for New Materials at Saarbrücken make specific use of the sol-gel process in which liquid starting materials are utilized at low temperatures for nanoscale metal, ceramic, glass, and semiconductor nanoparticles (Clasen 1990). The advantages, besides temperature, include the isolation of high purity powders. The institute has focused on four basic areas for spin-off and adaptation towards commercialization:

1. New functional surfaces with nanomers. Properties such as corrosion protection, wettability, coloration, micropatterned surfaces, porosity, and the ability for selective absorption of molecules produce multifunctionality.
2. New materials for optical applications. This area combines properties of lasers and ceramics with polymers. Such features as optical filters, transparent conducting layers, materials for optical telecommunications, photochromic layers, and holographic image storage are under investigation.
3. Ceramic technologies. In this area, a simple precipitation process such as sol-gel provides for pilot-scale production of agglomerate-free powder.
4. Glass technologies. Chemical incorporation of metal colloids with intelligent properties into glasslike structures is clearly possible.

During the WTEC team's site visits to the Industrial Research Institute of Nagoya, S. Kanzaki and M. Sando described methods for preparing synergy ceramics using nanoporous silica particles that are used to fabricate thin films with one-dimensional throughput channels (Kanzaki et al. 1994). The channels had 10-20 nm sizes. High temperature Fe2SiO4 oxides were prepared as both molecular sieves and particulate fibers. Japan plans to pursue the preparation of nanoporous materials for absorbing oil and identified particulates. The quest for low temperature nanoparticle preparation methods has spanned a wide range of systems. One that has been in existence for decades but has not been put into use in other industries is the method of preparing silver halide particles. Eastman Kodak in France, England, and the United States has utilized solution precipitation technology with well-controlled mixing and nucleation control to produce a wide range of grain sizes. "Lippmann"-type grains have a size of about 50 nm. Some of the properties of these fine-grain systems are discussed by Trivelli and Smith (1939). There are other methods of creating nanoparticles of organic materials such as filter dye applications in photographic films and spectral sensitizing dyes for use in silver halide grains. Ultrafine grinding media are used to almost sandpaper organic crystals to nanoparticle ranges of 20-80 nm (Czekai et al.

1994). Similar technology by Bishóp has been utilized in both pharmaceutical preparations and ink jet applications with good success (Bishóp et al. 1990). One other exciting area is in polymer science, where dendrimer molecules, often 10 nanometers in diameter, are prepared synthetically. These molecules have been studied in gene therapy, as aids in helping to detect chemical or biological agents in the air, and as a means to deliver therapeutic genes in cancer cells (Henderson 1996). The preparation of dendritic starburst molecules is described by Salamone (1996, 1814). One can imagine applications for coatings in which these molecules with their highly reactive surfaces participate in either classical (sub-nanoscopic chemistry) or novel nanoscopic conversions.

Nanoparticles Pass Muster as Vectors for Gene Therapy

Gene therapy, in which a viral vector is used to modify defective genes or replace missing ones, has shown significant potential as a way of treating disease in animal models. But its use in humans has been hampered by safety concerns, including some fatalities in clinical trials. Researchers have thus been looking into the possibility of using nonviral vectors, which should carry fewer inherent risks, to deliver therapeutic genes. In a paper published online this week by the *Proceedings of the National Academy of Sciences,* scientists report that silicon nanoparticles can perform this task successfully in mice.

Paras N. Prasad of the State University of New York (SUNY) at Buffalo and his colleagues manufactured nanoparticles using organically modified silicon. The surface of these particles can be tailored to target specific cells. The team used the tiny units to transport a fluorescent marker gene to dopamine neurons in the brains of mice. After injecting the nanoparticles, the researchers observed brain cells fluorescing using a new imaging technique that works on live animals. According to the report, the study is the first in which a nonviral vector has shown efficacy comparable to that of a viral delivery system in an animal model. What is more, a month later none of the animals had experienced adverse effects from the procedure.

The researchers also investigated the possibility of manipulating the behavior of specific brain cells, instead of solely tagging their presence. In so doing, they discovered that the nanoparticles can be used to reactivate adult stem cells by altering a nuclear growth factor receptor. The team will next test the approach on larger animals. "In the future," says study co-author Earl J. Bergey, also at SUNY Buffalo, "this technology may make it possible to repair neurological damage caused by disease, trauma or stroke." —*Sarah Graham*

Biotechnology

The first broad category of enabling technology for molecular manufacturing in the mechanical tradition is biotechnology. Molecular biologists study and modify systems of molecular machines, and genetic engineers reprogram these systems, sometimes to build novel molecular objects having complex functions. By 1998, biotechnologists could make almost any DNA sequence or poly-peptide chain desired (though not extremely long aperiodic ones), assemble "devices" such as artificial chromosomes and viruses, and were actively pursuing gene therapy, all of which are examples of molecular engineering. While earlier writers had suggested that biomolecules could be used as mechanical components, Drexler evidently was the first to point out, in 1981, that complex devices resembling biomolecular motors, actuators, bearings, and structural components could be combined to build versatile molecular machine systems analogous to machine systems in the macroscopic world. "Development of the ability to design protein molecules," Drexler wrote, "will open a path to the fabrication of devices to complex atomic specifications, thus sidestepping obstacles facing conventional microtechnology. This path will involve construction of molecular machinery able to position reactive groups to atomic precision."

Perhaps the best-known biological example of such molecular machinery is the ribosome, the only freely programmable nanoscale assembler already in existence. In nature, ~8000 nm^3 ribosomes act as general-purpose factories building diverse varieties of proteins by bonding amino acids together in precise sequences under instructions provided by a strand of messenger RNA (mRNA) copied from the host DNA, powered by the decomposition of adenosine triphosphate (ATP) to adenosine diphosphate (ADP). Each ribosome is a compact ribonucleoprotein particle consisting of two subunits, with each subunit consisting of several proteins associated with a long RNA molecule (rRNA). Biotechnology can employ the ribosomal machinery of bacteria to produce novel proteins, which proteins then might serve as components of larger molecular structures.

Of course, such biological mechanisms have been disparaged as "slipshod devices working in spite of haphazard design." Dan Heidel, a University of Washington biomimetics engineer, agrees that biological systems are sloppy but observes that they are designed to tolerate it, and offers high praise for the lowly ribosome:

"The average ribosome sits in an aqueous environment surrounded by thousands of cosolutes at concentrations just a hair's breadth from precipitating out of solution. The ribosome itself consists of three RNA molecules that spontaneously fold into floppy spaghetti noodle piles of chemically active groups surrounded by >50 proteins of similarly dubious structure. The other solutes are bombarding this motley assemblage at tens to hundreds of miles an hour and the solvating water molecules are a constantly shifting and unpredictable network of hydrogen bonds and polar ionic charges constantly impinging upon the ribosome. The ribosome must recognize one particular cognate transfer RNA (tRNA) for the codon in its A-site out of ~60 other nearly identical tRNAs. This particular tRNA must be discriminated from all its tRNA brethren along with whatever other solutes can fit into the A-site pocket using only three base pairs for discrimination. In fact, most codons use only two of the three base pairs for recognition so some tRNAs are recognized by as few as 4 hydrogen bonds. All of this requires sub-Angstrom precision in the active sites but since the ribosome is a massive complex of subunits held together by only ionic, hydrogen, hydrophobic and Van der Waals bonds, it's like getting submillimeter accuracy from points on a bunch of taped-together water balloons being rolled down a staircase."

"This ridiculous machine can assemble proteins at a 20 Hz subunit-incorporation frequency with a 10^3 error rate. Ribosomes are even more impressive when one considers that the number of amino acids in each ribosome's own protein subunits combined with the 10^3 protein assembly error rate ensures that each ribosome has several errors in it and that each cell almost certainly has no two ribosomes exactly alike. I dare any engineer on this planet to rationally design an assembler at any scale that is able to operate at a comparable level of performance under similarly adverse operating conditions."

Thus ribosomes can manufacture, with >99.9% fidelity to arbitrary specifications, linear strings of amino acids of great length which may then fold up to produce three-dimensional protein structures performing a broad range of highly differentiated functions, providing good analogs for working nanomachines capable of forming self-assembling systems. Much of the great diversity of protein activity flows from the selectivity with which they bind to specific receptor sites on other molecules. Some of these binding mechanisms might be copied to construct molecular robotic arms, while other devices could be made by adapting naturally-occurring proteins, or by synthesizing new, custom-built ones. Protein engineering might be used to create proteins with desired qualities as building blocks for constructing the first primitive assemblers.

The field of protein engineering has its own journal of the same name, and has already achieved remarkable results in the synthesis of novel structures. Drexler cites the creation of the first de novo structure by DeGrado, the development of synthetic branched protein-like structures that depart substantially from protein models, and the engineering of a branched, nonbiological protein with enzymatic activity. Since then, numerous alpha-helical peptides have been designed de novo, a 20-residue artificial peptide exhibiting designed beta-sheet secondary structure has been created, and various peptide analogs have also been synthesized. By 1998, advances in computational techniques allowed the design of precise sidechain packing in proteins with naturally occurring backbone structures, and the study of de novo backbone structures had begun. The catalytic task space was being laboriously investigated at the molecular level — for example, one study of comparative enzyme structures revealed that changing as few as four amino acids converted an oleate 12-desaturase to an oleate hydroxylase, and as few as six substitutions could convert an oleate hydroxylase into an oleate desaturase. To further expand the toolkit, Mills and Fahy proposed exploiting the degeneracy of the genetic code (wherein amino acids are specified by up to 6 different codons; by assigning some codons to unnatural amino acids, allowing the creation of artificial proteins with unprecedented structural or catalytic properties. With 61 usable codons available, there is room for up to 41 novel amino acids in addition to the natural 20 amino acids; unnatural amino acids have been introduced into beta-lactamase and T4 lysozyme enzymes in a site-directed fashion, using artificial tRNAs and mRNAs in an in vitro translation system, and into an engineered peptide as a reporter group. Additionally, artificial novel base pairs can be incorporated enzymatically into DNA and other schemes are possible. By incorporating one additional base pair, the number of possible codons is expanded from $4^3 = 64$ codons to $6^3 = 216$ codons, potentially allowing up to 193 novel amino acids to be incorporated into artificial protein structures — or at least 48 new amino acids assuming that present levels of redundancy are retained and the existing 4-letter code is kept as a subset for "upward compatibility". If the genetic alphabet is increased to eight bases ($8^3 = 512$ codons), the number of available codons for novel amino acid-like molecules increases to as many as 489.

One of the greatest challenges in protein engineering (where one objective is to create functional, atomically-precise 3-D aperiodic structures) has been the difficulty of protein fold design, because individual amino acids have no strong, natural complementarity. In 1998, Duan and Kollman successfully folded a solvated 36-residue (~12,000-atom) protein fragment (i.e., a peptide) by molecular dynamics simulation into a structure that resembles an intermediate native state. During a 1-microsec simulation, the chain folded during 150 nanosec into a compact structure resembling the intermediate native state, as known by NMR, then unfolded and refolded again for a shorter period. In 1998, whole-protein folding to the final stable native state had not yet been computed deterministically using molecular dynamics — the real protein will fold and refold hundreds of times before it stumbles into the stable conformation with the lowest free energy — but research pursuing this objective was active and ongoing. Structural changes that occur during protein function (e.g., enzymatic action) were also being avidly studied.

A different protein-based approach to near-atomically-precise 3-D structures avoids the folding problem by making use of semi-rigid protein nanoshells of deterministic size and shape to guide the ordered assembly of inorganic particles. For example, the capsid of the cowpea chlorotic mottle virus, a protein container with a 28-nm external diameter and an 18-nm internal cavity diameter, is composed of 180 identical coat protein subunits arranged on an icosahedral lattice. Each subunit presents at least nine basic residues (arginine and lysine) to the interior of the cavity, creating a positively charged interior surface which provides an interface for inorganic crystal nucleation and growth, while the outer capsid surface is not highly charged. In one experiment, the empty capsid shell was found to act as a spatially selective nucleation catalyst in paratungstate mineralization, in addition to its role as a size- and shape-constrained reaction vessel. Noted the authors: "The range of viral morphology and size allow great flexibility for adapting this methodology to control the size and shape of the entrapped material, which is limited only by access to the protein interior and could include inorganic and organic species. The electrostatic environment within the cavity could be altered by site-directed mutagenesis to induce additional specific interactions."

It may also be possible to borrow various existing protein devicesand apply them to new uses. The bacteriophage DNA injection system is self-assembling and could be engineered for other applications. Vogel has attached kinesin motors to flat surfaces in straight grooves; when ATP was added, the motors were activated, passing 25-nm wide microtubules hand over hand down the line in the manner of a ciliary array. Montemagno genetically engineered the 12-nm wide ATPase molecular rotary motor so that one end would adhere to a metal surface and the other end would provide an attachment site for 1-micron fluorescent streptavidin-coated bead payloads. When ATP was added, each bead individually attached to a biomotor in the motor array began twirling at ~10 Hz, generating a >100 pN force. Bead rotation was maintained continuously for more than 2 hours, indicating the high reliability of these ~100% efficient self-assembling biomotors. Montemagno explains that his laboratory's "long-term goal is the integration of the F_1ATPase biological motor with nanoelectromechanical systems (NEMS) to create a new class of hybrid nanomechanical devices." Other natural nanomotors such as the self-assembling bacterial flagellar motor and the hexagonal "packaging RNA" or pRNA pumping mechanism that packs DNA into the capsid shell of the bacteriophage Phi 29 are also being intensively studied.

Yet another material which allows moderately well-defined three-dimensional nanoassemblies is DNA. The ideas behind DNA nanotechnology have been around since 1980, but activity in this field accelerated in the 1990s after numerous experimental difficulties were surmounted. Horn and Urdea reported branched and forked DNA polymers. Niemeyer and Smith exploited DNA specificity to generate regular protein arrays, and both suggested using self-assembled DNA as an early material for molecular nanotechnology. Shi and Bergstrom attached DNA single strands to rigid organic linkers, showing that cyclical forms of various sizes could be formed with these molecules. Mirkin's group attached DNA molecules to 13-nm colloidal gold particles with ~6 nm particle spacings, with the goal of assembling nanoparticles into macroscopic materials; Alivisatos, Schultz and colleagues used DNA to organize 1.4-nm gold nanocrystals in arrays with 2-10 nm spacings. In either case, the specificity of DNA pairing should allow the construction of complex geometries since the use of colloidal particles can potentially add structural elements (which are more rigid than any polymer strand) to the set of building blocks for nanometer-scale structures — although to take full advantage of this capability would require particles which are atomically precise and which possess several chemically distinct anchoring points on each particle. Damha has synthesized V-shaped and Y-shaped branched RNA molecules, branching RNA dendrimers with "forked" and "lariat" shaped RNA intermediates, and even trihelical DNA. Henderson and coworkers have designed a simple DNA decamer that can form an extended linear staggered quadruplex array reaching lengths of >1000 nm, and have made branched oligonucleotides that template the synthesis of their branched "G-wires"; depending on the ratio of linear to branched building blocks, extensive DNA arrays with differing connectivities but irregular interstices can be created.

In 1998 the most intensive work on three-dimensional engineered DNA structures was taking place in Nadrian Seeman's laboratory in the New York University Department of Chemistry. Seeman originally conceived the idea of rigid 3-D DNA structures in the early 1980s, while examining DNA strands that had arranged themselves into unusual four-armed Holliday junctions. Seeman recognized that DNA had many advantages as a construction material for nanomechanical structures. First, each double-strand DNA with a single-strand overhang has a "sticky end," so the intermolecular interaction between two strands with sticky ends is readily programmed (due to base-pair specificity) and reliably predicted, and the local structure at the interface is known (sticky ends associate to form B-DNA). Second, arbitrary sequences are readily manufactured using conventional biotechnological techniques. Third, DNA can be manipulated and modified by a large variety of enzymes, including DNA ligase, restriction endonucleases, kinases and exonucleases. Fourth, DNA is a stiff polymer in 1-3 turn lengths and has an external code that can be read by proteins and nucleic acids.

During the 1980s, Seeman worked to develop strands of DNA that would zip themselves up into more and more complex shapes. Seeman made junctions with five and six arms, then squares, stick-figure cubes comprised of 480 nucleotides, and a truncated octahedron containing 2550 nucleotides and a molecular weight of ~790,000 daltons. The cubes were synthesized in solution, but Seeman switched to a solid-support-based methodology in 1992, greatly improving control by allowing construction of one edge at a time and isolating the growing objects from one another, allowing massively parallel construction of objects with far greater control of the synthesis sequence. By the mid-1990s, most Platonic (tetrahedron, cube, octahedron, dodecahedron, and icosahedron), Archimedean (e.g., truncated Platonics, semiregular prisms and prismoids, cuboctahedron, etc.), Catalan (linked rings and complex knots), and irregular polyhedra could be constructed as nanoscale DNA stick figures.

Seeman's DNA strands that formed the frame figures were strong enough to serve as girders in a molecular framework, but the junctions were too floppy. In 1993 Seeman discovered the more rigid anti-parallel DNA "double crossover" motif, which in 1996 he used to design and build a stiff double junction to keep his structures from sagging. The next goal was to bring together a large number of stick figures to form large arrays or cage-shaped DNA crystals that could then be used as frameworks for the assembly of other molecules into pre-established patterns. These DNA molecules would serve as the scaffolding upon which new materials having precise molecular structure could be assembled.

In 1998, Erik Winfree and colleagues in Seeman's laboratory reported the design and construction of two-dimensional DNA crystals using self-assembling double crossover molecules. Repeating array units were approximately 2 nm x 4 nm x 16 nm in size, and examination of the array with an Atomic Force Microscope (AFM) revealed domains of up to 500,000 interconnected units, showing that the self-assembly process could be very reliable under ideal conditions. While the work described the construction of two- and four-unit lattices, the number of component tiles that could be used in the repeat unit did not appear to be limited to such small numbers, suggesting that complex patterns could be assembled into periodic arrays, which might also serve as templates for nanomechanical assembly. Noted the authors: "Because oligonucleotide synthesis can readily incorporate modified bases at arbitrary positions, it should be possible to control the structure within the periodic group by decoration with chemical groups, catalysts, enzymes and other proteins, metallic nanoclusters, conducting silver clusters, DNA enzymes, or other DNA nanostructures such as polyhedra." In these experiments, DNA hairpin turns incorporated within selected units inside their structures were visible as topographic features under AFM imaging, proving the ability to build predictable, atomically precise 2-D crystals with design control over every lattice point. Since the density of lattice points was much sparser than atomic spacings, this DNA self-assembly technology could be paired with atomically precise synthesis of diverse nanoscale subassemblies so that each unique double crossover unit in a pattern could be decorated with a unique subassembly of comparable size, possibly allowing the construction of mechanical nanocomputers and other nanomachinery.

In early 1999, Seeman reported yet another breakthrough — the construction of a mechanical DNA-based device that is a possible prototype for a nanoscale robotic arm. The mechanism has two rigid arms a few nanometers long that can be rotated between fixed positions by introducing a positively charged cobalt compound into the solution surrounding the molecules, causing the bridge region to be converted from the normal B-DNA structure to the unusual Z-DNA structure. The free ends of the arms shift position by ~2.0 nm during the structural conversion. Explained Seeman: "Using synthetic DNA as a building material, we have constructed a controllable molecular mechanical system. In the long-term, the work will have implications for the development of nanoscale robots and for molecular manufacturing."

DNA can also serve as an assembly jig in solution phase. Bruce Smith and colleagues are devising a method for the assembly and covalent linkage of proteins into specific orientations and arrangements as determined by the hybridization of DNA attached to the proteins, called DNA-Guided Assembly of Proteins (DGAP). In this method, multiple DNA sequences would be attached to specific positions on the surface of each protein, and complementary sequences would bind, forcing protein building blocks (possibly including biomolecular motors, structural protein fibers, antibodies, enzymes, or other existing functional proteins) together in specific desired combinations and configurations, which would then be stabilized by covalent interprotein linkages. This technique could also be applied to nonprotein components that can be functionalized at multiple sites with site-specific DNA sequences, although proteins, at least initially, may be more convenient building blocks due to their size, their surface chemistry, the wide variety of functions and mechanical properties they can confer on the resulting assemblies, and the many existing techniques for introducing designed or artificially evolved modifications into natural proteins of known structure. (In 1998, custom DNA and peptide sequences could be ordered online.) Methods for covalently attaching functional proteins to a DNA backbone in a specified manner at ~8.5 nm (25 base-pair) intervals, addressable protein targeting in macro-molecular assembly, and "protein stitchery" were being explored by others. Drexler notes that evolution has not maximized the stability of natural proteins, and that substantially greater stability may be engineered by various means (e.g., increasing folding stability by >100 Kcal/mole.

Molecular and Supramolecular Chemistry

A second broad category of enabling technology for molecular manufacturing includes molecular chemistry — the conscious design of completely artificial, nonbiological chemical structures that could potentially serve as molecular "parts" or which have specific devicelike functionality. For example, catalysts may very loosely "be thought of as rudimentary assemblers that are slightly programmable through changes in the reaction milieu (i.e., changes in pH, temperature, etc.)". The number of natural molecular "parts" is immense — perhaps a few hundred thousand bioorganic compounds have been separated, purified and identified — but by 15 February 1999 chemists had already registered 19,245,458 well-characterized artificial molecules or other "substances" in the CAS Registry. Synthesis of natural molecules containing ~100 atoms (e.g., a ~1 nm^3 molecular "part"), using methods of classical organic synthesis often requiring ~1 synthesis step per atom with ~90% yield per step, was state-of-the-art in 1998.

"What is exciting about modern nanotechnology," says Nobel chemist Roald Hoffmann, "is (a) the marriage of chemical synthetic talent with a direction provided by 'device-driven' ingenuity coming from engineering, and (b) a certain kind of courage provided by those incentives, to make arrays of atoms and molecules that ordinary, no, extraordinary chemists just wouldn't have thought of trying. Now they're pushed to do so. And of course they will. They can do anything."

Molecular chemists study how small numbers of atoms combine covalently with each other to produce molecules with a wide variety of different properties. Supramolecular chemists are primarily concerned with the much weaker noncovalent forces that can arise between molecules — such as hydrogen bonds and van der Waals interactions — that can be sufficient to bind collections of molecules tightly enough to form functional nanometer-sized structures. Supramolecular chemistry studies "chemistry beyond the molecule" or "the chemistry of the noncovalent bond". Its subject matter includes atomically precise supermolecules (well-defined discrete oligomolecular species resulting from the intermolecular association of a few components, such as a receptor and its substrate) and polymerlike supramolecular assemblies (polymolecular entities that result from the spontaneous association of a large undefined number of components).

Self-assembly is a key concept in supramolecular chemistry as it is in nature, allowing the manufacture of large numbers of compound objects simultaneously and in parallel, rather than sequentially. Molecular self-assembly is a strategy for nanofabrication that involves designing molecules and supra-molecular entities so that complementarity causes them to aggregate into desired structures. Self-assembly of atomically precise supermolecules thus demands well-defined adhesion between selected molecules, which may require steric (shape and size) complementarity, interactional complementarity, large contact areas, multiple interaction sites, and strong overall binding.

G.M. Whitesides explains that self-assembly has a number of advantages as a strategy. First, it carries out many of the most difficult steps in nanofabrication — those involving the smallest atomic-level modifications of structure — using the very highly developed techniques of synthetic chemistry. Second, it draws from the enormous wealth of examples in biology for inspiration; self-assembly is one of the most important strategies used in biology for the development of complex, functional structures, such as the well-studied examples of complex biomachine self-assembly of whole bacteriophage virions and flagellar rotor motors from their smaller molecular "parts". Third, self-assembly can incorporate biological structures directly as components in the final systems. Fourth, self-assembly requires target structures to be among the most thermodynamically stable available to the system, thus tends to produce structures that are relatively defect-free and self-healing. As Lehn points out: "By increasing the size of its entities, nanochemistry works its way upward towards microlithography and microphysical engineering, which, by further and further miniaturization, strive to produce ever smaller elements". Hierarchical self-assembly of many billions of micron-scale spherical components into a periodic honeycomb-structured photonic crystal optical device ~30 microns thick and ~1 cm^2 in area was demonstrated in 1998.

There is a wide range of different molecular systems that can self-assemble, including those which form ordered monomolecular structures by the coordination of molecules to surfaces, called self-assembled monolayers (SAMs), self-assembling thin films and Langmuir-Blodgett films, and self-organizing nanostructures. In these systems, a single layer of molecules affixed to a surface allows both thickness and composition in the vertical axis to be adjusted to 0.1 nm by controlling the structure of the molecules comprising the monolayer, although control of in-plane dimensions to <100 nm is very difficult. Fluidic self-assembly of microscale parts and the dynamics of Brownian self-assembly have also been described, and the theory of designable self-assembling molecular machine structures is beginning to be addressed.

Another approach using solid-phase peptide synthesis and a massively convergent self-assembly process was employed by M.R. Ghadiri and colleagues, who designed and synthesized a number of self-assembling peptide-based nanotubes using cyclic peptides with an even number of alternating D- and L-amino acids for the building blocks of the nanotubes. The alternating stereo-chemistry of the cyclic peptides allowed all the side chains of the amino acids to be pointing outwards which would not be possible in an ordinary all L-cyclic peptide. In this conformation, the amide backbone can H-bond in a direction perpendicular to the plane of the cyclic peptide. The stacking of two cyclic peptides forms an H-bonding network resembling an antiparallel beta-sheet as is commonly found in natural proteins. The H-bonding lattice quickly propagates perpendicular to the plane of the cyclic peptide, forming a tubular microcrystalline structure with 0.75-nm pores, or, in another experiment, 1.3-nm pores. Another group of nanotubes was designed with a highly hydrophobic outer surface and a hydrophilic inner pore. These nanotubes were easily inserted into a lipid bilayer and have been shown to be highly efficient ion channels; nanotubes with slightly larger pores transport small molecules such as glucose as well. Stable flat nanodisks with diameters continuously (chemically) adjustable from 30-3000 nm have been self-assembled in surfactant solutions.

Self-assembled crystalline solids such as zeolitesincompletely fill space, leaving substantial voids that may be occupied by solvent molecules or other guest molecules, producing solid-state host-guest complexes known as clathrates (from the Latin *clathratus*, meaning "enclosed by the bars of a grating"). MacNicol and coworkers reported the first rationally designed clathrate host in 1978, but the first true de novo design of an organic clathrate was in 1991 and involved the formation of an extensive three-dimensional diamondlike porous lattice built from a single Tinkertoy-type subunit containing four tetrahedrally arrayed pyridone groups that acted as connectors to assemble the units together in a well-defined geometry. However, the crystal was held together only by weak hydrogen bonds and the links had a rotational degree of freedom, rendering the exact crystal molecular arrangement unpredictable. Engineered nanoporous molecular crystals can provide molecular-scale voids with controlled sizes, shapes, and embedded chemical environments at resolutions of 0.3-4.0 nm. Other self-assembling crystal structures have been described including the crystal engineering of diamondoid networks, template-directed colloidal crystallization or colloidal epitaxy, organic templating to form crystalline zeolite-type structures with ordering lengths <3 nm, 3-D polymer channels with chemically functionalizable channel linings, and micromolding combined with templating and cooperative assembly of block copolymers to produce hierarchical ordering over discrete and tunable characteristic length scales of ~10 nm, ~100 nm, and ~1000 nm in a single body.

Dendrimers, also known as arborols or fractal polymers, are a well-known set of self-assembling structures that possibly could be used as tools to assist in the assembly of early nanomachines. Dendrimers are large regularly-branching macro-molecules resembling fractal patterns, made by an iterative process in which small linear molecules are allowed to bind to each other at a certain number of sites along their length, building up branches upon branches with each iteration working outward from a core molecule having at least two chemically reactive arms, somewhat mimicking tree growth. Different core molecules or building-block chains produce macromolecules with different shapes, and the outer surface can be terminated with chemical groups of specific functionality, producing hundreds of differently-surfaced branched molecules that have already found use in medical research. Growth is regulated, so size can be accurately controlled — dendrimers are typically a few nanometers wide but have been constructed with masses exceeding a million protons (~10^5 atoms) and with diameters larger than 30 nm. In 1999, convergent assembly of specified 6-mer dendrimers, via trimerization of precursor dimer "parts," was demonstrated.

Autocatalysis or self-replication is yet another approach to the chemical self-assembly of subunits. In 1989, J. Rebek reported the synthesis of super-molecules that could generate copies of themselves when placed in a sea of simpler molecular parts, with each component part consisting of up to a few dozen atoms. Molecular self-replication has continued to be studied by others. Ghadiri reported a self-replicating peptide in 1996, and by 1998 his laboratory had moved on to studies of more complex autocatalytic cycle networks. It is unknown how complex such assembly cycles may be made, but Ghadiri was investigating a self-replicating 256-component molecular ecosystem incorporating ~32,000 possible binary interactions, in 1998.

By 1998, chemists had already synthesized a vast variety of simple molecular parts of which only a tiny sampling can be mentioned here. For instance, carbon rings have been linked edge-on or by corners to create propeller-shaped molecules called propellanes or rotanes, or stacked with short hydrocarbon links to make gear-looking molecules such as cyclophane and superphane. Hydrocarbon molecular polyhedra in the shape of triangular (e.g., triangulane), cubic (e.g., cubane, first made by P. Eaton in 1964), pentagonal and hexagonal prisms, collectively known as prismanes, have been created, along with more unusual geometrical forms such as paddlane, housane, basketane, churchane, pagodane, and bivalvane. Cryptands, spherands, cryptaspherands and carcerands, corannulene baskets, and calixarenes are bowl-shaped molecules whose benzene ring walls hold their cavities rigid. The bowl rim of a calixarene can be lined with chemical groups to determine which guests it will accept, or two bowls can be bonded together, making a hollow nanoscale reaction vessel, and various molecular cages and capsules with precise sizes and characteristics have been assembled. A variety of nonchiral and chiral square molecules have been synthesized, as well as molecular "bottlebrushes"; DNA-polymer constant-force springs; the ferric wheel ; molecular barrels and molecular tennis balls;various helical, boxlike, and other hollow organic structures; molecular "turnstiles" and "molecular tweezers"; simple and complex trefoil knot molecules; and stable carbyne molecular rods with chain lengths up to ~300 carbon atoms having alternating single and triple covalent bonds. Chemists are justifiably awed by the creation of structures like the ferric wheel; "for me," observed Nobel chemist Roald Hoffmann, "this molecule provides a spiritual high....The molecule is beautiful because its symmetry reaches directly into the soul, [playing] a note on a Platonic ideal."

Molecular gear systems first became a popular object of chemical study during the 1980s. For example, G. Yamamoto described "compounds that exist in conformations which are regarded as static meshed gears with two-toothed and three-toothed wheels and some of them behave as dynamic gears." H. Iwamura prepared one system that formed a chain of beveled molecular gears with ~GHz rotation rates, and another "doubly connected bevel gear system...[Transfer] of information from one end of the molecular system to the other end could take place in large molecules via cooperativity of the torsional motions of the chain." Kurt Mislow gave numerous examples of molecular gear systems that he had synthesized using the methods of traditional solution chemistry, which "resemble to an astonishing degree the coupled rotations of macroscopic mechanical gears." He added: "It is possible to imagine a role for these and similar mechanical devices, molecules with tiny gears, motors, levers, etc., in the nanotechnology of the future."

In 1998, Gimzewski and coworkers synthesized and observed a 1.5-nm diameter gearlike single molecule (hexatertbutyl decacyclene) rotating within a supramolecular bearing (a void in a 2-D lattice) at room temperature on the surface of a Cu{100} lattice. The authors suggest that added nonthermal noise (e.g., temperature differences produced by tunnel current heating of the rotor) could be rectified by asymmetries in the rotor/neighbor potential energy curve, allowing the rotor to turn unidirectionally. Says Gimzewski: "Our wheel is frictionless in the conventional sense of the word, and it doesn't wear out." Gimzewski's experimental work confirms Drexler's calculations that van der Waals bearings can operate in molecular systems, that multiatomic bearing surfaces with higher load capacities than sigma bond bearings can be built, that these bearings can run with no lubricants, and that such bearings have sufficiently low energy barriers to allow turning by thermal vibrations alone.

By 1998, limited control of the mechanical action of molecular parts had been demonstrated. For example, in 1994 T. Ross Kelly's group created a "paddlewheel" molecule (a spinning propeller-shaped wheel) with a built-in chemically-controlled brake. In solution, the wheel spins freely, but when Hg^{++} ions are added, a blocking ligand physically rotates into a new position (tripping the brake) and stopping the rotation; removal of the mercury causes the wheel to spinning. In 1997, Kelly's group used similar techniques to synthesize the first molecular ratchet, a propeller with three benzene ring "blades" that serve as gear teeth. A row of four rings (the pawl) sits between two of the blades, such that the propeller cannot turn without pushing it aside; because of a twist in the pawl, it is easier to turn clockwise than counterclockwise under thermal agitation, providing mechanical rectification without net motion or energy extraction. Shinkai and colleagues fabricated a pair of molecular tongs for grasping metal ions that uses two crown ether rings as the jaws, linked by two benzene rings joined through two double-bonded nitrogen atoms. Irradiation with ultraviolet light induces photoisomerization of the double bond between the nitrogen atoms, closing the jaws; heat reverses the process, opening the jaws. In related systems, collectively called "butterfly molecules," irradiation with visible light or pH changes can also induce opening. Seeman's chemically-operated mechanical DNA system provides similar control of molecular mechanical movement. Numerous controllable molecular electronics devices and possible nanocomputer components are reviewed.

Noncovalent self-assembly of molecular parts has also been demonstrated. For example, Fraser Stoddart and colleagues have extensively investigated the rotaxanes, molecules in which one part is threaded through a hole or loop in another, and kept from unthreading by end-groups, like a ring trapped on a barbell. In one system, a ring-shaped molecule slides freely along a shaft-like chain molecule, moving back and forth between stations at either end with a frequency of ~500 Hz, making an oscillating "molecular shuttle". This shuttling behavior can be controlled by a series of different chemical, electrochemical, or photochemical external stimuli. Using a different approach, G. Wenz threaded ~120 molecular beads onto a single, long, poly-iminooligomethylene polymer chain, making a "molecular necklace." Other groups are pursuing related research regarding [n]rotaxanes and pseudorotaxanes (mechanically-threaded molecules), while still others are using the necklace technique as a means, for example, of self-assembling ~2 nm wide cyclodextrin nanotubes.

Another interesting molecular system that can be made entirely by self-assembly is the catenanes, which have two or more closed rings joined like the links of a chain. The rings are mechanically linked — there are no covalent bonds between separate links. The first [2]catenane (2 linked rings) was constructed by Edel Wasserman of AT&T Bell Laboratories in 1960, using simple hydrocarbon rings; the first [3]catenane was synthesized in 1977. In 1994, Fraser Stoddart and David Amabilino constructed the first [5]catenane, having ~372 atoms, dubbed "olympiadane" because of its resemblance to the Olympic rings, and in 1997 the same group announced the first deterministically-ordered 7-ring heptacatenane; another group later reported the spontaneous self-assembly of a 10-component catenane. By 1998, much research on larger polymeric chains of linked rings (e.g., oligocatenanes), supramolecular "daisy chains", fullerene-containing catenanes, and supramolecular weaving was in progress. The smallest [2]catenane constructed to date has dimensions 0.4 nm x 0.6 nm. Self-assembled mechanically-interlocked 2-dimensional and 3-dimensional "infinite" arrays of single molecular ring-like species are known. Interestingly, in 1998 it was discovered that many viral coats (which also self-assemble) appear to be a 2-dimensional "chain-mail" weave of mechanically-interlocked protein rings — for instance, the spherical bacteriophage HK97 capsid shell consists of exactly 72 interlinked protein rings, specifically 60 hexamers and 12 pentamers.

Complex molecular parts also can be built up from simpler molecular parts by covalent bonding as low-symmetry shape-invariant molecular object polymers, or via processes variously known as modular chemistry, heterosupramolecular chemistry, chemical assembly, or structure-directed synthesis — including the well-known Diels-

Alder reactions which may be employed repetitively to fabricate a series of atomically-precise molecular rods, rings, and nanocages (e.g., "beltenes" and "collarenes") composed of polyacenequinone units with nanometer-scale dimensions. Such efforts have been described as steps toward "molecular LEGO", "molecular Meccano", "molecular Tinkertoys", or "molecular building blocks". Some structure-directed assembly has been achieved with simple DNA forms. Other simple techniques for synthesizing nanowires, nanorods, nanotubes, and nanocages, or for selecting polymers for length, are well-known.

For instance, Jean-Marie Lehn has created rectangular grid complexes by mixing rodlike ligands that make use of different numbers of binding sites — one example is a 4 x 5 array consisting of nine ligands and 20 metal ions, nanoscale grids which he claims might one day find use "as components within a futuristic information storage and processing nanotechnology". C.M. Drain has described a 21-component (5 nm)^2 square array of nine porphyrins tethered together by 12 palladium ions shaped like a four-pane window. The structure is built from three different kinds of porphyrins — (1) an X-shaped unit that coordinates to four metal ions, forming the center of the array, (2) a T-shaped unit that coordinates to three metals and forms each side of the array, and (3) an L-shaped unit that coordinates to only two metals and forms each corner of the array. When these components are placed in solution in the correct ratio, they form the square array within a half hour at room temperature with about 90% yield. The same porphyrin units can be combined in different ratios and induced to form wires or tapes. The properties of such arrays can be fine-tuned by choosing the appropriate metal ion linker and functionalized porphyrin unit.

Jeffrey Moore has investigated a three-dimensional nanoscaffolding comprised of a molecular lattice that could serve as a framework for catalysts, photosynthetic molecules, or more complex molecular devices. Moore's original objective was to synthesize modular building blocks with physical and chemical properties that would dictate a "programmed assembly" protocol for "nanoarchitectures". The characteristics of each modular unit shaped the weak intermolecular attractions — electrostatic, van der Waals, and hydrogen-bonding forces — between it and its neighbors so that units would array themselves in a larger structure with the desired spacing and geometry. Moore's group first linked phenylacetylene subunits into oligomers of various lengths which bend and twist like wire sculptures until their ends meet to form closed polyhedral molecules — the basic scaffold components (of which there were at least half a dozen shapes). These building blocks then self-assembled or folded into geometries at least partly controlled by designed hydrogen-bonding interactions. The molecular polyhedra each possessed several phenylene groups, "chemical handles" upon which other molecular groups could be hung. Moore then focused on ~10-nm dendrimeric forms and later began attaching photosensitive antenna molecules to these structures. By 1997, Moore had begun investigating norbornadiene on Si{100} in an effort to develop ways to covalently attach molecules onto surfaces with subnanometer spatial precision, and had suggested a packing model for interpenetrated diamondoid structures. Robson's group has also investigated possible scaffolding molecules.

Josef Michl is attempting to build a molecular construction kit using molecular rods and connectors, pursuing an explicit vision of "molecular Tinkertoys" by working with simple molecular structures that form stiff, flexible rods. Michl has assembled rods from a mixture of carbon-boron molecules (e.g., 10-vertex or 12-vertex carboranes) and carbon-hydrogen molecules, providing fine control over the total rod length. The rods are built up from more primitive molecular parts, such as propellane (a strained form of C_5H_6) and cubane (a strained form of C_8H_8), to make "staffanes" which are a series of cage units in a linear series. (Strained molecules are constructed with bonds that are forced out of their normal angles, as for example 90° in the case of cubane, compared with carbon's normal orientation of 109.5°.) Michl has fabricated rods whose lengths vary from 0.5 nm to 2.5 nm, in precise 0.1-nm steps. There are other ways of making rodlike molecules, but Michl's are highly inert, do not absorb visible or UV light, are stable up to at least 200°C, and do not react with the oxygen in air even at high temperatures.

Related work is underway in Michl's laboratory on connectors to join the rods together. Metal atoms would be the simplest solution, offering the useful quality of strong joints that can be easily disassembled. Different metals give different binding geometries — square, octahedral, and so on. Many of the connectors are metal-containing species; some are symmetrically trisubstituted or hexasubstituted benzenes or tetrasubstituted cyclobutadiene complexes. Michl must build the right chemical groups onto the rod termini to make them adhere to the connectors as desired. In one example, tiny crosses were built by attaching carboxylate groups to one end of the staffanes, allowing them to bind to a connector made of two rhodium atoms. The groups at the other end of the rods were converted to an ester group, which doesn't bind to rhodium, and so the crosses self-assembled correctly. Michl has also fabricated "star connectors" in which the rods are built into the molecules as covalently bonded arms of a star. For instance, one 3-arm star connector uses three large carboranes coupled to a central benzene ring, allowing a pedestal to be attached vertically to the benzene ring via a ruthenium "sandwich" bond. Additional devices could be installed on this molecular scaffolding, such as optically active molecules or even molecular mechanical "windmills". Nanoscale square and hexagonal planar grids have been fabricated. Extensive ab initio quantum mechanical calculations accompany the experimental work and aid in the interpretation of the results. Michl's ultimate goal is "the production of thin layers of solids of completely controlled aperiodic structure consisting of an inert covalent scaffolding carrying selected active groups."

G. Leach and colleagues used computers to simulate various diamond and graphite nanostruts using molecular dynamics calculations, analyzing solid rectangular struts with varying aspect ratios, cross-sections, terminating atoms (for diamond), potential energy functions, and temperatures. The most dramatic differences were seen between struts with 100:1 aspect ratios and struts with lower (10:1 or 1:1) aspect ratios — the diamond strut with a 100:1 aspect ratio and a unit cell (~0.4 nm)^2 cross-section begins to curl (with end-to-end distance decreasing 1.2 nm after 20 picosec) due to thermal vibration (both at 150 K and 300 K), while a unit strut with an aspect ratio of 10:1 or a 100:1 strut with a (1 nm)^2 cross-section showed end-to-end length fluctuations of only ~0.1 nm. The authors concluded that support struts with a cross-section of at least (1 nm)^2 would be sufficient to keep both end-to-end distance and transverse fluctuations below ~0.1 nm in struts with 10:1 aspect ratios. Other work by the same group claims that many nanomachine designs may actually perform better than molecular dynamics results might suggest, or may require fewer atoms for the same positional stability. The elastic and wear characteristics of diamond are being studied computationally by others.

No discussion of molecular "parts" could be complete without mention of the fullerenes, first discovered in 1985 and vigorously investigated ever since. Fullerenes are one of the three known allotropic forms of carbon. In graphite, the most common allotrope, carbon atoms are arranged in hexagonal rings and strongly bonded into parallel planar sheets, with much weaker bonding between the sheets, giving graphite its excellent lubricating properties. In diamond, the second allotropic form, carbon atoms are arranged in a symmetric, tetrahedral structure, giving immense strength. Exposed to air, both diamond surfaces and graphite edges are quickly coated or "passivated" with hydrogen or other atoms that tie up the dangling bonds. In fullerenes, the third allotropic form, carbon atoms form large and hollow cage-like structures, often roughly spherical or tubular, made up of closed curved one-atom-thick sheets of carbon atoms arranged in a number of five-, six-, and higher-membered rings. Interestingly, the fullerenes need no passivating hydrogens or other atoms to satisfy their chemical bonding requirements at the surface — in this sense, "fullerenes are the first and only stable forms of pure, finite carbon".

A new journal, *Fullerene Science and Technology*, appeared in 1993, and by the mid-1990s ~3000 papers on fullerenes and their properties had been published and 149 fullerene-related patents had been issued in the U.S. alone (through 1996). By 1998, an incredible variety of fullerenes of many shapes and sizes had been synthesized.

The original fullerene, C_{60}, consists of exactly 60 carbon atoms arranged in a soccerball structure with 20 hexagons and 12 pentagons (with each pentagon entirely surrounded by hexagons). In theory, small molecules such as H_2 or CO, and probably CH_4, would fit inside C_{60}. A slightly larger variant, C_{70}, is the same except that an extra belt of carbon atoms has been inserted around the equator. Smaller and larger variants have been observed; as cages get bigger, the corners (where the 12 pentagons needed for closure reside) get sharper. The topologically smallest possible fullerene is a dodecahedron consisting of 12 pentagons and 20 carbon atoms, but the ring isomer of C_{20} (a hoop of single carbons) is apparently energetically favored over the bowl or spherical (fullerene) isomers; in 1998, C_{28} was the smallest fullerene that had been observed experimentally, and C_{36} could be made in significant quantities by arc-discharge. The approximate diameter (measured to atomic centers) of a spheroidal fullerene of formula C_n is $D_{ball} \sim 1.1\ (n/60)^{1/2}$ nanometers. In 1998, purified C_{60} cost \$27.50/gm (99.5% pure) or \$60/gm (99.9%), C_{70} \$250/gm (98%), and C_{84} \$3,750/gm (90%) from Dynamic Enterprises Ltd. of London.

Insertion of many additional equatorial belts of carbon atoms (incorporated as hexagons) into a spheroidal fullerene results in a long cylinder with graphite-like ("graphene") walls and spherical endcaps, making a class of fullerenes known as the capsular fullerenes or single-walled carbon nanotubes (SWNT) that come in many sizes and chiral forms. The molecular form of carbon nanotubes has been likened to rolled chicken wire. SWNTs may be synthesized by vaporizing graphite spiked with 1% catalyst from the nickel, cobalt and iron group above 3000°C, then allowing the vapor to slowly condense. Tubes form because the metal atoms interact with dangling bonds at the end of a tube, favoring tube extension over hemispherical capping by newly arriving carbon vapor atoms. (For any given mass number, ball-shaped fullerenes are energetically favored over tube-shaped fullerenes during high-temperature synthesis in the absence of catalyst.) SWNTs typically self-assemble in 1.1 nm wide tubes, although smaller single-walled nanotubes and larger single-walled and multi-walled nested nanotubes exist. SWNT carbon atoms are bonded in virtually flawless hexagonal arrays. Simulations show that isolated flaws migrate to the ends of the tube and are eliminated, a phenomenon known as "self-healing". Single-molecule nanotubes of $C_{1,000,000}$ or larger, with lengths >100,000 times longer than their widths (e.g., ~1 mm long), had been synthesized by 1998. While graphite is a very brittle material, the graphene walls of a carbon nanotube are quite resilient. Computer simulations and experiments confirm that nanotubes kink when bent, then snap back when released. Nanotube diameter is $D_{tube} = 0.078\ (n^2 + nm + m^2)^{1/2}$ nanometers, where (n, m) is the "rollup vector" defined by the number of steps required for a repeat pattern along two crystallographic cylinder wall axes. The cohesive energy per atom required to curve a flat graphene sheet into a cylinder is $E_{rollup} = (13\ zJ\text{-}nm^2 / D_{tube}{}^2)$. (1 zeptojoule (zJ) = 10^{21} joules.) As an example, a (10,10) nanotube has $D_{tube} = 1.36$ nm and $E_{rollup} = 7$ zJ/atom. Young's modulus for individual SWNTs, assuming a hollow (open-ring) cross-section, has been estimated as high as $\sim 5.5 \times 10^{12}\ N/m^2$; elastic bending modulus (measured as 1×10^{12} to $0.1 \times 10^{12}\ N/m^2$) decreases sharply with increasing diameter (from 8-40 nm). In 1998, purified carbon singlewalled nanotubes were available via the Internet for \$1400/gm from Tubes@Rice, or for \$200/gm at lesser purity from CarboLex.

From spatial geometry, it is known that hexagonal tiling produces flat sections (like cylinder walls), and inserting pentagons into an hexagonal array produces positive curvature (like spheres), making endcaps. But 7-sided heptagons can also be inserted into hexagonal arrays to induce negative curvature, thus permitting concave surface deformations such as saddle-shaped fullerenes or possibly helical tubular structures. Indeed, note Colbert and Smalley, "the use of hexagonal, pentagonal, and heptagonal substructures are sufficient to produce caged carbon structures of any topology". Nature makes use of this same geometrical principle in the radiolarians, tiny protozoans with fullerene-like siliceous skeletons, and in the conical nucleoprotein core particle of the HIV-1 virus. R. Smalley observes that a graphene sheet has the highest tensile strength of any known 2-dimensional network, and the packing density of atoms in the sheet (e.g., atoms/m^2) is higher than any other network made of any atoms in the periodic table and higher than the packing density in any 2-dimensional slice through any 3-dimensional object — even diamond which has the highest known 3-D packing density. Hence the graphene sheet is effectively impermeable under normal chemical conditions.

Carbon nanotubes have also been observed with a variety of endcap shapes and with tubes that reduced or increased their diameter for some distance before terminating. Toroidal fullerenes can be produced experimentally; conelike, spindlelike, and helical fullerene objects have also been examined. By 1998, the mechanical properties of fullerenes and carbon nanotubes were being extensively investigated and there were experimental demonstrations of:

1. a limited ability to cut fullerene nanotubes to specific desired lengths;
2. dissolving derivatized SWNTs in common organic solvents;
3. the synthesis of self-oriented nanotube arrays; and
4. the trapping of individual C_{60} molecules in a perforated Langmuir-Blodgett film "workpiece holder" (though an even simpler means for producing 2-nm hole arrays with 78 nm hole spacings was later reported).

Nanotubes were also being investigated as nanoscale sensor components.

Given an ability to synthesize a wide variety of fullerene shapes and sizes, the next task is to find ways to join them together in specific desired molecular architectures. Two hydrogens were readily added to fullerene carbon atoms, making specific isomers of $C_{60}H_2$ and $C_{70}H_2$, and in the 1990s fullerene chemistry began to be explored. A little more than a decade after the discovery of C_{60}, gram quantities of buckyballs became widely available for chemical experimentation. Progress in covalent fullerene chemistry exploded. By the late 1990s, fullerene surfaces could be regioselectively functionalized in many interesting ways.

Theory and experiment suggest that SWNTs may be joined at 30° angles to create complex structures, including helices and three-way nanotube junctions as suggested.

In 1998, the idea of using fullerenes as "Tinkertoys" for molecular construction was just starting to be considered. For example, given the synthesis of C_{60} exohedral monoadducts and multiple adducts described above, it may be possible to fabricate simple gears by bonding rigid ligands onto the external surfaces of carbon nanotubes in the manner of gear teeth, although the chemical synthesis of nanotubes with precisely positioned teeth will not be easy. Given the success in 1996 of IBM scientists in positioning individual organic molecules (each having a total of 173 atoms and a 1.5-nm diameter) at room temperature by purely mechanical means, it might also be possible to align and maintain these molecular gear teeth in atomically precise meshed positions. If this could be done, would such devices actually work in the manner of macroscale gears?

J. Han and colleagues at NASA/Ames performed a detailed 2000-atom molecular dynamics simulation to investigate the properties of molecular gears fashioned from carbon nanotubes with teeth added via a benzyne reaction known to occur with C_{60}. Computationally, one gear is powered by forcing the atoms near the end of the nanotube to rotate, and a second gear is allowed to rotate by keeping the atoms near the end of its nanotube constrained to a cylinder (i.e., the ends of the shaft were constrained to not elongate but were allowed to move within a plane transverse to the tube symmetry axis). The meshing aromatic gear teeth transfer angular momentum from the powered gear to the driven gear. Each gear is made of a 1.1-nm diameter (14, 0) nanotube with seven benzyne teeth. The spacing between two nanotubes is 1.8 nm and the smallest distance between a tooth atom and a tube atom is ~0.4 nm. The results show that the gears can operate up to 70 GHz in vacuo at room temperature without overheating or slipping. As rotational speed rises above 150 GHz, the gears overheat and begin to slip with tooth tilting up to 20°, but no bond or tooth breaking occurs up to at least 3000 K and slipping gears can always be returned to proper operation by lowering the temperature or the rotation rate. A related nanotube gear system (58 sprockets, 290-464 atoms) simulated by Robertson and colleagues at the Naval Research Laboratory (NRL) showed similar overheating at 500 GHz but was stable when accelerated to only 20 GHz.

The NASA group also simulated several other types of nanotube-based gear systems. For instance, a rack and pinion system was designed using a gear made from a (14, 0) nanotube with teeth separated by two hexagon rings and a shaft made from a (9, 9) tube with teeth separated by three rings. Gear and shaft are 1.94 nm apart, with tooth face normal to the radial direction of nanotube (14, 0) for the gear, but in the axis direction of nanotube (9, 9) for the shaft. The gear could receive power, driving the shaft, or vice versa, and worked well for shaft translational velocities up to ~100 m/sec. (Most devices described in Drexler typically move at 1 m/sec.) Since shaft mass is almost twice gear mass, it takes more power for the gear to drive the shaft. In another simulation involving a large 1.4-nm gear coupled to a smaller 0.8-nm gear, the large gear drives the smaller gear smoothly, but if power is instead applied to the smaller gear at a sufficiently large acceleration, then the smaller gear does not drive the larger one but instead "bounces back and forth several times, like elastic collisions of a small ball between two boards."

Another research group at Oak Ridge National Laboratory (ORNL) used classical molecular dynamics to investigate the properties of molecular bearings consisting of an inner and an outer carbon nanotube. The graphite bearings ranged in size from inner shafts between 0.4-1.6 nm in diameter up to 12 nm long, and outer cylinders between 1.0-2.3 nm in diameter up to 4 nm long. The original simulations found excessive vibrational motion, but subsequent work using a more complete quantum approach found that under certain conditions the nanobearing is "frictionless" and undergoes superrotation, a classical dynamical behavior reminiscent of superfluidity. The regime of superrotary motion is somewhat restricted when the nanobearing is under load, which suggests that a very careful design is required to ensure optimum performance.

The ORNL group simulated a fullerene motor consisting of two concentric graphite cylinders (shaft and sleeve) with one positive and one negative electric charge attached to the shaft. Rotational motion of the shaft was induced by applying one, or sometimes two, oscillating laser fields. The shaft cycled between periods of undesirable rotational pendulum-like behavior and good unidirectional motor-like behavior. The NASA/ Ames group simulated a pulsed-laser-powered carbon nanotube gear-motor system which rotated consistently in one direction, although that direction could be either clockwise or counterclockwise.

Building Blocks & Nanotechnology

Although it seems at first that Nature has provided a limited number of basic building blocks—amino acids, lipids, and nucleic acids—the chemical diversity of these molecules and the different ways they can be polymerized or assembled provide an enormous range of possible structures. Furthermore, advances in chemical synthesis and biotechnology enable one to combine these building blocks, almost at will, to produce new materials and structures that have not yet been made in Nature. These self-assembled materials often have enhanced properties as well as unique applications. The selected examples below show ways in which clever synthetic methodologies are being harnessed to provide novel biological building blocks for nanotechnology. The protein polymers produced by Tirrell and coworkers (1994) are examples of this new methodology. In one set of experiments, proteins were

Top: a 36-mer protein polymer with the repeat sequence (ulanine-glycine)3 – glutamic acid – glycine. Bottom: idealized folding of this protein polymer, where the glutamic acid sidechains (+) are on the surface of the folds. designed from first principles to have folds in specific locations and surfacereactive groups in other places (Krejchi et al. 1994; 1997). One of the target sequences was -((AG)3EG)- 36. The hypothesis was that the AG regions would form hydrogen-bonded networks of beta sheets and that the glutamic acid would provide a functional group for surface modification. Synthetic DNAs coding for these proteins were produced, inserted into an *E. coli* expression system, and the desired proteins were produced and harvested. These biopolymers formed chain-folded lamellar crystals with the anticipated folds. In addition to serving as a source of totally new materials, this type of research also enables us to test our understanding of amino acid interactions and our ability to predict chain folding. Biopolymers produced via biotechnology are monodisperse; that is, they have precisely defined and controlled chain lengths; on the other hand, it is virtually impossible to produce a monodisperse synthetic polymer. It has recently been shown that polymers with well-defined chain lengths can have unusual liquid crystalline properties.

For example, Yu et al. (1997) have shown that bacterial methods for polymer synthesis can be used to produce poly(gamma-benzyl alpha L-glutamate) that exhibits smectic ordering in solution and in films. The distribution in chain length normally found for synthetic polymers makes it unusual to find them in smectic phases. This work is important in that it suggests that we now have a route to new smectic phases whose layer spacings can be controlled on the scale of tens of nanometers. The biotechnology-based synthetic approaches described above generally require that the final product be made from the natural, or L-amino acids. Progress is now being made so that biological machinery (e.g., *E. coli*), can be co-opted to incorporate non-natural amino acids such as b-alanine or dehydroproline or fluorotyrosine, or ones with alkene or alkyne functionality (Deming et al. 1997). Research along these lines opens new avenues for producing controlled-length polymers with controllable surface properties, as well as biosynthetic polymers that demonstrate electrical phenomena like conductivity. Such molecules could be used in nanotechnology applications. Novel chemical synthesis methods are also being developed to produce "chimeric" molecules that contain organic turn units and hydrogen-bonding networks of amino acids (Winningham and Sogah 1997). Another approach includes incorporating all tools of chemistry into the synthesis of proteins, making it possible to produce, for example, mirror-image proteins. These proteins, by virtue of their D-amino acid composition, resist biodegradation and could have important pharmaceutical applications (Muir et al. 1997). Arnold and coworkers are using a totally different approach to produce proteins with enhanced properties such as catalytic activity or binding affinity. Called "directed evolution," this method uses random mutagenesis and multiple generations to produce new proteins with enhanced properties. Directed evolution, which involves DNA shuffling, has been used to obtain esterases with five- to six-fold enhanced activity against *p*-nitrobenzyl esters (Moore et al. 1997).

The ability of biological molecules to undergo highly controlled and hierarchical assembly makes them ideal for applications in nanotechnology. The self-assembly hierarchy of biological materials begins with monomer molecules (e.g., nucleotides and nucleosides, amino acids, lipids), which form polymers (e.g., DNA, RNA, proteins, polysaccharides), then assemblies (e.g., membranes, organelles), and finally cells, organs, organisms, and even populations (Rousseau and Jelinski 1991, 571-608). Consequently, biological materials assembly on a very broad range of organizational length scales, and in both hierarchical and nested manners (Aksay et al. 1996; Aksay 1998). Research frontiers that exploit the capacity of biomolecules and cellular systems to undergo self-assembly have been identified in two recent National Research Council reports (NRC 1994 and 1996). Examples of self-assembled systems include monolayers and multilayers, biocompatible layers, decorated membranes, organized structures such as microtubules and biomineralization, and the intracellular assembly of CdSe semiconductors and chains of magnetite. A number of researchers have been exploiting the predictable basepairing of DNA to build molecular-sized, complex, three-dimensional objects. For example, Seeman and coworkers (Seeman 1998) have been investigating these properties of DNA molecules with the goal of forming complex 2-D and 3-D periodic structures with defined topologies.

DNA is ideal for building molecular nanotechnology objects, as it offers synthetic control, predictability of interactions, and well-controlled "sticky ends" that assemble in highly specific fashion. Furthermore, the existence of stable branched DNA molecules permits complex and interlocking shapes to be formed. Using such technology, a number of topologies have been prepared, including cubes (Chen and Seeman 1991), truncated octahedra (Zhang and Seeman 1994), and Borromean rings (Mao et al. 1997). Other researchers are using the capacity of DNA to self-organize to develop photonic array devices and other molecular photonic components (Sosnowski et al. 1997).

This approach uses DNA-derived structures and a microelectronic template device that produces controlled electric fields. The electric fields regulate transport, hybridization, and denaturation of oligonucleotides. Because these electric fields direct the assembly and transport of the devices on the template surface, this method offers a versatile way to control assembly. There is a large body of literature on the self-assembly on monolayers of lipid and lipid-like molecules (Allara 1996, 97-102; Bishop and Nuzzo 1996). Devices using self-assembled monolayers are now available for analyzing the binding of biological molecules, as well as for spatially tailoring. Idealized truncated octahedron assembled from DNA. This view is down the four-fold axis of the squares. Each edge of the octahedron contains two double-helical turns of DNA surface activity. The technology to make self-assembled monolayers (SAMs) is now so well developed that it should be possible to use them for complex electronic structures and molecular-scale devices.

Research stemming from the study of SAMs (e.g., alkylthiols and other biomembrane mimics on gold) led to the discovery of "stamping" (Kumar and Whitesides 1993). This method, in which an elastomeric stamp is used for rapid pattern transfer, has now been driven to < 50 nanometer scales and extended to nonflat surfaces. It is also called "soft lithography" and offers exciting possibilities for producing devices with unusual shapes or geometries. Self-assembled organic materials such as proteins and/or lipids can be used to form the scaffolding for the deposition of inorganic material to form ceramics such as hydroxyapatite, calcium carbonate, silicon dioxide, and iron oxide. Although the formation of special ceramics is bio-inspired, the organic material need not be of biological origin. An example is production of template-assisted nanostructured ceramic thin films (Aksay et al. 1996). A particularly interesting example of bio-inspired self-assembly has been described in a recent article by Stupp and coworkers (Stupp et al. 1997). This work, in which organic "rod-coil" molecules were induced to selfassemble, is significant in that the molecules orient themselves and selfassemble over a wide range of length scales, including mushroom-shaped clusters; sheets of the clusters packed side-by-side; and thick films, where the sheets pack in a head-to-tail fashion. The interplay between hydrophobic and hydrophilic forces is thought to be partially responsible for the controlled assembly.

Dispersion and Coatings

Nanoparticles and nanospheres have considerable utility as controlled drug delivery systems (Hanes et al. 1997). When suitably encapsulated, a pharmaceutical can be delivered to the appropriate site, its concentration can prevented from undergoing premature degradation. Nanoparticles (as opposed to micron-sized particles) have the advantage that they are small enough that they can be injected into the circulatory system. Highly porous materials are also ideal candidates for controlled drug delivery (Schnur et al. 1994) and for tissue engineering (Hubbell and Langer 1995). An example of a controlled drug delivery system comes from the area of microtubules. Phospholipid bilayers can self-assemble into long cylindrical tubes with diameters usually below a micron and lengths up to hundreds of microns (Schnur 1993). During synthesis, drugs can be entrained original patterned silicon *elastomeric stamp elastomeric stamp* substrate. An elastomeric stamp (top left) is made from an original master (bottom left). The stamp is dipped into the biological material (top right) and the pattern is transferred to the substrate (bottom right). Mushroom-shaped clusters formed from self-assembly of rod-coil molecules; these clusters can undergo further packing to form sheets. in these nanotubes, and the final product can be used as a controlled delivery system. Tubules prepared from phospholipid bilayers are ideal for such applications because they are biocompatible. Dendrimers can be prepared so they are of discrete size and contain specific functional groups (Karak and Maiti 1997). They can be functionalized and used in biomedicine. Examples include gene transfer agents for gene therapy, made to carry and control the relaxivity of paramagnetic MRI (magnetic resonance imaging) contrast agents (Toth et al. 1996), and to deliver drugs on a controlled release basis.

Role of Nanoparticles in Health and Pollution

Although beyond the scope of this review, it is important to keep in mind the potential role of atmospheric nanoparticles in photocatalytic and thermal production of atmospheric pollutants. Atmospheric aerosols in heavily polluted areas have the potential to accelerate ozone formation reactions. Furthermore, because they are respirable, they could represent a health hazard. Two controversial studies (the Harvard University six-city study and the American Cancer Society study) have linked the presence of fine particular matter to premature mortality (*Chemical and*

Engineering News 1997). Atmospheric aerosols generally contain two major components: one is composed of amorphous carbon that has fullerene-like particles dispersed in it; the second is inorganic and consists of oxides and sulfides supported on clay minerals. In particular, the iron oxide, manganese oxide, and iron sulfide nanoparticles have band-gaps that could enhance the photocatalytic adsorption of solar radiation. In addition, these materials are acidic and may be coated with water, which would enhance their catalytic ability to crack hydrocarbons and create free radicals (Chianelli 1998). At present this is an underexplored area of research that bears scrutiny. It is interesting to note that some microorganisms produce and sequester CdSe and CdS nanoparticles in response to high levels of toxic Cd++ in their environment (Brus 1996). A large number of organisms also have the ability to precipitate ferrimagnetic minerals Fe_3O_4 and Fe_3S_4.

High Surface Area Materisl

Supported polymeric membranes can be used to remove low molecular weight organics from aqueous solution (Ho et al. 1996). They work by filling the pores of microfiltration or ultrafiltration membranes with functional, polymeric, or oligomeric liquids that have an affinity for the compound of interest. These ideas can be extended to microemulsions and used for separations of large biological molecules. For example, non-ionic microemulsions (i.e., oil-water systems) have been used for hemoglobin extraction. These results augur well for the separation of other biological materials (Qutubuddin et al. 1994). Molecular imprinting can also be used to produce high surface area adsorbents that are enantioselective for amino acids and other biological molecules (Vidyasankar et al. 1997).

Crystalline bacterial cell surface layers (S-layers) are composed of repeating protein units. These layers self-assemble and have a high binding capacity. They have been explored as patterning elements for molecular nanotechnology. For example, they have been used to pattern cadmium sulfide superlattices (Shenton et al. 1997).

Functional Nanostructures

The smallest possible computer would ideally be able to perform computations on a molecular scale. Even though the computation may be carried out on that scale, the issue becomes one of having enough molecules to obtain sufficient signal-to-noise ratio to read out the answer. Consequently, at least at present, these computations must be carried out in bulk or with extremes in temperature. There has been a recent development along this line. Adleman (1994) has shown how the rules of DNA selfassembly, coupled with polymerase chain reaction (PCR) amplification of DNA, can lead to a molecular computer of sorts. The system was used to solve the Hamiltonian path problem, a classic and difficult (NP complete) mathematical problem that involves finding a path between vertices of a graph. The starting and ending vertices of each edge of the graph were encoded as the first and second halves of a strand of DNA. A solution to the problem (what is the path between two specific vertices?) was obtained by using PCR primers for the two vertices. Others have extended these ideas and shown that it is possible to make DNA add (Guarnieri et al. 1996). Although these are exciting demonstrations, at present this technology has a number of drawbacks, including the labor and time it takes to analyze the results of the computation, and the uncertainties associated with wet chemistry.

Optoelectronic Devices

Biological molecules and assemblies, such as the photochemical reaction center, are capable of capturing light with good quantum efficiency and transforming it into chemical energy. If properly exploited, such assemblies have potential applications as biomolecule information processing units. Bacteriorhodopsin, from the purple membrane bacterium *Halobacterium halobium*, is one such system that has been studied extensively and has been commercialized into optical holographic memories (Birge 1995). In the bacterium, the protein bacteriorhodopsin self-assembles into ordered lipid patches. The protein absorbs light and undergoes a cycle involving a complex series of intermediates, resulting in a proton being pumped across the membrane. It was information developed from understanding the basic science behind the way that bacteriorhodopsin works that led to the use of bacteriorhodopsin as a biomolecule information processing unit. To be used for information storage, the protein is placed under nonbiological conditions. It is dispersed and immobilized into a matrix (e.g., collagen or another polymeric substance) and held at liquid nitrogen temperatures. At this low temperature, the protein acts as an optically-driven bistable switch. One form, the light-adapted form of bacteriorhodopsin, absorbs light at 570 nm. When irradiated with green light at 77 K, it switches to a different stable form that absorbs at 630 nm. Using light of different colors enables one to read and write images onto these memories. By subtracting one memory from the other, these memories are especially useful for realtime, rapid comparison of images.

Molecular Motors

The molecular motors found in biology provide for bacterial locomotion, as well as for the active transport and delivery of molecules. For example, the bacterial flagellar motor is about 20 nm in diameter, and is comprised of a complex assembly of more than 10 different proteins (Imae and Atsumi 1989). The role of the motor is to rotate the helical flagella of the bacterium so that it is able to swim. Chemical energy (in this case protons or sodium) is transduced into mechanical energy. Other examples of molecular motors include RNA polymerase (Yin et al. 1995), F1-ATPase (Noji et al. 1997), myosin, and kinesin (Seventh Biophysical Discussions 1995). The fuel that powers these motors is ATP (adenosine triphosphate). A number of researchers have proposed schemes by which such motors could be used to deliver molecules, one at a time, for the purpose of the ground-up assembly of nanoscale devices (NRC 1996). It is envisioned that the highways could be actin or tubulin, which would need to be immobilized onto a surface. Myosin or kinesin, which naturally travel along these highways, could be used to deliver molecular "packages" to a specific assembly site.

Other Forms of Biological Transport Using Nanoparticles

In the human body, the function of the high density lipoproteins is to transport cholesterol. The ~ 7.5 nm discoidal lipoprotein assemblies are sandwiched between discs of phospholipids and stacked, poker-chip style.

The lipoproteins stabilize the cholesterol particles and assist in their transport. Current research involves manipulating and fusing the assemblies and particles with an atomic force microscope (ATM) tip (Sligar 1998).

Bioelectronics

There has been considerable research activity in molecular electronics and bioelectronics, particularly in Japan (Aizawa 1994). Although this area of nanotechnology is still not as well developed as others in this report, it bears watching. For example, Shionoya and coworkers at the Institute for Molecular Science of the Okazaki National Research Institutes in Japan have proposed that novel combinations of DNA, metal ligands, DNA templating, and proteins could produce molecular wires; molecular hoops through which DNA could be threaded; and doublestranded peptides whose helix pitch could be controlled by an entrained metal. The active site containing the metal could be induced to go from Cu(I)tetrahedral to CU(II)square planar, perhaps by electrons delivered by a scanning tunneling microscope (STM) tip.

Consolidated Materials

Ceramics

Biomolecule/inorganic interactions can be used to produce ceramics with increased toughness. Fundamental studies of biomineralization, in which an organic substance (usually protein or peptide or lipid) interacts with an inorganic phase (e.g., calcium carbonate or hydroxyapatite) have led to the bioinspired synthesis of composite materials. Novel combinations of DNA, metal ligands, DNA templating, and proteins are being investigated for molecular wires, inductors, and switches (photo courtesy of Shionoya and coworkers, Inst. for Molecular Science).

The structure and porosity of the inorganic phase can be controlled by templating with an organic surfactant, vesicular arrays, or liquid crystalline materials. Micelle-templated synthesis can produce ceramics with 20-100 Å pore dimensions (Ying 1998). These tailored pores can be used as catalysts and absorbents, and for gas/liquid separations and thermal and acoustic insulation. Their selectivity makes them very useful for biochemical and pharmaceutical separations. Bioceramics can also be made that are more compatible with teeth and bone. An interesting example of an organic/inorganic composite is the new packaging material that has been developed to replace the polystyrene "clam-shell" for fast food products. Composed of potato starch and calcium carbonate, this foam combines the advantages of good thermal insulation properties and light weight with biodegradability (Stucky 1978).

Biological Bar Magnets

It has been well documented that a very large number of organisms have the ability to precipitate ferrimagnetic minerals such as Fe3O4 and Fe3S4. In addition, linear chains of membrane-bound crystals of magnetite, called magnetosomes, have been found in microorganisms and fish (Kirschvink et al. 1992). For example, the Fe3O4 domain size in the organism *A. magnetotactum* is about 500 Å, and a chain of 22 of those domains has a magnetic moment of 1.3 x 10-15. It is not immediately clear how these particles could be exploited for nonbiological applications.

Opportunities and Challenges

This section of the review describes current limits on the biological aspects of nanotechnology. It outlines selected broad areas of research opportunity, and it sets forth challenges for the intermediate and long-term future.

Enabling Technologies

We owe much of the recent progress in the biological aspects of nanotechnology and nanoparticles to important enabling technologies. Many of these enabling technologies are described in other parts of this report. Those that have had a significant impact on biological measurements include techniques and instrumentation such as optical traps, laser tweezers, and "nano-pokers" to measure femtoNewton forces (Svoboda and Block 1994), and AFM and STM (Wildoer et al. 1998).

There is also a need to move detection away from the ensemble, to the single molecule scale. This includes driving instrumentation toward singlemolecule spectroscopy; single-spin detection (Rugar et al. 1994); chemical analysis of nanoliter volumes (Hietpas et al. 1996, 139-158); nuclear magnetic resonance microcoils; nanoscale electrode arrays and chemical sensing and detecting technology (McConnell 1996, 97-102); separations technology; chemical analysis of single cells (Hietpas et al. 1996); new biological transformations; and chemical probes of nanostructures. Enhanced computational infrastructure, both hardware and algorithm development, will also be required to investigate supramolecular assemblies and to understand interactions that occur over a wide variety of time and length scales. At present, we are fairly well equipped to handle quantum mechanical calculations. At the next level of computational complexity, the molecular dynamics force fields and atomic charges can presently handle up to about a million particles and still retain their accuracy. The next scale of complexity, mesoscale modeling, currently requires the use of pseudo atoms. Much more research is required to improve the accuracy of finite element calculations and the modeling of materials applications (Goddard 1998).

Surface Interactions and the Interface Between Biomolecules and Substrates

Much of the progress in the biological aspects of nanotechnology has come from research on surface interactions. It will be important to continue to develop a better understanding of the interface between biomolecules and surfaces. Of particular interest are questions about whether surfaces cause biomolecules to denature, the optimum length of linker groups, and ways to communicate from the biomolecule, through the linker, to the substrate. Another challenge is to produce nanometer ultrathin films that have stable order (Jaworek et al. 1998).

Robustness of Biomolecules and Their Interactions in Aqueous Solution

Although much progress has been made on ways to confer additional robustness on biomolecules, it will be necessary to continue to improve the stability and reliability of biomolecular assemblies. Such research may be along the lines of understanding the thermal stability of biomolecules, perhaps by examining extremophiles and molecules produced by directed evolution. It will also be necessary to produce molecules that can work in the absence of water, or to devise ways in which aqueous solutions can be used reliably.

Assemblers or Templating

The hierarchical self-assembly of biomolecules is often touted as one of their key attributes for nanotechnology applications. However, there remains much to be learned about how to make periodic arrays of biomolecular assemblies, how to use biological templating in an efficient manner, how to mimic biological self-assembly with nonbiological molecules, and how to exploit differences between biological and nonbiological self-assembly.

Combinatorial Approaches to Nanotechnology

Combinatorial chemistry has had an enormous impact on drug discovery and development in the past several years. Ability to develop molecularly sensitive screening techniques is one of the limiting factors of combinatorial chemistry. It would be intriguing to discover screening methods that are particularly sensitive for picking up new phenomena exhibited by biological nanoparticles.

Biomimetics and Polymer-Biopolymer Chimeras

The largest impact of biological sciences on nanotechnology may prove to be through the lessons one can learn from biology, rather than from directly employing the biological molecules themselves. To this end, it is important to support research on various aspects of biomimetics, including activities in which synthetic and biologically-based molecules are combined in chimeric fashion.

Nanotechnology for Bio Materials and Tissues

Technology Focus

Historically, synthetic materials have not served as sufficient implants.For example, the current average lifetime of an orthopedic implant, such as hip, knee, ankle, etc., is only 15 years. Similarly, for the vascular community, small diameter vascular grafts are only functional 25% of time past 5 years of use. The same limited lifetimes for a number of failing organs could be stated for cartilage, bladder, and the peripheral and central nervous systems. Clearly, conventional materials, or those materials with constituent dimensions greater than 1 micron have not invoked proper cellular responses to regenerate tissue that allows for these devices to be successful for long periods of times.

In contrast, due to their ability to mimic the dimensions of constituent components of natural tissues, like proteins, nanophase materials may be an exciting successful alternative. Nanophase materials are defined as materials with constituent dimension less than 100 nm in at least one direction. Materials investigated to date include nanophase ceramics, metals, polymers, and composites. Nanophase materials have revolutionized many traditional engineering fields such as those involving catalysis, electronics, optics, magnetics, etc. For example, in the catalytic field, higher surface reactivity of nanophase materials allows for faster reaction times.

To date,there has also been the emergence of data that nanophase materials may be optimal materials for tissue engineering applications. This is not only due to their ability to simulate dimensions of proteins that comprise tissues, but also because of their higher reactivity for interactions of proteins that control cell adhesion and, thus, the ability to regenerate tissues.

There are numerous reasons to consider using nanophase materials as better tissue engineering materials.While most of the promise is conjecture, limited scientific evidence does exist and is generating wide support.

Course Objectives

The first objective of this course is to define current problems with implants. You will be familiar with issues related to implant failure particular for orthopedic,dental, cartilage,vascular,bladder,and the nervous systems.

The second objective is to understand how cells interact with implanted materials to generate a biological response.We will describe how proteins interact with materials to control cell function leading to either a desirable or undesirable biological response to an implanted material.

The third objective is to understand how nanophase materials could be used as better implants. What are their promising properties and what experimental evidence exists? We will sort through the experimental data and separate promise from "hype". The discussion will make you be able to indicate what are the promising properties of nanophase materials and how to make them to control initial protein interactions leading to successful regeneration of tissue.

The last objective is to understand what further information will be needed to elucidate if nanophase materials will make better biomaterials for these organ systems. While there is no golden standard chemistry for various implant applications, this course will discuss the possibility of a standard topography that may be useful to regenerate tissue for a wide range of materials including ceramics, metals, and polymers. You will be able to identify future experiments needed for the further consideration of nanophase materials to regenerate tissue

Course Content

This course will begin with a general overview of current problems with orthopedic, dental, cartilage, vascular, bladder, and central as well as peripheral nervous system devices. These organs will be chosen since they are the ones in which the most implantable devices have been used to date. One day will be spent describing situations in which these devices are needed and why we have not been able to develop sufficient materials to successfully regenerate these tissues. To do this, we will cover the basic anatomy, physiology, and cellular structures of these organs. This will help us understand the complexity of our organs and consequently to design materials that match the most important properties of tissue to promote tissue regeneration.

Next,we will cover how the body responds to implanted materials by first adsorbing a thin layer of proteins and then through select cellular responses. Topics such as how proteins adsorb to surfaces,how cells recognize proteins, how cells adhere and function on materials, and how tissues may or may not regrow will be the focus of this day.The biological response section will be complemented with results from experiments that have delineated fundamental material properties that control such responses. By understanding how the body originally responds to implanted materials, we can begin to develop materials to obtain a desirable response, such a tissue regrowth.

We will then interject a new classification of materials that are beginning to be considered to serve as better biomaterials: nanophase materials. One day will be spent defining these materials and highlighting their special properties pertinent for implants. This includes their mechanical, optical, electrical, and surface properties. We will also discuss various routes to synthesize nanophase materials. Although, to date, the special properties of nanophase materials have been used in non-biological applications, we will emphasize their role in biological applications. We will constantly compare theory will experimental evidence in this section. The day will end with the most promising results obtained thus far in terms of increased tissue regeneration on nanophase materials.

Lastly, we will discuss the most promising future directions for the use of nanophase materials in biological applications. This day will focus on key unanswered questions if nanophase materials are to be used in the body. Central to this question is the immunological response to nanoparticles, which will comprise a majority of the day.

To enhance learning as well as stimulate discussion, the course will be interactive with participation from the learners throughout.

Current Implant Failures

Orthopedic

- Current Materials Used
- Current Modes of Failure: Poor Bonding to Surrounding Bone – Generation of Wear Debris – Stress and Strain Imbalances at the Tissue Implant Interface
- Current Statistics on Orthopedic Implant Failure

Dental

- Current Materials Used
- Current Modes of Failure: Poor Bonding to Surrounding Bone – Generation of Wear Debris – Stress and Strain Imbalances at the Tissue Implant Interface
- Current Statistics on Dental Implant Failure

Cartilage

- Current Treatment Methods and Materials Used
- Current Modes of Failure: Poor Regeneration of Cartilage Tissue – Generation of Wear Debris – Stress and Strain Imbalances at the Tissue Implant Interface
- Current Statistics on Cartilage Implant Failure

Vascular

- Current Materials Used
- Current Modes of Failure: Poor Bonding to Surrounding Vascular – Endothelium Removal Due to Shear Stress – Generation of Stenosis - Stress and Strain Imbalances at the Tissue Implant Interface
- Current Statistics on Vascular Implant Failure

Bladder

- Current Materials Used
- Current Modes of Failure: Poor Bonding to Surrounding Tissue – Stress and Strain Imbalances at the Tissue Implant Interface
- Current Statistics on Bladder Implant Failure

Central and Peripheral Nervous System

- Current Materials Used
- Current Modes of Failure: Poor Bonding to Surrounding Tissue – Build-up of Glial Scar Tissue – Poor Maintenance of Electrical Properties – Stress and Strain Imbalances at the Tissue Implant Interface
- Current Statistics on Central and Peripheral Nervous System Implant Failure

Biological Response of Implanted Materials

Protein Interactions with Implanted Materials

- Properties of Proteins
- Adsorption
- Conformation/Bioactivity

Cellular Recognition of Proteins Adsorbed on Materials Surfaces

- Adhesion
- Migration
- Differentiation

Cellular Extracellular Matrix Deposition Leading to Tissue Regeneration

- Bone
- Dental Tissue
- Cartilage
- Vascular
- Bladder
- Central and Peripheral Nervous System.

REFERENCES

Baird, D.: 1993, 'Analytical Chemistry and the Big Scientific Instrumentation Revolution', *Annals of Science*, 50, 267-290.

Baird, D.; Nordmann, A. & Schummer, J. (eds.): 2004, *Discovering the Nanoscale*, Amsterdam: IOS Press.

Baker, R.T.K. Synthesis, properties and applications of graphite nanofibers. In *R&D status and trends*, ed. Siegel et al.

Ball, P.: 2002, 'Tutorial: Natural Strategies for the Molecular Engineer', *Nanotechnology*, 13, 15-28.

Bate, R., Frazier, G., Frensley, W., Reed., M., "An Overview of Nanoelectronics," *Texas Instruments Technical Journal*, July-August 1989, pp. 13-20.

Baum, R.; Drexler, K.E.; Smalley, R.: 2003, 'Nanotechnology: Drexler and Smalley Make the Case For and Against 'Molecular Assemblers', *Chemical and Engineering News*, 81, 37-42.

Beck, J.S., J.C. Vartuli, W.J. Roth, M.E. Leonowicz, C.T. Kresge, K.D. Schmitt, C.T.-W. Chu, D.H. Olsen, E.W. Shepard, S.B. McCullen, J.B. Higgins, and J.L. Schlenker. 1992. *J. Am. Chem. Soc.* 114:10834.

Bensaude-Vincent, B.: 2001, 'The Construction of a Discipline: Materials Science in the United States', *Historical Studies in the Physical and Biological Sciences*, 31, 223-248.

Bensaude-Vincent, B.: 2004, 'Two Cultures of Nanotechnology?', *Hyle: International Journal for Philosophy of Chemistry*, 10(2), 65-82.

Berube, D.M.: 2004, 'The Rhetoric of Nanotechnology', in: D. Baird, A. Nordmann & J. Schummer (eds.), *Discovering the Nanoscale*, Amsterdam: IOS Press, pp. 173-192.

Bijker, W.E.: 1995a, *Of Bicycles, Bakelite, and Bulbs: Toward a Theory of Sociotechnical Change*, MIT Press, Cambridge, MA.

Bijker, W.E.: 1995b, 'Sociohistorical Technology Studies', in: S. Jasanoff, G.E. Markle, J. Petersen, T. Pinch (eds.), *Handbook of Science and Technology Studies*, Sage, London, pp. 229-256.

Bijker, W.E.; Law, J.: 1992, 'Do Technologies Have Trajectories?', in: W.E. Bijker, J. Law (eds.), *Shaping Technology/Building Society: Studies in Sociotechnical Change*, MIT Press, Cambridge, MA, pp. 1-16.

Bijker, W.E.; Pinch, T.: 1987, 'The Social Construction of Facts and Artifacts: Or How the Sociology of Science and the Sociology of Technology Might Benefit Each Other', in: W.E. Bijker, T.P. Hughes, T. Pinch (eds.), *The Social Construction of Technological Systems: New Directions in the Sociology and History of Technology*, MIT Press, Cambridge, MA, pp. 17-50.

Bimber, B.: 1994, 'Three Faces of Technological Determinism', in: M.R. Smith, L. Marx (eds.), *Does Technology Drive History?: The Dilemma of Technological Determinism*, MIT Press, Cambridge, MA, pp. 79-100.

Bowes, C.L., A. Malek, and G.A. Ozin. 1996. *Chem. Vap. Deposition* 2:97.

Braun, P.V., P. Osenar, and S.I. Stupp. 1996. *Nature*. 368: 2.

Braun, T.; Schubert, A. & Zsindely, S.: 1997, 'Nanoscience and nanotechnology on the balance', *Scientometrics*, 38, 321-325.

Brinker, C.J. 1996. *Curr. Opin. Solid State Mater. Sci.* 1:798.

Brooks, L.J.: 2003, *The Institute for the Future: A Site for Mapping, Maintaining, and Transforming Images of the Future*, (Ph.D. dissertation) University of California, San Diego, La Jolla.

Brown, N.: 2004, 'Needle on the Real: Technoscience and Poetry at the Limits of Fabrication', in: N.K. Hayles (ed.), *Nanoculture: Implications of the New Technoscience*, Bristol, UK: Intellect Books, pp. 173-190.

Bueno, O.: 2004, 'The Drexler-Smalley Debated on Nanotechnology: Incommensurability at Work?', *Hyle: International Journal for Philosophy of Chemistry*, 10(2), 83-98.

Bueno, O.: 2004, 'Von Neumann, Self-Reproduction and the Constitution of Nanophenomena', in: D. Baird, A. Nordmann & J. Schummer (eds.), *Discovering the Nanoscale*, Amsterdam: IOS Press, pp. 101-115.

Bush, V.: 1945, *Science, the Endless Frontier*, US Government Printing Office, Washington, DC.

Carroll, J.S.: 2001, 'Social Science Research Methods for Assessing Societal Implications of Nanotechnology', in: M.C. Roco and W.S. Bainbridge (eds.), *Societal Implications of Nanoscience and Nanotechnology*, Dordrecht: Kluwer, pp. 188-192.

Chen, C-Y., S.L. Burkett, H.-X. Li, and M.E. Davis. 1993. *Microporous Mater*. 2:27.

Chianelli, R.R. 1998. Synthesis, fundamental properties and applications of nanocrystals, sheets, and fullerenes based on layered transition metal chalcogenides. In *R&D status and trends,* ed. Siegel et al.

Chianelli, R.R., M. Daage, and M.J. Ledoux. 1994. *Advances in Catalysis* 40:177.

Claeson, T., and Likharev, K.," Single Electronics," *Scientific American,* June 1992, pp. 80-85.

Clausen-Schaumann, H.; Seitz, M.; Krautbauer, R.; Gaub, H.E.: 2000, 'Force Spectroscopy with Single Bio-Molecules', *Current Opinion in Chemical Biology*, 4, 524-530.

Coenen, C.: 2004, 'Nanofuturismus: Anmerkungen zu seiner Relevanz, Analyse und Bewertung', *Technikfolgenabschätzung ' Theorie und Praxis*, 13 (2), 78-85.

Constant, E.W., II: 1980, *The Origins of the Turbojet Revolution*, Johns Hopkins University Press, Baltimore.

Crandall, B.: 1999, 'Molecular Engineering', in: B. Crandall (ed.), *Nanotechnology: Molecular Speculations on Global Abundance*, MIT Press, Cambridge, MA, pp. 1-46.

Crandall, B.C. and Lewis, J., *Nanotechnology: Research and Perspectives*, MIT Press, Cambridge, 1992.

Crow, M. & Sarewitz, D.: 2001, 'Nanotechnology and Societal Transformation', in: M.C. Roco and W.S. Bainbridge (eds.), *Societal Implications of Nanoscience and Nanotechnology*, Dordrecht: Kluwer, pp. 45-54.

Decker, M.; Fiedeler, U.; Fleischer, Th.: 2004, 'Ich sehe was, was Du nicht siehst... zur Definition von Nanotechnologie', *Technikfolgenabschätzung ' Theorie und Praxis*, 13 (2), 10-16.

Dresselhaus, M., and G. Dresselhaus. 1995. *Ann. Rev. Mat. Sci*. 25:487.

Dresselhaus, M.S. 1998. Carbon-based nanostructures. In *R&D status and trends*, ed. Siegel et al.

Dresselhaus, M.S., G. Dresselhaus, and P. Eklund. 1996. *Science of fullerenes and carbon nanotubes*. San Diego: Academic Press.

Drexler, K.E.: 1981, 'Molecular Engineering – an Approach to the Development of General Capabilities for Molecular Manipulation', *Proceedings of the National Academy of Sciences of the United States of America-Physical Sciences*, 78, 5275-5278.

Drexler, K.E.: 1992, *Nanosystems: Molecular Machinery, Manufacturing, and Computation*, Wiley, New York.

Drexler, K.E.: 2004, 'Nanotechnology: From Feynman to Funding', *Bulletin of Science, Technology & Society*, 24, 21-27.

Dupuy, J.-P.: 2004, 'Complexity and Uncertainty: A Prudential Approach To Nanotechnology', in: European Commission (Community Health and Consumer Protection): *Nanotechnologies: A Preliminary Risk Analysis on the Basis of a Workshop, Brussels,* 1-2 March 2004, pp. 71-93.

Dupuy, J.P.: 2004, 'Pour une évaluation normative du programme nanotechnologique', *Annales des Mines*, (February 2004), 27-32.

Eigler, D.M.; Schweizer, E.K.: 1990, 'Positioning Single Atoms With a Scanning Tunneling Microscope', *Nature*, 344, 524-526.

Einsiedel, E.F.; Goldenberg, L.: 2004, 'Dwarfing the Social? Nanotechnology Lessons from the Biotechnology Front', *Bulletin of Science, Technology & Society*, 24, 28-33.

ETC: 2003a, *The Big Down: Atom Tech ' Technologies Converging at the Atomic Scale*. Winnipeg, Canada: Action Group on Erosion, Technology and Concentration [www.etcgroup.org/documents/TheBigDown.pdf].

ETC: 2003b, 'No Small Matter II: The Case for a Global Moratorium ' Size Matters!' *Occasional Paper Series* 7(1) [www.etcgroup.org/documents/Occ.Paper_Nanosafety.pdf].

Etzkowitz, H., 2001, 'Nano-Science and Society: Finding a Social Basis for Science Policy', in: M.C. Roco and W.S. Bainbridge (eds.), *Societal Implications of Nanoscience and Nanotechnology*, Dordrecht: Kluwer, pp. 121-128.

European Comission: *European Workshop on Social and Economic Research on Nanotechnologies and Nanosciences, Brussels, 14-15 April 2004* [www.stage-research.net/STAGE/PAGES/Nano.html].

European Commission (Community Health and Consumer Protection): 2004, *Nanotechnologies: A Preliminary Risk Analysis on theBasis of a Workshop, Brussels, 1-2 March 2004.*

European Commission: 2004, *Converging Technologies: Shaping the Future of European Societies (Report of the High Level Expert Group "Foresighting the New Technology Wave")*, Brussels: European Commission Research [http://europa.eu.int/comm/research/conferences/2004/ntw/index_en.html].

F. Buot, "Mesoscopic Physics and Nanoelectronics: Nanoscience and Nanotechnology," *Physics Reports*, pp.73-174, 1993.

F. Capasso and S. Datta, "Quantum Electron Devices," *Physics Today*, pp.74-82, February 1990.

Feynman, R., "There's Plenty of Room at the Bottom: An invitation to Enter a New Field of Physics," Talk at the Annual Meeting of the American Physical Society, 29 December 1959.

Feynman, R.: 1999 [1959], 'There's Plenty of Room at the Bottom: An Invitation to Enter a New Field of Physics', in: *The Pleasure of Finding Things Out*, Perseus, Cambridge, MA, pp. 117-139.

Fielder, F.A.; Reynolds, G.H.: 1994, 'Legal Problems of Nanotechnology: An Overview', *Southern California Interdisciplinary Law Journal*, 3, 593-629.

Fleischer, Th.; Decker, M. & Fiedeler, U. (eds.): 2004, *Große Aufmerksamkeit für kleine Welten ' Nanotechnologie und ihre Folgen*, special issue of *Technikfolgenabschätzung ' Theorie und Praxis*, 13 (2), 5-85 [www.itas.fzk.de/tatup/042/inhalt.htm].

Fogelberg, H. & Glimell, H.: 2003, 'Molecular Matters: In Search of the Real Stuff', in: H. Fogelberg & H. Glimell, *Bringing Visibility to the Invisible: Towards A Social Understanding of Nanotechnology*, Göteborg: Göteborg University, pp. 5-32 [www.sts.gu.se/publications/STS_report_6.pdf].

Fogelberg, H.: 2003, 'The Material Culture of Nanotechnology', in: H. Fogelberg & H. Glimell, *Bringing Visibility to the Invisible: Towards A Social Understanding of Nanotechnology*, Göteborg: Göteborg University, pp. 99-114.

Francoeur, E.: 1997, 'The Forgotten Tool: The Design and Use of Molecular Models', *Social Studies of Science*, 27, 7-40.

Frazier, G., "An Ideology For Nanoelectronics," in *Concurrent Computations: Algorithms, Architecture, and Technology*, Plenum Press, New York, 1988.

Galison, P.: 1996, 'Computer Simulations and the Trading Zone', in: P. Galison, D.J. Stump (eds.), *The Disunity of Science: Boundaries, Contexts, and Power*, Stanford University Press, Stanford, pp. 118-157.

Galison, P.: 1997, *Image and Logic: A Material Culture of Microphysics*, University of Chicago Press, Chicago.

Galison, P.; Stump, D.J. (eds.): 1996, *The Disunity of Science: Boundaries, Contexts, and Power*, Stanford University Press, Stanford.

Gieryn, T.F.: 1999, *Cultural Boundaries of Science: Credibility on the Line*, University of Chicago Press, Chicago.

Gieryn, T.F.; Figert, A.E.: 1986, 'Scientists Protect Their Cognitive Authority: The Status Degradation Ceremony of Sir Cyril Burt', in: G. Böhme, N. Stehr (eds.), *The Knowledge Society*, Reidel, Dordrecht, pp. 67-89.

Glimell, H.: 2001, 'Challenging Limits' Excerpts from an Emerging Ethnography of Nano Physicists', in: H. Glimell & O. Johlin (eds.), *The Social Production of Technology: On the Everyday Life with Things*, Göteburg, SE: BAS Publisher, chapter 7, pp. 111-131 (reprinted in: H. Fogelberg & H. Glimell, *Bringing Visibility to the Invisible: Towards A Social Understanding of Nanotechnology*, Göteborg: Göteborg University, pp. 115-139.

Glimell, H.: 2001, 'Dynamics of the Emerging Field of Nanoscience', in: M.C. Roco and W.S. Bainbridge (eds.), *Societal Implications of Nanoscience and Nanotechnology*, Dordrecht: Kluwer, pp. 156-160.

Glimell, H.: 2003, 'Dynamics of the Emerging Field of Nanoscience', in: H. Fogelberg & H. Glimell, *Bringing Visibility to the Invisible: Towards A Social Understanding of Nanotechnology*, Göteborg: Göteborg University, pp. 79-85 [www.sts.gu.se/publications/STS_report_6.pdf].

Glimell, H.: 2004, 'Grand Visions and Lilliput Politics: Staging the Exploration of the 'Endless Frontier'', in: D. Baird, A. Nordmann & J. Schummer (eds.), *Discovering the Nanoscale*, Amsterdam: IOS Press, pp. 231-246.

Goldhaber-Gordon, D., Montemerlo, M.S., Love, J.C., Opiteck, G.J., and Ellenbogen, J.C., "Overview of Nanoelectronic Devices," submitted to the *Proceedings of the IEEE*, February 1997. For more information, please send e-mail to nanotech@mitre.org.

Gorman, M.E.; Groves, J.F. & Shrager, J.: 2004, 'Societal Dimensions of Nanotechnology as a Trading Zone: Results from a Pilot Project', in: D. Baird, A. Nordmann & J. Schummer (eds.), *Discovering the Nanoscale*, Amsterdam: IOS Press, pp. 63-73.

Grinbaum, A. & Dupuy, J.-P.: 2004, 'Living with Uncertainty: Toward the Ongoing Normative Assessment of Nanotechnology', *Techne: Research in Philosophy and Technology*, 8(3) (forthcoming).

Grinbaum, A.: 2004, 'La condition de l'homme moderne et les nanotechnologies', in: G. Nivat (ed.),

Les limites de l'humain. 39èmes Rencontres Internationales de Genève, L'Age d'Homme, Genève, p. 141.

Gross, M.: 1999, *Travels to the Nanoworld: Miniature Machinery in Nature and Technology*, Plenum, New York.

Grunwald, A.: 2004, 'Ethische Aspekte der Nanotechnologie. Eine Felderkundung', *Technikfolgenabschätzung ' Theorie und Praxis*, 13 (2), 71-78.

Gupta, V.K. & Pangannaya, N.B.: 2000, 'Carbon nanotubes: bibliometric analysis of patents', *World Patent Information*, 22, 185-189.

Guthold, M.; Falvo, M.R.; Matthews, W.G.; Paulson, S.; Washburn, S.; Erie, D.A.; Superfine, R.; Brooks, F.P.; Taylor, R.M.: 2000, 'Controlled Manipulation of Molecular Samples with the NanoManipulator', *IEEE-ASME Transactions on Mechatronics*, 5, 189-198.

Hacking, I.: 1983, *Representing and Intervening: Introductory Topics in the Philosophy of Natural Science*, Cambridge University Press, Cambridge, UK.

Hacking, I.: 1992, 'The Self-Vindication of the Laboratory Sciences', in: A. Pickering (ed.), *Science as Practice and Culture*, University of Chicago Press, Chicago, pp. 29-64.

Hansson, S.O.: 2004, 'Great Uncertainty about Small Things', *Techne: Research in Philosophy and Technology*, 8(3) (forthcoming).

Haruta, M. 1997. *Catalyst surveys of Japan* 1:61 and references therein.

Hayles, N.K. (ed.): 2004, *Nanoculture: Implications of the New Technoscience*, Bristol, UK: Intellect Books.

Hayles, N.K.: 2004, 'Connecting the Quantum Dots: Nanotechscience and Culture', in: N.K. Hayles (ed.), *Nanoculture: Implications of the New Technoscience*, Bristol, UK: Intellect Books, pp. 11-26.

Hecht, G., Allen, M.T.: 2001, 'Introduction: Authority, Political Machines, and Technology's History', in: M.T. Allen, G. Hecht (eds.), *Technologies of Power: Essays in Honor of Thomas Parke Hughes and Agatha Chipley Hughes*, MIT Press, Cambridge, MA, pp. 1-23.

Hecht, G.: 1998, *The Radiance of France: Technology, Politics, and National Identity after World War II*, MIT Press, Cambridge, MA.

Heidegger, M.: 1977, *The Question Concerning Technology and Other Essays*, Harper Colophon Books, New York.

Helmreich, S.: 1998, *Silicon Second Nature: Culturing Artificial Life in a Digital World*, University of California Press, Berkeley, CA.

Hemerén, G.: 2004, 'Nano Ethics Primer', in: European Commission (Community Health and Consumer Protection): *Nanotechnologies: A Preliminary Risk Analysis on theBasis of a Workshop, Brussels, 1-2 March 2004* (www.europa.eu.int/comm/health/ ph_risk/documents/ ev_20040301_en.pdf), pp. 95-102.

Hennig, J.: 2004, 'Changes in the Design of Scanning Tunneling Microscopic Images from 1980 to 1990', *Techne: Research in Philosophy and Technology*, 8(3) (forthcoming).

Hessenbruch, A.: 2004, 'Nanotechnology and the Negotiation of Novelty', in: D. Baird, A. Nordmann & J. Schummer (eds.), *Discovering the Nanoscale*, Amsterdam: IOS Press, pp. 135-144.

Hughes, T.P.: 1983, *Networks of Power: Electrification in Western Society, 1880-1930*, Johns Hopkins University Press, Baltimore.

Hughes, T.P.: 1994, 'Technological Momentum', in: M.R. Smith, L. Marx (eds.), *Does Technology Drive History? The Dilemma of Technological Determinism*, MIT Press, Cambridge, MA, pp. 101-114.

Hullmann, A. & Meyer, M.: 2003, 'Publications and Patents in Nanotechnology. An overview of previous studies and the state of the art', *Scientometrics*, 58 (3) 507-527.

Huo, Q., D.I. Margolese, U. Ciesla, P. Feng, T.E. Gier, P. Sieger, R. Leon, P.M. Petroff, F. Schuth, and G.D. Stucky. 1994. *Nature* 368:317.

Ihde, D.: 1991, *Instrumental Realism: The Interface between Philosophy of Science and Philosophy of Technology*, Indiana University Press, Bloomington.

Jena, P., S.N. Khanna, and B.K. Rao. 1996. In *Science and technology of atomically engineered materials*, ed. P. Jena. River Edge, NJ: World Scientific.

Johansson, M.: 2003, 'Plenty of room at the bottom: Towards an anthropology of nanoscience', *Anthropology Today*, 19 (No. 6), 3-6.

Johnson, A.: 2004, 'The End of Pure Science: Science Policy from Bayh-Dole to the NNI', in: D. Baird, A. Nordmann & J. Schummer (eds.), *Discovering the Nanoscale*, Amsterdam: IOS Press, pp. 217-230.

Jones, C.W., and W.J. Koros. 1994. *Carbon* 32:1419.

Jounet, C., W.K. Maser, P. Bernier, A. Loiseau, M. Lamy de la Chapelle, S. Lefrant, P. Deniard, R. Lee, and J.E. Fischer. 1997. *Nature* 388:756.

K.E. Drexler, *Nanosystems: Molecular Machinery, Manufacturing, and Computation*, Wiley, New York, 1992.

Karger, J., and D.M. Ruthven. 1992. *Diffusion in zeolites*. New York: J. Wiley.

Kay, L.: 2000, *Who Wrote the Book of Life? A History of the Genetic Code*, Stanford University Press, Stanford.

Keiper, A.: 2003, 'The Nanotechnology Revolution', *The New Atlantis*, 2, 17-34.

Khushf, G.: 2004, 'A Hierarchical Architecture for Nano-scale Science and Technology: Taking Stock of the Claims About Science Made By Advocates of NBIC Convergence', in: D. Baird, A. Nordmann & J. Schummer (eds.), *Discovering the Nanoscale*, Amsterdam: IOS Press, pp. 21-33.

Khushf, G.: 2004, 'Systems Theory and the Ethics of Human Enhancement: A Framework for NBIC Convergence', *Annals of the New York Academy of Sciences*, 1013, 124-149.

Kline, R.: 1992, *Steinmetz: Engineer and Socialist*, Johns Hopkins University Press, Baltimore.

Kline, R.: 1995, 'Construing 'Technology' as 'Applied Science'': Public Rhetoric of Scientists and Engineers in the United States, *Isis*, 86, 194-221.

Kline, R.: 2000, 'The Paradox of 'Engineering Science': A Cold War Debate about Education in the United States', *IEEE Technology and Society Magazine*, 19, 19-25.

Kline, R.; Pinch, T.: 1996, 'Users as Agents of Technological Change: The Social Construction of the Automobile in the Rural United States', *Technology and Culture*, 37, 763-95.

Knorr-Cetina, K.: 1992, 'The Couch, the Cathedral, and the Laboratory: On the Relationship between Experiment and Laboratory in Science', in: A. Pickering (ed.), *Science as Practice and Culture*, University of Chicago Press, Chicago, pp. 113-138.

Knorr-Cetina, K.: 1999, *Epistemic Cultures: How the Sciences Make Knowledge*, Harvard University Press, Cambridge, MA.

Kranakis, E.: 1997, *Constructing a Bridge: An Exploration of Engineering Culture, Design, and Research in Nineteenth-Century France and America*, MIT Press, Cambridge, MA.

Krätschmer, W., L.D. Lamb, K. Fostiropoulos, and D.R. Huffman. 1990. *Nature* 347:354.

Kresge, C.T., M.E. Leonowicz, W.J. Roth, J.C. Vartuli, and J.S. Beck. 1992. *Nature* 359:710.

Kuhn, T.S.: 1996, *The Structure of Scientific Revolutions*, University of Chicago Press, Chicago.

Kupperman, A., S. Nadimi, S. Oliver, G. Ozin, J. Garcés, and M. Olken. 1993. *Nature* 365:239.

Kuusi, O.; Meyer, M.: 2002, 'Technological generalizations and leitbilder ' the anticipation of technological opportunities', *Technological Forecasting & Social Change*, 69, 625-639.

Landon, B.: 2004, 'Less is More: Much Less is Much More: The Insistent Allure of Nanotechnology Narratives in Science Fiction', in: N.K. Hayles (ed.), *Nanoculture: Implications of the New Technoscience*, Bristol, UK: Intellect Books, pp. 131-146.

Laszlo, P.: 2004, 'Is There Life After Partington?', *Hyle: International Journal for Philosophy of Chemistry*, 10(2), 169-178.

Latour, B.: 1983, 'Give Me a Laboratory and I Will Raise the World', in: K. Knorr-Cetina, M. Mulkay (eds.), *Science Observed*, Sage, London, pp. 141-170.

Latour, B.: 1996, *Aramis or The Love of Technology*, Harvard University Press, Cambridge, MA.

Layton, E.T., Jr.: 1971, 'Mirror-Image Twins: The Communities of Science and Technology in 19th-Century America', *Technology and Culture*, 12, 562-580.

Lenhard, J.: 2004, 'Nanoscience and the Janus-Faced Character of Simulations', in: D. Baird, A. Nordmann & J. Schummer (eds.), *Discovering the Nanoscale*, Amsterdam: IOS Press, pp. 93-100.

Lenhard, J.: 2004, 'Nanoscience and the Janus-Faced Character of Simulations', in: D. Baird, A. Nordmann, J. Schummer (eds.), *Discovering the Nanoscale*, IOS Press, Amsterdam, pp. 93-100.

Lent, C.S., Tougaw, P.D., Porod, W., Bernstein, G.H., "Quantum Cellular Automata," *Nanotechnology*, Vol. 4, p. 49, 1993.

Lewak, S.E.: 2004, 'What's the Buzz? Tell Me What's A-Happening: Wonder, Nanotechnology, and Alice's Adventures in Wonderland', in: N.K. Hayles (ed.), *Nanoculture: Implications of the New Technoscience*, Bristol, UK: Intellect Books, pp. 201-.

Lide, D.R., ed. 1993-1994. *CRC Handbook of Chemistry and Physics*, 74th ed.

Lin-Easton, P.C.: 2001, 'It's Time for Environmentalists to Think Small ' Real Small: A Call for the Involvement of Environmental Lawyers in Developing Precautionary Policies Molecular Nanotechnology', *Georgetown International Law Review*, 14, 106-134.

López, J.: 2004, 'Bridging the Gaps: Science Fiction in Nanotechnology', *Hyle: International Journal for Philosophy of Chemistry*, 10(2), 129-152.

Lösch, A.: 2004, 'Nanomedicine and Space: Discursive Orders of Mediating Innovations', in: D. Baird, A. Nordmann & J. Schummer (eds.), *Discovering the Nanoscale*, Amsterdam: IOS Press, pp. 193-202.

Mackenzie, D.: 1996a, 'Economic and Sociological Explanations of Technological Change', in: *Knowing Machines: Essays on Technical Change*, MIT Press, Cambridge, MA, pp. 49-65.

Mackenzie, D.: 1996b, 'Marx and the Machine', in: *Knowing Machines: Essays on Technical Change*, MIT Press, Cambridge, MA, pp. 23-47.

Marshall, K.: 2004, 'Atomizing the Risk Technology', in: N.K. Hayles (ed.), *Nanoculture: Implications of the New Technoscience*, Bristol, UK: Intellect Books, pp. 147-160.

Martin, T.P., N. Malinowski, U. Zimmerman, U. Naher, and H. Schaber. 1993. *J. Chem. Phys.* 99:4210.

Martin, T.P., U. Naher, H. Schaber, U. Zimmerman. 1993. *Phys. Rev. Lett.* 70:3079.

Marty, A.: 2003, *Testimony before the Committee on Science on HR 766, Nanotechnology Research and Development Act*, US Government Printing Office, Washington, DC.

Marx, K.: 1963 [1847], *The Poverty of Philosophy*, International Publishers, New York.

Mayer, S.: 2002, 'From genetic modification to nanotechnology: the dangers of 'sound science'', in: T. Gilland (ed.), *Science: Can We Trust the Experts?*, London: Hodder and Stoughton, pp. 1-15.

Mehta, M.D.: 2002, 'Nanoscience and Nanotechnology: Assessing the Nature of Innovation in These Fields', *Bulletin of Science, Technology & Society*, 22(4), 269-273.

Mehta, M.D.: 2004, 'From Biotechnology to Nanotechnology: What Can We Learn from Earlier Technologies?, *Bulletin of Science, Technology & Society*, 24, 34-39.

Meyer, M. & Kuusi, O.: 2004, 'Nanotechnology: Generalizations in an Interdisciplinary Field of Science and Technology', *Hyle: International Journal for Philosophy of Chemistry*, 10(2), 153-168.

Meyer, M.: 2000a, 'Does science push technology? Patents citing scientific literature', *Research Policy*, 29, 409-434.

Meyer, M.: 2000b, 'Patent citations in a novel field of technology: What can they tell about interactions of emerging communities of science and technology?', *Scientometrics*, 48, 151-178.

Meyer, M.: 2001a, 'Patent citations in a novel field of technology: An exploration of nano-science and nano-technology', *Scientometrics*, 51, 163-183.

Meyer, M.: 2001b, 'Socio-economic Research on Nanoscale Science and Technology: A European Overview and Illustration', in: M.C. Roco and W.S. Bainbridge (eds.), *Societal Implications of Nanoscience and Nanotechnology*, Dordrecht: Kluwer, pp. 217-241.

Meyer, M.; Persson, O.: 1998, 'Nanotechnology ' interdisciplinarity, patterns of collaboration and differences in application', *Scientometrics*, 47, 195'205.

Milburn, C.: 2002, 'Nanotechnology in the Age of Posthuman Engineering: Science Fiction as Science', *Configurations*, 10, 261-295.

Milburn, C.: 2004, 'Nano/Splatter: Disintegrating the Postbiological Body', *New Literary History*, 35 (in print).

Misa, T.: 1988, 'How Machines Make History and How Historians (and Others) Help Them to Do So', *Science, Technology, and Human Values*, 13, 308-331.

Mnyusiwalla, A.; Abdallah, S.D.; Singer, P.A.: 2003, 'Mind the gap: science and ethics in nanotechnology', *Nanotechnology*, 14, R9-R13.

Mody, C.C.M.: 2000, "A New Way of Flying': Différance, Rhetoric, and the Autogiro in Interwar Aviation', *Social Studies of Science*, 30, 513-543.

Mody, C.C.M.: 2001, 'A Little Dirt Never Hurt Anyone: Knowledge-Making and Contamination in Materials Science', *Social Studies of Science*, 31, 7-36.

Mody, C.C.M.: 2004, 'How Probe Microscopists Became Nanotechnologists', in: D. Baird, A. Nordmann & J. Schummer (eds.), *Discovering the Nanoscale*, Amsterdam: IOS Press, pp. 119-133.

Mody, C.C.M.: 2004, 'How Probe Microscopists Became Nanotechnologists', in: D. Baird, A. Nordmann, J. Schummer (eds.), *Discovering the Nanoscale*, IOS Press, Amsterdam, pp. 119-133.

Mody, C.C.M.: 2004, 'Instruments in Training: The Growth of American Probe Microscopy in the 1980s', in: D. Kaiser (ed.), *Pedagogy and the Practice of Science: Producing Physical Scientists, 1800-2000*, Cambridge, MA: MIT Press (forthcoming).

Mody, C.C.M.: 2004, 'Small, but Determined: Technological Determinism in Nanoscience', *Hyle: International Journal for Philosophy of Chemistry*, 10(2), 99-128.

Mody, C.C.M.: 2004, *Crafting the Tools of Knowledge: The Invention, Spread, and Commercialization of Probe Microscopy, 1960-2000*, Ph.D. dissertation, Cornell University.

Montemerlo, M.S., Love, J.C., Opiteck, G.J., Goldhaber, D. J., and Ellenbogen, J.C., "Technologies and Designs for Electronic Nanocomputers," MITRE Technical Report 96W0000044, The MITRE Corporation, McLean, VA, July 1996. For more information, send e-mail to nanotech@mitre.org.

Moor, J.H. & Weckert, J.: 2004, 'Nanoethics: Assessing the Nanoscale From an Ethical Point of View', in: D. Baird, A. Nordmann & J. Schummer (eds.), *Discovering the Nanoscale*, Amsterdam: IOS Press, pp. 301-310.

Munn Sanchez, E.: 2004, 'The Expert's Role in Nanoscience and Technology', in: D. Baird, A. Nordmann & J. Schummer (eds.), *Discovering the Nanoscale*, Amsterdam: IOS Press, pp. 257-266.

Nordmann, A.: 2003, 'Shaping the World Atom by Atom: Eine nanowissenschaftliche WeltBildanalyse', in: A. Grunwald (ed.), *Technikgestaltung zwischen Wunsch und Wirklichkeit*, Berlin: Springer, pp. 191-199.

Nordmann, A.: 2004, 'Molecular Disjunctions: Staking Claims at the Nanoscale', in: D. Baird, A. Nordmann & J. Schummer (eds.), *Discovering the Nanoscale*, Amsterdam: IOS Press, pp. 51-62.

Nordmann, A.: 2004, 'Molecular Disjunctions', in: D. Baird, A. Nordmann, J. Schummer (eds.), *Discovering the Nanoscale*, IOS Press, Amsterdam, pp. 51-62.

Nordmann, A.: 2004, 'Nanotechnology: Convergence and Integration', Presentation at *EuroNanoForum, Trieste, December 10, 2003* (Proceedings in preparation).

Nordmann, A.: 2004, 'Nanotechnology's WorldView: New Space for Old Cosmologies', *IEEE Technology and Society Magazine*, 23 (forthcoming).

Nordmann, A.: 2004, 'Social Imagination for Nanotechnology', in: European Commission (Community Health and Consumer Protection): *Nanotechnologies: A Preliminary Risk Analysis on the Basis of a Workshop, Brussels, 1-2 March 2004*, pp. 111-113 [www.europa.eu.int/comm/health/ph_risk/documents/ev_20040301_en.pdf].

Nordmann, A.: 2004, 'Was ist TechnoWissenschaft? ' Zum Wandel der Wissenschaftskultur am Beispiel von Nanoforschung und Bionik', in: T. Rossmann & C. Tropea (eds.), *Bionik ' Neue Forschungsergebnisse aus Natur-, Ingenieur- und Geisteswissenschaften*, Berlin: Springer, 2004 (forthcoming)

Nye, D.E.: 1994, *American Technological Sublime*, MIT Press, Cambridge, MA.

Nye, D.E.: 2003, *America as Second Creation: Technology and Narratives of New Beginnings*, MIT Press, Cambridge, MA.

Paschen, H.; Coenen, C.; Fleischer, T.; Grünwald, R.; Oertel, D. & Revermann, C.: 2004, *Nanotechnologie: Forschung, Entwicklung, Anwendung*, Berlin: Springer (*Nanotechnologie; TAB-Arbeitsbericht 92*, Berlin: Büro für Technikfolgen-Abschätzung beim Deutschen Bundestag).

Pinch, T.: 1986, *Confronting Nature*, Reidel, Dordrecht.

Pinch, T.: 1993, "Testing – One, Two, Three ... Testing!': Toward a Sociology of Testing', *Science, Technology, and Human Values*, 18, 25-41.

Pitt, J.C.: 2004, 'The Epistemology of the Very Small', in: D. Baird, A. Nordmann & J. Schummer (eds.), *Discovering the Nanoscale*, Amsterdam: IOS Press, pp. 157-163.

Polanyi, M.: 1962, *Personal Knowledge: Towards a Post-Critical Philosophy*, Harper Torchbooks, New York.

Pressman, J.: 2004, 'Nano Narrative: A Parable from Electronic Literature', in: N.K. Hayles (ed.), *Nanoculture: Implications of the New Technoscience*, Bristol, UK: Intellect Books, pp. 191-200.

Prigogine, I., and S. Rice. 1988. *Advances in chemical physics*, Vol. 70, Parts 1 & 2. New York: J. Wiley.

R. T. Bate, "Nanoelectronics," *Nanotechnology*, Vol. 1, pp. 1-7, 1990.

Rademann, K., B. Kaiser, U. Even, F. Hensel. 1987. *Phys. Rev. Lett.* 59:2319.

Rao, C.N.R., B.C. Satishkumar, and A. Govindaraj. 1997. *Chem. Commun.* 1581.

Rao, M.B., and S. Sircar. 1993. *Gas Separation and Purification* 7:279.

Ratner, M.; Ratner, D.: 2003, *Nanotechnology: A Gentle Introduction to the Next Big Idea,* Prentice Hall, Upper Saddle River, NJ.

Reed, M.A., "Quantum Dots," *Scientific American,* January 1993, pp. 118-123.

Regis, E.: 1995, *Nano: The Emerging Science of Nanotechnology*, Little, Brown, Boston.

Reinhardt, C.: 2004, 'Chemistry in a Physical Mode: Molecular Spectroscopy and the Emergence of NMR', *Annals of Science*, 61, 1-32.

Rheinberger, H.-J.: 1997, *Toward a History of Epistemic Things: Synthesizing Proteins in the Test Tube*, Stanford University Press, Stanford.

Roberts, J.A.: 2004, 'Deciding the Future of Nanotechnologies: Legal Perspectives on Issues of Democracy and Technology', in: D. Baird, A. Nordmann & J. Schummer (eds.), *Discovering the Nanoscale*, Amsterdam: IOS Press, pp. 247-255.

Robinson, C.: 2004, 'Images in NanoScience/Technology', in: D. Baird, A. Nordmann & J. Schummer (eds.), *Discovering the Nanoscale*, Amsterdam: IOS Press, pp. 165-169.

Robison, W.L.: 2004, 'Nano-Ethics', in: D. Baird, A. Nordmann & J. Schummer (eds.), *Discovering the Nanoscale*, Amsterdam: IOS Press, pp. 285-300.

Roco, M.C. & Bainbridge, W.S. (eds.): 2001, *Societal implications of nanoscience and nanotechnology*, (Proceedings of a workshop organized by the National Science Foundation, September 28-29, 2000), Kluwer: Dordrecht [available online at http://itri.loyola.edu/nano/societalimpact/nanosi.pdf]

Roco, M.C. & Tomellini, R. (eds.): 2002, *Nanotechnology: Revolutionary Opportunities and Societal Implications, Workshop, Lecce (Italy),* 31 January-1 February 2002, Luxemburg: Office for Official Publications of the European Communities, [200 pp.]

Roco, M.C.: 2003, 'Broader Societal Issues of Nanotechnology', *Journal of Nanoparticle Research*, 5, 181-189.

Roco, M.C.; Bainbridge, W.S. (eds.): 2002, *Converging Technologies for Improving Human Performance: Nanotechnology, Biotechnology, Information Technology and the Cognitive Science*, Arlington, VA: National Science Foundation.

Rohlfing, E.A., D.M. Cox, and A. Kaldor. 1984. *J. Chem. Phys.* 81:3846.

Rohrer, H.: 1995, 'The Nanometer Age – Challenge and Chance', *Microelectronic Engineering*, 27, 3-15.

Rosen, P.: 1993, 'The Social Construction of Mountain Bikes: Technology and Postmodernity in the Cycle Industry', *Social Studies of Science*, 23, 479-513.

Ruthven, D.M., S. Farooq, K.S. Knaebel. 1994. *Pressure swing adsorption*. New York: VCH Publishers.

Sarewitz, D.; Woodhouse, E.: 2003, 'Small is Powerful', in: A. Lightman, D. Sarewitz & Chr. Desser, (eds.), *Living with the Genie: Essays on Technology and the Quest for Human Mastery*, Washington, DC: Island Press, pp. 63-83.

Schiemann, G.: 2004, 'Dissolution of the Nature-Technology Dichotomy? Perspectives on Nanotechnology from an Everyday Understanding of Nature', in: D. Baird, A. Nordmann & J. Schummer (eds.), *Discovering the Nanoscale*, Amsterdam: IOS Press, pp. 209-213.

Schmidt, J.C.: 2004, 'Unbounded Technologies: Working Through Technological Reductionism of Nanotechnology', in: D. Baird, A. Nordmann & J. Schummer (eds.), *Discovering the Nanoscale*, Amsterdam: IOS Press, pp. 35-50.

Schuler, E.: 2004, 'Perception of Risks and Nanotechnology', in: D. Baird, A. Nordmann & J. Schummer (eds.), *Discovering the Nanoscale*, Amsterdam: IOS Press, pp. 279-284.

Schummer, J.: 2001, 'Ethics of Chemical Synthesis', *Hyle: International Journal for Philosophy of Chemistry*, 7, 103-124 [online: www.hyle.org/journal/issues/7/schummer.htm].

Schummer, J.: 2003, 'Aesthetics of Chemical Products: Materials, Molecules, and Molecular Models', *Hyle: International Journal for Philosophy of Chemistry*, 9, 73-104 [online: www.hyle.org/journal/issues/9-1/schummer.htm].

Schummer, J.: 2003, 'The Notion of Nature in Chemistry', *Studies in History and Philosophy of Science*, 34 (2003), 705-736.

Schummer, J.: 2004, 'Interdisciplinary Issues in Nanoscale Research', in: D. Baird, A. Nordmann & J. Schummer (eds.), *Discovering the Nanoscale*, Amsterdam: IOS Press, pp. 9-20.

Schummer, J.: 2004, 'Interdisciplinary Issues in Nanoscale Research', in: D. Baird, A. Nordmann, J. Schummer (eds.), *Discovering the Nanoscale*, IOS Press, Amsterdam, pp. 9-20.

Schummer, J.: 2004, 'Multidisciplinarity, Interdisciplinarity, and Patterns of Research Collaboration in Nanoscience and Nanotechnology', *Scientometrics*, 59, 425-465.

Schummer, J.: 2004, 'Naturverhältnisse in der modernen Wirkstoff-Forschung', in: K. Kornwachs (ed.), *Technik ' System ' Verantwortung*, Münster: LIT, pp. 629-638.

Schummer, J.: 2004, 'Why do Chemists Perform Experiments?', in: D. Sobczynska, P. Zeidler, E.

Zielonacka-Lis (eds.), *Chemistry in the Philosophical Melting Pot*, Peter Lang (Frankfurt/ M.), 2004, pp. 395-410.

Schummer, J.: 2005, 'Reading Nano: The Public Interest in Nanotechnology as Reflected in Book Purchase Patterns", *Public Understanding of Science*, 14 (2) (forthcoming).

Schwarz, A.E.: 2004, 'Shrinking the 'Ecological Footprint' with NanoTechnoScience?', in: D. Baird, A. Nordmann & J. Schummer (eds.), *Discovering the Nanoscale*, Amsterdam: IOS Press, pp. 203-208.

Schwarz, R.B. 1998. Storage of hydrogen in powders with nanosized crystalline domains. In *R&D status and trends*, ed. Siegel et al.

Siegel, R.W., E. Hu, and M.C. Roco, eds. 1998. *R&D status and trends in nanoparticles, nanostructured materials, and nanodevices in the United States*. Baltimore: Loyola College, International Technology Research Institute. NTIS #PB98-117914.

Smith, M.R.; Marx, L. (eds.): 1994, *Does Technology Drive History? The Dilemma of Technological Determinism*, MIT Press, Cambridge, MA.

Soffer, A., J.E. Koresh, S. Saggy. 1987. United States Patent 4,685,940.

Stix, G.: 2001, 'Little Big Science: Nanotechnology Is All the Rage. But Will It Meet Its Ambitious Goals? And What The Heck Is It?', *Scientific American* 285, 32-37.

Stuckless, J.T, D.E. Starr, D.J. Bald, C.T. Campbell. 1997. *J. Chem. Phys*. 107:5547.

Suchman, M.: 2001, 'Envisioning Life on the Nano-Frontier', in: M.C. Roco and W.S. Bainbridge (eds.), *Societal Implications of Nanoscience and Nanotechnology*, Dordrecht: Kluwer, pp. 211-216.

Suchman, M.: 2002, 'Social Science and Nanotechnologies', M. Roco & R. Tomellini (eds.), *Nanotechnology ' Revolutionary Opportunities and Societal Implications*, Luxembourg: European Communities, pp. 95-99.

Sullivan, D.: 1994, 'Exclusionary Epideictic: *NOVA's* Narrative Excommunication of Fleischmann and Pons', *Science, Technology, and Human Values*, 19, 283-306.

Swami, N.: 2002, *Testimony before Senate Committee on Commerce, Science, and Transportation*, US Government Printing Office, Washington, DC.

Sweeney, A.E.; Seal, S. & Vaidyanathan, P.: 2003, 'The promises and perils of nanoscience and nanotechnology: Exploring emerging social and ethical issues', *Bulletin of Science, Technology & Society*, 23 (4), 236-245.

Tschope, A.S., W. Liu, M. Flyzani-Stephanopoulos, and J.Y. Ying 1995. *J. Catal*. 157:42.

Turner, F.: forthcoming, 'When the Counterculture Met Computer Networks and the New Economy: Revisiting the WELL and the Origins of Virtual Community', *Technology and Culture*.

Vermaas, P.E.: 2004, 'Nanoscale Technology: A Two-Sided Challenge for Interpretations of Quantum Mechanics', in: D. Baird, A. Nordmann & J. Schummer (eds.), *Discovering the Nanoscale*, Amsterdam: IOS Press, pp. 77-91.

Wald, C.A.: 2004, 'Working Boundaries on the NANO Exhibition', in: N.K. Hayles (ed.), *Nanoculture: Implications of the New Technoscience*, Bristol, UK: Intellect Books, pp. 83-108.

Weckert, J.: 2001, 'The Control of Scientific Research: The Case of Nanotechnology', *Australian Journal of Professional and Applied Ethics*, 3, 29-44.

4

Nanocomposites & Nanostructred Materials: Properties & Applications

Nanocomposites: An Introduction

The definition of nanocomposite material has broadened significantly to encompass a large variety of systems such as one-dimensional, two-dimensional, three-dimensional and amorphous materials, made of distinctly dissimilar components and mixed at the nanometer scale. The general class of nanocomposite organic or inorganic materials is a fast growing area of research. Significant effort is focused on the ability to obtain control of the nanoscale structures via innovative synthetic approaches. The properties of nanocomposite materials depend not only on the properties of their individual parents but also on their morphology and interfacial characteristics. This rapidly expanding field is generating many exciting new materials with novel properties. The latter can derive by combining properties from the parent constituents into a single material. There is also the possibility of new properties which are unknown in the parent constituent materials. The inorganic components can be three-dimensional framework systems such as zeolites, two-dimensional layered materials such as clays, metal oxides, metal phosphates, chalcogenides, and even one-dimensional and zero-dimensional materials such as $(Mo_3Se_3\text{-})_n$ chains and clusters. Experimental work has generally shown that virtually all types and classes of nanocomposite materials lead to new and improved properties when compared to their macrocomposite counterparts. Therefore, nanocomposites promise new applications in many fields such as mechanically reinforced lightweight components, non-linear optics, battery cathodes and ionics, nanowires, sensors and other systems.

The general class of organic/inorganic nanocomposites may also be of relevance to issues of bio-ceramics and biomineralization in which *in-situ* growth and polymerization of biopolymer and inorganic matrix is occurring. Finally, lamellar nanocomposites represent an extreme case of a composite in which interface interactions between the two phases are maximized. Since the remarkable properties of conventional composites are mainly due to interface interactions, the materials dealt with here could provide good model systems in which such interactions can be studied in detail using conventional bulk sample (as opposed to surface) techniques. By judiciously engineering the polymer-host interactions, nanocomposites may be produced with a broad range of properties. Inorganic layered materials exist in great variety. They possess well defined, ordered intralamellar space potentially accessible by foreign species. This ability enables them to act as matrices or hosts for polymers, yielding interesting hybrid nano-composite materials. Lamellar nanocomposites can be divided into two distinct classes, intercalated and exfoliated. In the former, the polymer chains alternate with the inorganic layers in a fixed compositional ratio and have a well defined number of polymer layers in the intralamellar space. In exfoliated nano-composites the number of polymer chains between the layers is almost continuously variable and the layers stand >100 Å apart. The intercalated nanocomposites are also more compound-like because of the fixed polymer/layer ratio, and they are interesting for their electronic and charge transport properties. On the other hand, exfoliated nanocomposites are more interesting for their superior mechanical properties. A subclass of nanocomposites which is the lamellar class of intercalated organic/inorganic nanocomposites and namely those systems that exhibit electronic properties in at least one of the components offers the possibility of obtaining well ordered systems some of which may lead to unusual electrical and mechanical properties. Selected members of this class may be amenable to direct structural characterization by standard crystallographic methods. An important issue in this area is the few structural details that are available, therefore, any system which can be subjected to such analysis is of interest. Nanocomposites also offer the possibility to combine diverse properties which are impossible within a single material, e.g. flexible mechanical properties and superconducting properties. This work is now in its infancy and extensive investigations are underway. Another exciting aspect is the possibility of creating heterostructures composed of different kinds of inorganic layers, which could lead to new multifuntional materials. Several general synthetic routes have been developed for inserting polymer chains into host structures and many novel nanocomposites have been designed.

These include:

— *In-situ* intercalative polymerization (ISIP) of a monomer using the host itself as the oxidant. The rationale behind intercalative polymerization is that host matrices with high electron affinity can oxidatively polymerize appropriate monomers in their interior.

— Monomer intercalation followed by topotactic intralamellar solid state polymerization. This route creates conjugated polymers inside non-oxidizing hosts.

— Direct precipitative encapsulation of polymer chains by colloidally dispersed single layers of a host. This approach gives access to a large variety of nanocomposites with many kinds of polymers and hosts.

Materials with features on the scale of nanometers (10^{-9} meter) often have properties dramatically different from their bulk-scale counterparts. For example, nanocrystalline copper is five times harder than ordinary copper with its micrometer-sized crystalline structure. The development of such materials is currently a research area of great interest. Important among these nanoscale materials are nanocomposites, in which the constituents are mixed on a nanometer-length scale. They often have properties that are superior to conventional microscale composites and can be synthesized using surprisingly simple and inexpensive techniques. A nanoscale dispersion of sheet-like inorganic silicate particles in a polymer matrix, for example, is superior to either constituent in such properties as optical clarity, strength, stiffness, thermal stability, reduced permeability, and flame retardancy. The key to the synthesis of nanocomposites is how the silicates are made to disperse in the polymer. Since the organic and inorganic components are typically immiscible (like water and oil), the silicate surface is modified by attaching surfactant molecules (like adding soap to water). Polymer-based nanocomposites are also being developed for electronics applications such as thin-film capacitors in integrated circuits and solid polymer electrolyes for batteries. Nature also makes fabulous nanocomposites, and scientists are trying to emulate such processes. The abalone shell, for example, has alternating layers of calcium carbonate and a rubbery biopolymer; it is twice as hard as and a thousand times tougher than its components. The study of nanocomposite materials requires a multidisciplinary approach, involving novel chemical techniques and an understanding of physics and surface science. It is a field of broad scientific interest with impressive technological promise. Studies have revealed clearly the property advantages that nanomaterial additives can provide in comparison to both their conventional filler counterparts and base polymer. Properties which have been shown to undergo substantial improvements include:

- Mechanical properties e.g. strength, modulus and dimensional stability.
- Decreased permeability to gases, water and hydrocarbons.
- Thermal stability and heat distortion temperature.
- Flame retardancy and reduced smoke emissions.
- Chemical resistance.
- Surface appearance.
- Electrical conductivity.
- Optical clarity in comparison to conventionally filled polymers.

To date one of the few disadvantages associated with nanoparticle incorporation has concerned toughness and impact performance. Some of the data presented has suggested that nanoclay modification of polymers such as polyamides, could reduce impact performance. Clearly this is an issue which would require consideration for applications where impact loading events are likely. In addition, further research will be necessary to, for example, develop a better understanding of formulation/structure/property relationships, better routes to platelet exfoliation and dispersion etc. In addition it is important to recognise that nanoparticulate/fibrous loading confers significant property improvements with very low loading levels, traditional microparticle additives requiring much higher loading levels to achieve similar performance. This in turn can result in significant weight reductions (of obvious importance for various military and aerospace applications) for similar performance, greater strength for similar structural dimensions and, for barrier applications, increased barrier performance for similar material thickness.

Data provided by Hartmut Fischer of TNO in the Netherlands relating to polyamide – montmorillonite nanocomposites indicates tensile strength improvements of approximately 40 and 20 per cent at temperatures of 23°C and 120°C respectively and modulus improvements of 70 per cent and a very impressive 220 per cent at the same temperatures. In addition Heat Distortion Temperature was shown to increase from 65°C for the unmodified polyamide to 152°C for the nanoclay-modified material, all the above being achieved with just a 5 per cent loading of montmorillonite clay. Similar mechanical property improvements were presented for polymethyl methacrylate—clay hybrids.

Further data provided by Akkepeddi of Honeywell relating to polyamide-6 polymers confirms these property trends. In addition, the further benefits of short/long glass fibre incorporation, together with nanoclay incorporation, are clearly revealed.

Nanocomposite materials with an inorganic glass and an organic polymer constitute a relatively new and unique area in material science. The term "ormocers", "ormosils" and "ceramers" are often utilized to describe this class of nanocomposite. By combining at the molecular level inorganic and organic polymeric material a blending of unique physical properties can be achieved. The value in these materials is apparent, from fiber optics to paints these materials may provide the requisite physical properties to achieve the next technological advance. There are several different ways of synthesizing this class of nanocomposite; therefore a means of classification is necessary. Most developed nomenclature is based on synthetic techniques; Wilkes has a relatively recent and exhaustive categorization. However, we chose to classify these materials upon a simpler system first suggested by Novak. Five categories cover the majority of composites synthesized with more recent techniques being modifications or combinations from this list.

Type I: Organic polymer embedded in an inorganic matrix without covalent bonding between the components.

Type II: Organic polymer embedded in an inorganic matrix with sites of covalent bonding between the components

Type III: Co-formed interpenetrating networks of inorganic and organic polymers without covalent bonds between phases.

Type IV: Co-formed interpenetrating networks of inorganic and organic polymers with covalent bonds between phases.

Type V: Non-shrinking simultaneous polymerization of inorganic and organic polymers.

Before discussing different kinds of nanocomposites it is appropriate to discuss the sol-gel process. Although zirconium, titanium, aluminum and boron oxides have been utilized as the inorganic component, the great majority of nanocomposites incorporate silica from tetraethoxysilane (TEOS). The formation of the inorganic component involves two steps, hydrolysis and condensation. The important insight in the formation of this part of the composite is based upon the relative kinetic rates of each step. For example, if the rate of hydrolysis is fast compared to condensation, then simple particles or highly branched silicate matrices are formed. Conversely, if the condensation step is quicker than hydrolysis, then string-like filaments are formed. These changes in morphology of the silicate matrix can be manipulated by the choice of sol-gel catalyst with dramatic effect on the physical properties of the nanocomposite.

Perhaps the most common and straightforward nanocomposites found in the literature are the Type I composites. Typically, TEOS is used to form an inorganic network around an organic polymer component. The goal in the process is to form a completely interpenetrating network (IPN) of both inorganic and organic phases. Homogeneous nanocomposites with good IPNs are often stronger, more resilient, and optically transparent, whereas heterogeneous composites are often mechanically weaker and opaque. Choice of co-solvents is critical throughout the formation of the inorganic component. As the hydrolysis and/or condensation reaction occurs, changes in the polarity of the solvent mixture can result in undesired phase separation of the inorganic and organic components. Phase separations can produce a range of results from small zones of one component dominance to total inorganic/organic separations. However, the result is that the same sites of instability are created. The timing and physical conditions in which phase separations occur can produce isolated silicate formations within composite formulations.

Type II composites utilize modified organic polymers with sites capable of covalently bonding directly to the forming inorganic phase of the nanocomposite. Most often an organosilane with a reactive pendant group is used to form the covalent bond to the organic polymer component. These pendant groups can be isocyanates, which easily react with polymers containing alcohols or amines. Another option that has been explored is hydrosilation reactions with polymers containing a terminal alkene. Once bound, the silane's remaining pendant ethoxide or chloride groups will be available to bond with such sites on other silane species, forming the inorganic component.

Type V composites are unique examples of these materials because of the physical conditions in which they are formed. Novak and Grubbs developa method of simultaneously polymerizing both inorganic and organic components without shrinkage of the material. The insight was that by replacing the ethoxide found in TEOS with an organic monomer oxide, solvent loss created during the hydrolysis step is eliminated. It is the loss of ethanol (or other solvents) in the sol-gel process, which creates material shrinkage and leads to cracking from capillary pressures. Although extremely innovative, unfortunately a limited number of composites have been formed successfully with this process. This is presumably because of the difficulties in precisely matching the rates of polymerization of the organic component with both the rates of hydrolysis and condensation of the ceramic component to prevent disastrous phase separations.

A general synthesis for a base, acid, or salt catalyzed polyphosphazene, polyethylene oxide (PEO), and polyethylene oxide/polypropylene oxide (PPO/PEO) block nanocomposite is as follows: 300 mg of polymer is dissolved into 10 mL of a 50/50 by volume tetrahydrofuran (THF)/ethanol mixed solvent in a capped vial. To this solution is added TEOS (336 mg). A catalyst is then introduced as an aqueous solution (150 µl) and the mixture is capped and sonicated at 50°C for 30 minutes. The solution is aged from hours to days depending upon the catalyst used in a sealed vial and poured into a teflon mold and loosely covered at room temperature. The nanocomposite self assembles as the volatile solvent slowly escapes during the condensation process. The synthesis of polyvinyl acetate (PVAc)/silicate nanocomposites requires a different approach from the other nanocomposites. PVAc (300 mg) is dissolved into an 50/50 by volume acetic acid/methanol (10 mL) mixed solvent in a capped vial. To this solution is added TEOS (373 mg). The solution is then sonicated for 5 minutes in a sealed vial at room temperature and poured into a teflon mould and loosely covered at room temperature. The nanocomposite self assembles during the curing process, which typically lasts up to 24 hours. Additional heating at 100°C for 30 minutes aids in removing lingering acetic acid from the nanocomposite.

Various membrane samples were tested on a Dynamic Mechanical Analyzer (TA Instruments, Model 2980, New Castle, DE) fitted with a tension (film) or penetration (compression) clamp. A simple linear ramp-force protocol was developed to evaluate the membrane strength in tension and compression. Both tension and compression protocols consist of the establishment of 0.03 Newton (N) static force and equilibration to 25°C. A force is then applied to the specimen at a specific rate, 0.5 N per minute for tension, and 1.0 N per minute for compression (18 N maximum) or until the maximum limit of travel is reached (24.7 mm). The membrane specimens are cut with a laboratory fixture consisting of two razor blades mounted on a parallel plate yielding a 6.20 mm by 39.0 mm sample. The specimen's thickness is measured with a caliper (Mitutoyo, Model ID-C112EBS, Japan) in several places along the length to obtain an average membrane thickness.

Various thin films were imaged as an unstressed, swelled, or stretched nanocomposite with a Philips XL30ESEM. A 10 to 20 kV electron beam was utilized. Wet mode analyses employed a 500 µm pressure-limiting aperture (PLA). To image the topography either a gaseous secondary electron detector (wet mode) or secondary electron detector (HiVac) was used.

The polyphosphazene family of polymers was chosen for the initial investigations into nanocomposites due to their diverse array of physical and solubility properties. The researcher was specifically interested in some heretofore inaccessible phosphazenes that lacked the mechanical stability to be useful for membrane and barrier applications. The polymer poly[*bis*-(2-(2-methoxyethoxy)ethoxy)phosphazene] (MEEP) is such a polymer - one where the chemical properties, such as selective transport of ions are extremely interesting, yet the polymer alone lacks the requisite mechanical integrity to form a stable membrane for evaluation and subsequent use. This water-soluble polyphosphazene is a viscous liquid with two phosphorus bound polyether ligands on an alternating nitrogen phosphorus backbone. In all of the nanocomposites TEOS was employed as the ceramic precursor. The condensation of the ceramic component was catalyzed in a solvent composed of various ratios of water, ethanol and THF. Depending upon the nature of the catalyst employed, differing properties were observed in the resulting nanocomposites. These catalysts are best separated into three categories; acid, base or ionic salt; and each will be discussed separately below. Acid catalyzed nanocomposites that were examined in this study were found to have the highest tensile strength of any of the catalyst types studied; yet they were also found to be glassy and brittle when compared to salt or base catalyzed analogues. While providing a maximum of physical stabilization for the liquid-like MEEP polymer, this translates directly into extremely poor durability when wetted, as exhibited in aqueous swelling tests. An ESEM image of an aqueous swelled/deswelled membrane, which has fractured from the internal hydrostatic pressure. The dominant reason for this lack of durability when wet is the nature of the ceramic morphology of these acid catalyzed nanocomposites. It is known from the ceramic literature that the general morphology of acid catalyzed sol-gel condensations forms long filament structures instead of star-like or particulate structures. This more rigid morphology imparts a high tensile strength to these nanocomposites, yet leaves them less elastic and highly susceptible to damage upon contact with liquids. Base catalyzed nanocomposites have the best elastic and swelling characteristics of all the nanocomposites studied. As seen in the acid catalyzed nanocomposites, the morphology of the condensing ceramic component dominates the physical properties of the material. An ESEM image of a MEEP nanocomposite after being soaked to equilibrium in water. Before swelling, the nanocomposite has a featureless uniform appearance, as it does after drying. During wetting, the hydrophilic polymer component expands as it absorbs the water yet the ceramic is not capable of expansion. This behaviour produces the creases seen in the micrograph. The villae-like features reveal the underlying structure of the ceramic component of the nanocomposite, i.e. areas of higher silicate composition. This is consistent with the morphology expected for a base catalyzed nanocomposite—one that contains interlocking star-like structures. In regions between the ceramic star structures, the polymer is capable of absorbing water and expanding, while regions within the ceramic star structures maintain their structural rigidity. Because a single material contains both of these types of different regions, this renders base catalyzed nanocomposites simultaneously both tough and flexible. Salt catalyzed nanocomposites exhibit properties that are between those found for acid and base catalyzed nanocomposites. Again, the nature of the ceramic dominates the physical properties of the nanocomposite. There is considerable variation in the properties of this class of nanocomposites as there are a variety of simple salts that have proven to be effective catalysts. To investigate the nature of ionic salt catalysts in MEEP nanocomposites, a course of fluoride salts with various alkali counter-cations was examined, as it is known that fluoride anion salts are superior ceramic condensation catalysts over other halide salts. As was expected, the mechanical properties of fluoride catalyzed nanocomposites more closely resembled their base catalyzed analogues because the condensation was driven by an anionic catalyst, similar to OH^-.

The initial results indicated salt catalysts that contain large cations like cesium performed better than small ones such as lithium. Intuitively, hard ions such as lithium, the closest of the alkali metal ions to the proton (an extremely effective catalyst) should outperform the softer ionic catalysts, yet exactly the opposite trend is observed. The reason behind the observed trend becomes apparent when one considers that the strongly electronegative fluoride ion has a poorer affinity towards large polarizable cations (like cesium) than harder cations (like lithium). Catalyst availability in these polar, protic organic solvent systems becomes the dominant factor in the total catalytic effectiveness of these systems. There is a linear correlation between the lattice energy of the alkali fluoride salt employed and the final strength of the resulting nanocomposite, with the lowest energy salt yielding the strongest material. This is due to higher availability of the "naked" fluoride anion in systems possessing the lower lattice energies. In these systems, the catalyst ions spend less time during synthesis associated as intimate ion pairs in solution and are consequently more effective in driving the ceramic condensation.

A better understanding of the formation, properties, and morphology of Type I nanocomposites was gained during studies with polyphosphazene polymers. This information was used to try to better understand the properties of nanocomposites employing more conventional organic polymers. Commercially available polymers were used in the preparation of these polymer/silicate materials. A block polymer consisting of PEO/PPO (M_w = 2K), PEO (2K), polyvinylalcohol (PVA, 30–70K), PVAc (83K and 167K) and poly (acrylonitrile) (PAN, 150K) were the primary organic polymers of interest. Drawing upon previous experience, a 50/50 (w/v) polymer to silicate ratio was used to form these nanocomposites. Acid catalyzed composites were the focus of most of these studies. Numerous attempts were made with low molecular weight PEO and PEO/PPO block polymers to prepare a nanocomposite with acceptable mechanical integrity. Unfortunately, all attempts resulted in materials that were too brittle to be lifted as a single unit suitable for mechanical analyses. This was surprising in light of the fact that acid catalyzed composites were previously found to yield the strongest nanocomposites. Explanation for this behaviour lies in the low molecular weight of the polymer component. Typical molecular weights for the MEEP polymer were in the 10^6-10^7 range, providing a directing effect for the condensing ceramic. The PEO polymer was sufficiently light that the ceramic exhibits a directing effect on the orientation of the polymer leading to a significantly different ceramic morphology, more base-like in nature. Polyvinyl alcohol (PVA) and polyvinyl acetate (PVAc) were found to readily form tractable nanocomposites with good mechanical integrity, good flexibility and interesting wetting characteristics. In general, the PVA nanocomposites yielded a significantly higher storage modulus then the PVAc nanocomposites. However, results of a simple destructive force ramp test, at the individual material's glass transition temperature (PVA 85°C, PVAc 30°C), suggests that the PVAc nanocomposite actually is mechanically stronger. Another interesting feature of these two nanocomposites that was noted is that they tend to behave somewhat as a thermoplastic. That is, once heated above the polymer component's glass transition temperature, the material can be easily deformed or bent, and the material will retain its shape once cooled. This is a rare example of where the polymer dictates the physical properties of the entire nanocomposite.

The behaviour of PAN nanocomposites was also explored. However, PAN has limited solubility in solvents commonly used to form this type of nanocomposite due to polarity differences between the polymer (lower polarity) and the solvents or TEOS (higher polarity). Attempts to form PAN nanocomposites through our usual synthetic protocols resulted in an unusual material unlike the other homogeneous nanocomposites previously produced. The polymer and ceramic components retreated into separate phase domains, forming an inhomogeneous composite material. As was expected, the mechanical properties of this material were poor due the poor interfacial characteristics between the polymer and ceramic phases. Additional attempts were made at producing a true homogeneous nanocomposite by using N-methylpyrrolidone as the solvent for the condensation. Though a homogeneous nanocomposite was formed with this method, difficulties in the condensation/curing process resulted in a highly fractured material such that a mechanically acceptable material was never obtained.

Applications

There are a host of applications of nanocomposites. One of the most interesting of these applications is as solid polymer electrolytes (SPE) for lithium batteries. The polyphosphazene MEEP is a well-known SPE with very high room temperature conductivity, however it lacks the mechanical stability to be used in a practical device. Traditional stabilization methods, such as deep UV or electron beam crosslinking methods do improve the physical stability of SPEs, however this crosslinking lowers ionic conductivity – tests performed in the laboratory revealed this to be a factor of 30-45 for MEEP-like phosphazene polymers. This reduction is due to the additional covalent linkages formed during the crosslinking process that inhibit chain segmental motion and ion transfer. Since the nanocomposites formed by the ceramic condensation process do not form bonds to the polymer component, (Type I nanocomposites) mechanical stabilization is achieved without a great loss of ionic conductivity. AThe ceramic component forms a "scaffolding" around the polymer providing the needed stability, while not interfering with the polymer's ability to easily transport ions. It is hoped that these nanocomposites will afford the development of all-solid lithium batteries that will be better, safer, and more environmentally friendly than those available today. Another set of applications for the nanocomposites is in separations membranes and other selective barriers. Polymers that are used in selective gas separations, and liquid pervaporation separations must usually be mechanically stabilized, typically by crosslinking, before they become practically usable. Since the nanocomposite is able to physically stabilize the polymer component in a manner very different than these traditional methods, a real advantage may be gained through nanocompositing rather than crosslinking. One final separation application that is being explored is using nanocomposites as subsurface permeable reactive barriers. The nanocomposite matrix is being used to provide a platform for selective chelating agents that remove contaminants from subsurface water. The design employs reproducibly swellable hydrophilic materials (such as the base catalyzed materials discussed above) that allow subsurface water to flow through the barrier, allowing the entrained chelators to selectively sequester the contaminants of concern. This type of *in-situ* remediation technology is of particular interest to the U.S. Department of Energy, but is also applicable to other contaminated sites. From auto parts to barrier packaging, the race is on to commercialize nanoclay thermoplastic composites. Just a pinch of these infinitesimally small particles can dramatically raise mechanical, thermal, barrier, and flame-retardant properties.

Why all the excitement about the new class of mineral-reinforced plastics known as nanocomposites? Because relatively small amounts (2-5 per cent) of nanometer-sized clay particles can provide large improvements in mechanical and thermal properties, as well as gas barrier and flame resistance. They also reduce shrinkage and warpage. And all of these benefits are available without significantly raising the density of the compound or reducing light transmission (nanoclays are in the same size range as visible-light wavelengths). Most of the current R&D work is focused on automotive parts and packaging. The goals are physical and thermal property enhancements and reduced permeability to gases, moisture, and hydrocarbons. Nylon-based nanocomposites are the first commercial materials to emerge, and there is now a frenzy of activity aimed at nano-reinforcing commodity thermoplastics such as polypropylene and PET. Two research consortia on nanocomposites have been formed in recent months. The Edison Polymer Innovation Corp. (EPIC) of Akron, Ohio, got one under way in March. Headed by staff scientist Jon Collister, this three-year, $3-million program will coordinate investigations by industry and academic researchers on thermoplastics (PP, PET, PE, and acrylic), advanced composites, coatings, and elastomers. Meanwhile, the National Institute of Standards and Technology (NIST) in Gaithersburg, Md., has formed a consortium of government and industry scientists to explore nanocomposites' potential for reducing the flammability of thermoplastics.

The current favorite nanoclay is montmorillonite, a layered alumino-silicate whose individual platelets measure on the order of one micron diameter, giving them an aspect ratio of about 1000:1. Because it is hydrophilic, montmorillonite is not compatible with most polymers and must be chemically modified to render its surface more hydrophobic. The most popular surface treatments are organic ammonium cations, which can be exchanged for the inorganic cations existing on the silicate surface. The treated clay is then incorporated into the resin matrix either during polymerization or by melt compounding. As an example of what can result, commercial nylon 6 nanocomposite materials provide tensile modulus, tensile strength, and HDT equivalent to 40 per cent mineral-filled nylon, while specific gravity remains at the level of unfilled nylon 6. With lower-modulus polymers such as polyolefins, nanocomposites have the potential to upgrade physical properties to levels competitive with costlier engineering resins. Nanocomposites already look attractive for molded car parts such as body panels and under-hood components, as well as electrical/electronic parts, power-tool housings, lawnmowers, aircraft interiors, and appliance components. On the packaging side, nanocomposites can slow transmission of gases and moisture vapor through plastics by creating a "tortuous path" for gas molecules to thread their way among the obstructing platelets. Bottles and food packaging aren't the only areas of interest here. According to EPIC, nanocomposites hold commercial benefits for reducing hydrocarbon emissions from hoses, seals, and auto fuel-system components. Nanocomposites of polypropylene might permit blow molding auto fuel tanks at lower cost than current multi-layer barrier structures.

The first nanocomposites on the market are based on nylon 6 and produced in the reactor. Toyota Central R& D Laboratories in Japan pioneered the work on nanocomposites a decade ago. Within the last few years, Toyota has licensed its patented in-reactor nylon nanocomposite technology to companies such as Nanocor and Japan's Ube Industries. Nanocor, in turn, sublicenses the technology to firms that want to make nylon nanocomposites. Ube was among the first to come out with a commercial nanocomposite—its "nylonclay hybrids" (NCH) of nylon 6 and nylon 6/66 copolymer for film and structural applications. Commercial applications of NCH include nylon 6 barrier film for food packaging and a timing-belt cover for Toyota Motors. Compared with unfilled nylon 6, Ube's NCH is said to have 68 per cent higher tensile modulus and 126 per cent higher flexural modulus. Ube also reports that cast films of NCH-2 grade (2 per cent clay) have half the oxygen permeability of straight nylon 6. At about the same time, Unitika Co. of Japan introduced a nylon 6 nanocomposite for injection molding, which is marketed and distributed in North America by Toyota Tsusho. Nylon M2350 is produced with Unitika's proprietary technology, which incorporates a synthetic silicate during polymerization. Nylon M2350 has been used by Mitsubishi Motors for an engine cover on its GDI models, where the nanocomposite is said to offer a 20 per cent weight reduction and excellent surface finish. Lighting covers and knife handles are other applications.

Sometime back, Bayer AG in Germany announced the development of nylon 6 nanocomposites for transparent barrier film packaging. The compounds are made in the reactor using Nanocor's nanoclay. Bayer is currently marketing nanocomposites only in Europe. It has two nylon 6-based nanocomposite grades for cast film: Durethan LPDU 601-1 and LPDU 601-2, which offer different degrees of barrier improvement. Relative to neat nylon 6, they reportedly cut oxygen transmission rate (OTR) by up to 50 per cent, plus offer increased barrier to other gases, chemicals, and flavors/odors. The nanofilms are also said to be stiffer than unfilled film. AlliedSignal Corp. has been exploring nylon 6 nanocomposites for potential applications in automotive, consumer, and packaging markets. According to Kris Akkapeddi, team leader in plastics R&D, the company is currently evaluating this technology for some niche applications with specific customers. Developmental products at AlliedSignal include toughened nylon nanocomposites for injection molding and high-viscosity nanocomposites for blow molding. Compared with unfilled nylon, Akkapeddi says the nanocomposites have 50-80 per cent higher stiffness and up to 126°F higher HDT without much increase in specific gravity.

Nylon nanocomposites produced by melt compounding that reportedly approach the performance of their in-reactor counterparts are now emerging. The trick is to exfoliate (delaminate) the clay particles sufficiently so that the ultimate level of reinforcement is achieved. Nanocor has introduced a new nanoclay grade, Nanomer I.24TC, which reportedly allows compounders to produce a nylon 6 compound with flexural modulus and HDT about 90 per cent of the levels achieved with in-reactor blends. RTP Co., a specialty thermoplastics compounder, announced in late April that it has produced what it believed to be the first commercially available compounded nylon 6 nanocomposite. The material is offered in injection molding grades and film grades with 3-5 per cent loadings. As shown in the table, RTP's injection grade shows 53 per cent higher HDT than unfilled resin—comparable to 20-30 per cent mineral-filled nylon—but with little change in specific gravity. Tensile strength is much better than either unfilled or mineral-filled nylon 6. Reduced flammability is also indicated by a 50 per cent reduction in rate of heat release. Toyota Central R&D Labs. recently succeeded in producing nylon 6 clay nanocomposites by melt processing in a twin-screw extruder. Dr. Arimitsu Usuki reports that the compounding approach improves productivity and opens the way for making nanocomposites of nylons 12, 66, and 6/66 copolymer. Showa Denko of Japan recently started commercial production of nylon 66 nanocomposites using a kneading extruder. The materials are said to show improved flame retardancy and rigidity. Two special FR grades contain no halogen or phosphorus. Systemer FE 30600 and 30602 reportedly provide UL 94V-0 and V-2 performance, respectively, at only 1/64-in. thickness. Their flex moduli are 30-80 per cent higher, and HDTs are 54-144°F higher, than those of standard nylon 66. Melt-compounded nylon 6 and 66 nanocomposites are also reported to be forthcoming from BASF AG and from Solutia, respectively. (Solutia markets through Dow Plastics.)

An agreement announced sometime ago by Eastman Chemical and Nanocor to jointly develop nanocomposites of PET and other packaging materials (e.g., polyethylene) is the first indication that commercialization of such materials may not be far off. Jim Lewis, v.p. of container plastics at Eastman, says his firm has worked with Nanocor since 1995, but several breakthroughs have occurred over the last year and a commercial product introduction is likely to come fairly soon. Eastman is developing its first PET nanocomposites via the in-reactor approach. Its initial focus is on rigid containers. "Nanocomposites are one of the strategic technology components we are developing to enable emerging markets in food packaging that require superior barrier, including beer bottles and single-serve juice bottles," Lewis says. He hints that these applications are likely to utilize a coinjected multi-layer preform with the nanocomposite barrier layer on the inside. Active projects on other barrier resins are also under way. Nanocor says it has made significant progress in joint development of EVOH nanocomposites for multi-layer packaging. Nanocor even offers a new nanoclay, Nanomer I.35L, designed for compounding with EVOH. Within the range of 50-80 per cent R.H., it reduces oxygen permeability of EVOH by 66-80 per cent. Besides packaging, EVOH nanocomposites are being investigated as a hydrocarbon barrier for automotive fuel tanks and fuel-system components.Nanocor has another new nanoclay, Nanomer I.28MC, for compounding with MXD6 aromatic nylon from Mitsubishi Chemical. Applications could include packaging films for moisture- and oxygen-sensitive foods and electronics.

Automotived-driven development of PP and TPO nanocomposites is one of the hottest areas of current interest. Because PP is a strongly nonpolar resin, it has been especially challenging to come up with an appropriate surface treatment for the clay. Nonetheless, late last year, three development partners—Montell North America, General Motors R& D in Warren, Mich., and Southern Clay—jointly announced the world's first TPO-based nanocomposite. According to Jim Keeler, advanced program manager at Montell's Automotive Business Group, the partners are testing several TPO nanocomposites for exterior and interior applications. The initial focus is on exterior door and rear quarter panels. Although the nanocomposites are produced by melt compounding, Montell is using its Catalloy in-reactor TPOs in order to simplify the overall compounding process. According to Keeler, previous PP/nanoclay compounds achieved only 10-15 per cent exfoliation of the layered clay particles, whereas the current work has achieved much greater exfoliation and thus enhanced property development. TPO nanocomposites with specific gravity as low as 0.91 have been produced with better dimensional stability than unfilled TPOs and comparable to talc-filled grades (CLTE of 5 × 10-5 mm/mm/°C). These developmental TPO nancomposites also have flexural moduli of 150,000 to 250,000 psi. In a 5-mph impact test, the TPO remains ductile down to –30°C. The partners also report that the nanocomposite has improved surface appearance due to the low filler loading. Work on nanocomposites is also well under way at Ford Motor Co., according to Jeff Helms, director of research and materials science. "We have been focusing on anything polypropylene-based, including TPOs, impact-modified PP, and PP alloys," he says. "We have explored several different methods via compounding with which we can make PP nanocomposites."

Ford's initial target is interior trim—instrument panels, A and B pillars, door panels, and consoles. The chief challenge, Helms says, is to maximize coupling between the silicate and the polymer matrix. He indicates that this could prove simpler with surface-treated synthetic clays. "With natural clays, you may need a surface treatment plus a compatibilizer." He says his group recently has achieved very good results in terms of stiffness/impact balance. The next step is to team up with material suppliers. Meanwhile, Toyota Central R&D Labs recently produced PP nanocomposites in a twin-screw extruder using 5 per cent clay and maleic anhydride-modified PP oligomer as a compatibilizer. According to Toyota, the "polypropylene clay hybrid" (PPCH) provides higher stiffness than the same loading of talc or glass.

Dow Plastics and automotive processor Magna International, Southfield, Mich., are in their second year of a joint effort to develop nanocomposites of PP and TPO. The five-year project, which is partially funded by the U.S. Commerce Dept.'s Advanced Technology Program, is aimed at developing methods for dispersing the nanoclay in such resins and demonstrating cost-effectiveness of nanocomposites for exterior front and rear car parts. For auto under-hood and electronics uses, Showa Denko just commercialized the first acetal nanocomposite. It reportedly shows low warpage and few sinks. Compared with unfilled acetal copolymer, it has 40 per cent higher flexural modulus and 45° F higher heat resistance.

Flame-retardant properties of nanocomposites are of interest on many fronts, as indicated by ongoing work at NIST and its new consortium. Over the last three years, NIST fire researchers demonstrated that the rate of heat release for nylon 6 nanocomposites with 5 per cent clay is 63 per cent lower than the RHR of pure nylon 6. There was also no increase in carbon monoxide or soot generated during combustion. Though the mechanism of nanocomposites' FR action is not known for sure, a durable carbon-silicate char was observed to form very rapidly in the burn tests. NIST's work last year showed that the reduced flammability of nanocomposites extends to other thermoplastics, including PP and PS. In one study, PP samples were grafted with 0.4 per cent maleic anhydride to make them more compatible with the clay. Fire researcher Jeffrey Gilman thinks that nanoclay in combination with reduced amounts of conventional FR fillers such as ATH and MgOH at lower use levels are likely to meet FR standards such as 94V-0 and V-2. Gilman notes that Showa-Denko has reported V-0 performance of nylon 6 nanocomposites containing melamine.

Flame-retardant properties of nanocomposites are of interest on many fronts, as indicated by ongoing work at NIST and its new consortium. Over the last three years, NIST fire researchers demonstrated that the rate of heat release for nylon 6 nanocomposites with 5 per cent clay is 63 per cent lower than the RHR of pure nylon 6. There was also no increase in carbon monoxide or soot generated during combustion. Though the mechanism of nanocomposites' FR action is not known for sure, a durable carbon-silicate char was observed to form very rapidly in the burn tests. NIST's work last year showed that the reduced flammability of nanocomposites extends to other thermoplastics, including PP and PS. In one study, PP samples were grafted with 0.4 per cent maleic anhydride to make them more compatible with the clay. Fire researcher Jeffrey Gilman thinks that nanoclay in combination with reduced amounts of conventional FR fillers such as ATH and MgOH at lower use levels are likely to meet FR standards such as 94V-0 and V-2. Gilman notes that Showa-Denko has reported V-0 performance of nylon 6 nanocomposites containing melamine.

Nanocomposites & Plastics

Nanosized particles have mega-potential in plastics because just a pinch does so much more than heavy loadings of other additives. Three recent conferences presented almost 200 papers on the feverish pace of 'nano' R&D on boosting plastics' mechanical and barrier properties, flame retardancy, and electrical conductivity. They are still in their infancy, but if the forecasts are right, nanocomposites could turn out to be the biggest little thing to hit plastics in decades. Polymers reinforced with as little as 2 per cent to 5 per cent of these particles via melt compounding or *in-situ* polymerization exhibit dramatic improvements in thermo-mechanical properties, barrier properties, and flame retardancy. They also can outperform standard fillers and fibers in raising heat resistance, dimensional stability, and electrical conductivity. Dispersions of nanoscale reinforcements in polymers are already entering the marketplace in automotive and packaging applications, albeit in a low-profile manner and slower than had been anticipated. But that pace is expected to speed up dramatically, as indicated by the enthusiasm of researchers and marketers shown in roughly 200 papers delivered at three technical conferences. These were Nanocomposites 2004 in San Francisco, the SPE Antec 2004 in Chicago, and Nanocomposites 2004 in Brussels, Belgium. A report from market-research firm pegs the total worldwide market for polymer nanocomposites at 24.5 million lb in 2003, valued at $90.8 million. It also projects the market to grow at an average annual rate of 18.4 per cent to reach $211.1 million by 2008. Even if nano developments hit some snags, BCC says some applications will grow faster than 20 per cent per year. The leading nanoscale fillers in R&D and commercial projects are layered silicate nanoclays and nanotalcs, plus carbon nanotubes and graphite platelets. But other candidates are being actively investigated, such as synthetic clays, polyhedral oligomeric silsesquioxane (POSS), and even natural fibers like flax and hemp.

The two types of nanofillers that have been most widely discussed and the first to break into commercial use are nanoclays and carbon nanotubes. Both must be chemically modified with surface treatments in order to achieve the fine dispersion and resin coupling that are required to derive maximum benefit. Both of these nanofillers have demonstrated improvements in structural, thermal, barrier, and flame-retardant properties of plastics. Carbon nanotubes also enhance electrical conductivity. So far, nanoclays have shown the broadest commercial viability due to their lower cost—$2.25 to $3.25/lb—and their utility in common thermoplastics like PP, TPO, PET, PE, PS, and nylon. The leading nanoclay is montmorillonite, a layered alumino-silicate whose individual platelets measure around 1 micron diam., giving them an aspect ratio of 1000:1. The two major US producers are Nanocor with its Nanomer line and Southern Clay Products with its Cloisite line. Both companies have formed alliances with suppliers of resins and surfactants, plus compounders and automotive OEMs and packaging firms. While much of their work is proprietary, they have disclosed several commercial successes.

General Motors has taken the lead in putting nanocomposites on the road. GM launched the first commercial auto exterior use of a nanocomposite in the step assist on the 2002 GMC Safari and Chevrolet Astro van. The part also appears on 2003 and 2004 models. More recently, a PP/nanoclay composite appeared on the body side molding of General Motors' highest-volume car, the 2004 Chevrolet Impala. The compound was developed by GM's R&D Center in Warren, Mich., in cooperation with Basell North America and Southern Clay Products. The latest application is on the 2005 GM Hummer H2 SUT. The vehicle's cargo bed uses about seven pounds of molded-in-colour nanocomposite parts for its center bridge, sail panel, and box-rail protector. The material is Basell's Profax CX-284 reactor TPO with nanoclay. While nanoclay adds muscle to plastics, carbon nanotubes impart electrical and thermal conductivity. Nanotubes' commercial potential has been limited by their high price tags—reportedly in the range of $100/gram, although they are available in masterbatches for $50/lb and up. Still, nearly every car produced in the U.S. since the late 1990s contains some carbon nanotubes, typically blended into nylon to protect against static electricity in the fuel system. Static-dissipative compounds containing nanotubes are also protecting computer read/write heads. Carbon nanotubes include both single-and multi-walled structures. The former have a typical outside diameter of 1 to 2 nm while the latter have an OD of 8 to 12 nm. They can range in length from the typical 10 microns to as much as 100 microns and have at least a 1000:1 aspect ratio. Carbon nanotubes have 50 times the tensile strength of stainless steel (100 GPa vs. 2 GPa) and five times the thermal conductivity of copper. When incorporated into a polymer matrix, they have the potential to boost electrical or thermal conductivity by orders of magnitude over the performance possible with traditional fillers such as carbon black or metal powders. US suppliers of nanotubes include Hyperion Catalysis with its Fibril multi-walled nanotubes and newcomer Zyvex Corp. with its Nanosolve single- or multi-walled tubes. Both suppliers now offer their products in masterbatches that typically contain 15 to 20 per cent nanotubes. A different but related category is vapor-grown carbon nanofibers from Pyrograf Products, a spin-off from Applied Sciences. Its Pyrograf III nanofibers reportedly can compete with nanotubes in providing thermal and electrical conductivity and dramatically enhancing mechanical properties and fire resistance (char formation). What's more, nanofibers cost significantly less—around \$100 to \$150/lb. Evaluations are under way in nylon, PP, and polyurethanes.

Alliances between Nanocor and two specialty compounders, have resulted in commercial-scale nanocomposite concentrates and compounds for structural and barrier applications. Noble Polymers new Forte PP nanocomposite compound made its commercial debut in the seat backs of the Honda Acura TL 2004 car. Forte replaced glass-filled PP, which caused processing difficulties, visual defects, and warping. Forte has a low density of 0.928 g/cc, superior mechanical properties, and improved surface quality and recyclability. Noble reports that the Forte nanocomposite will also be used to produce the center console for a 2006 model light truck. Other applications in the works include office furniture—replacing 20 per cent glass-filled PP—and appliance parts, where Forte reduces weight and possibly material cost.

PolyOne recently introduced the Maxxam LST line of PP homopolymer/nanoclay compounds that boast high stiffness and impact resistance. Through a patent-pending process, PolyOne reports that it has been able to overcome previous problems of incomplete exfoliation and dispersion of the nanoclay, resulting in performance that meets or exceeds many engineering thermoplastics. Lighter weight, aesthetic and processing advantages, and lower cost are also claimed. PolyOne also offers Nanoblend concentrates of up to 40 per cent nanoclay in homopolymer PP, modified PP, LLDPE, LDPE, HDPE, or an ethylene copolymer. Some grades are tailored specifically for barrier enhancement. PolyOne reports that applications nearing commercialization include pallets and dunnage, where Maxxam LST compounds are specified as alternatives to engineering resins due to their improved dimensional control, which is critical for robotic assembly. In addition, they boast good impact strength and lighter weight. Maxxam LST is also being considered for consumer disposable applications due to a combination of chemical resistance and stiffness, as well as dramatic cycle-time improvements. Meanwhile, the Nanoblend concentrates are being considered for auto interior and exterior TPO parts. Key drivers are dimensional stability, lighter weight, and stiffness without loss of impact. Nanoblend concentrates are being evaluated in films for enhancing barrier, stiffness, HDT, and controlled release or migration of additives such as biocides and dyes. In blow molded packaging, Nanoblend is being considered for improved barrier and the potential for thinwalling and faster cycles. Thin-walling and faster cycles are also attractions in injection molded containers and totes. Some industry sectors are evaluating the concentrates for improving flame retardancy.

Papers presented by General Motors and Southern Clay Products discussed numerous enhancements to automotive TPOs obtained with nanoclays. Those advances did not come easily: Early processing problems caused by clay agglomeration were ultimately resolved by optimizing the clay infeed position at the extruder, the screw design, screw speed, temperature, and pressure. Once processing issues were resolved, nanocomposite TPOs outperformed conventional talc-filled TPOs in consistency of properties, retention of low-temperature ductility, elimination of "tiger striping," reduced paint delamination, and improved knit-line appearance, colourability, grain patterns, scratch and mar resistance, and recyclability. What's more, the lower filler level means 3 per cent to 21 per cent lower density (0.92 vs. 0.96 to 1.13 g/cc). Lighter weight requires less adhesive for attachment, which cuts cost. Among the many auto exterior, interior, and under-hood applications for which nanocomposites appear suited are fascias, rocker covers, side trim, grilles, hood louvers, instrument panels, seat/IP foams, door inners, pillar covers, vertical and horizontal body and closure panels, engine shrouds, fan shrouds, air intakes, fuel tanks, and fuel lines. In addition to TPO/nanoclay composites, GM has explored using carbon nanotubes to replace current thermoset structural composites. GM is interested in reducing reinforcement levels in Class A applications by replacing continuous carbon fibers with nanotubes or short nanofibers. nanotubes also have the potential to reduce plastics' coefficient of thermal expansion more effectively.

Polymer barrier technology is also getting a boost from nanoclays. Mitsubishi Gas Chemical (MGC) and Honeywell Specialty Polymers both are using Nanocor's nanoclays in nylons as barrier layers in multi-layer PET bottles and films for food packaging. MGC's MXD6 nylon nanocomposite, called Imperm N, is used commercially in Europe in multi-layer PET bottles for beer and other alcoholic beverages. It is also being evaluated for small carbonated soft-drink bottles. Other Imperm applications that will debut in the next six months are multi-layer thermoformed containers for deli meats and cheeses and flexible multi-layer films for potato chips and ketchup. Honeywell has aimed its Aegis nylon 6 nanocomposites initially at PET beer bottles. In late 2003, a version containing an oxygen scavenger made a commercial splash with the introduction of the 1.6-liter Hite Pitcher beer bottle from Hite Brewery Co. in South Korea. Aegis is the barrier layer in this three-layer structure, which is said to provide a 26-week shelf life. Honeywell is aiming other Aegis nanocomposite grades (without oxygen scavenger) as replacements for EVOH in films and pouches. Such grades reportedly are lower in cost than EVOH, provide a better barrier that allows for lightweighting, and also have better puncture resistance and good clarity. (Because of their size, nanoparticles do not interfere with light transmission.) The U.S. military and NASA, in conjunction with Triton Systems, Inc., Chelmsford, Mass., are looking into nanoclay as a barrier enhancer for EVOH in long-shelflife packaging. An experimental thermoformed food tray was made from EVOH plus 3 per cent of Southern Clay's Cloisite in a layer sandwiched between two PP layers. It reportedly imparts three- to five-year shelf life without refrigeration, plus good clarity, processability, and recyclability. Alcoa CSI, Crawfordsville, Ind., is seeking a patent on coextruded barrier liners for plastic bottle caps for beer, juice, or carbonated soft drinks. The liners include a layer of nylon 6/ nanoclay composite plus one or two EVA layers with oxygen scavengers. This liner is said to outperform other barrier materials at very high humidity (95 per cent to 96 per cent RH). LG Chem Ltd. of South Korea has developed high-barrier, monolayer blow molded containers of HDPE with 3 per cent to 5 per cent nanoclay for handling toluene and light hydrocarbon fluids. LG reports that permeation of the hydrocarbon solvents is cut by a factor of 40 to 200 compared with neat HDPE.

Since the early 1990s, automotive fuel-line components such as quick connectors and filters have used inner barrier layers consisting of nylon 12 and carbon nanotubes. Hyperion Catalysis now aims to introduce nanotubes into other resins used in auto fuel systems, such modified nylons and fluoropolymers. A new fluoropolymer/nanotube compound is being used to make O-rings for automotive fuel connectors. In electronics, polycarbonate and polyetherimide (GE's Ultem) components of computer hard drives have been reinforced with nanotubes to render them conductive and very smooth. Over the last three years, a major automotive OEM in Europe has been using carbon nanotubes in GE's Noryl GTX nylon/PPO alloy to mold exterior fenders. This conductive nanocomposite allows for electrostatic painting.

Michigan State University's Composite Materials and Structures Center in East Lansing developed a new surface-treated graphite nano-platelet. Graphite has a modulus several times that of clay and also has excellent electrical and thermal properties. When incorporated into an epoxy, it results in superior mechanical properties and excellent electrical conductivity compared with standard carbon fibers and nanosized carbon black. MSU sees potential in ESD protection and EMI shielding. Plastic nano-graphite compounds are expected to sell for up to $5/lb, significantly less than compounds based on nanotubes or vapor-grown carbon fibers. Carbon nanotubes have more going for them than just conductivity. Researchers at the National Institute of Standards and Technology (NIST), Gaithersburg, Md., report that carbon nanotubes in PP not only enhance the material's strength and properties, but also dramatically change how the molten polymer flows, virtually eliminating die swell.

Extensive research at NIST has established nanoclays' effectiveness as flame-retardant synergists. Nanoclay levels of 2 per cent and 5 per cent in nylon 6 reduced the rate of heat release by 32 per cent and 63 per cent, respectively, NIST found. Specialty compounder Foster Corp. recently demonstrated that higher levels (13.9 per cent) of nanoclay can be added to nylon 12 elastomers to achieve UL 94V-0 rating at 1/8-in. thickness. Used as a char former, the nanoclay allows the typical 50 per cent loading of halogen/antimony oxide flame-retardant system to be cut in half, which significantly reduces detrimental effects on physical properties. The company first introduced nylon 12/nanoclay compounds for tubing and film in 2001. Germany's Sud-Chemie (U.S. office in Louisville, Ky.) offers modified nanoclays called Nanofil as flame retardants. It recently developed halogen-free EVA/PE wire and cable compounds containing 3 per cent to 5 per cent of new Nanofil SE 3000 plus 52 per cent to 55 per cent alumina trihydrate or magnesium hydroxide (typically used at 65 per cent levels). The result is said to be improved mechanical properties, smoother cable, and higher extrusion speeds. According to Hyperion Catalysis, two recent studies show that multi-walled carbon nanotubes may act as a flame retardant without use of halogen. In both EVA and maleic-anhydride-modified PP, 2.4 per cent to 4.8 per cent loadings of nanotubes show heat-release rates comparable to or better than those obtained with nanoclays.

Among its many virtues, nanoclay can work as a nucleating agent to control foam cell structure and enhance properties of polymeric foams for applications from insulation to packaging. The University of Toronto's Dept. of Mechanical and Industrial Engineering studied extrusion of chemically foamed LDPE/wood-fiber compounds. Addition of 5 per cent nanoclay to the mix decreased the cell size, increased the cell density and facilitated foam expansion. When burned, the foam showed good char formation. Similar results wer_2 gas. Researchers at Ohio State University's Dept. of Chemical Engineering (Columbus) in USA found that small amounts of nanoclay surface-grafted with PMMA can reduce cell size and increase cell density in microcellular PS foamed with CO_2. Another OSU study showed that smaller cell size and higher density can be achieved with 5 per cent nanoclay in polyurethane foams blown with pentane or water. Louisiana State University's Mechanical Engineering Dept. (Baton Rouge) reports that 4 per cent to 5 per cent nanoclay increases the flexural strength and elongation of epoxy syntactic foams used as core materials for sandwich composites in structural applications.

Automotive and Barrier Packaging

Nanocomposites are gradually gaining acceptance in the mainstream of global plastics processing. These polymer compounds, containing relatively low loadings (under 6 per cent by weight) of nanometer-sized mineral particles, are beginning to show up in polypropylene and TPO-based automotive exterior claddings (see sidebar), barrier beer bottles, nylon packaging films, polyethylene pipe and wire/cable coatings, and more. Optimism surrounding these novel materials has increased since they burst into industry consciousness two or three years ago. Exploratory effort has intensified as a growing body of data substantiates the potential of established nylon/clay nanocomposites, emerging polyolefin versions, and a range of other resin matrixes and nanofillers. Real-world applications are coming more slowly, in part because of the need to validate cost-effectiveness in the face of high price tags for nano-particle ingredients. Some early application-development programs have lapsed for cost reasons. Such casualties include an automotive timing-belt cover based on a nylon 6 nanocomposite from Japan's Unitika and an automotive mirror housing of conductive PPO/nylon alloy from GE Plastics. Yet the promise of nanocomposites is undiminished. They can improve polymers' stiffness, HDT, dimensional stability, gas barrier, electrical conductivity, and flame retardancy. Nanoparticles are so small and their aspect ratio (L/D) so high that properties improve with lower loadings and fewer penalties (such as higher density, brittleness, or loss of clarity) than with conventional reinforcers like talc or glass. Nanoclays are believed to increase barrier properties by creating a maze or "tortuous path" that slows the progress of gas molecules through the matrix resin. At the same time, these nanoplatelets are only 1 nm thick, less than the wavelength of light, so they do not impede light's passage.

Nanocomposites now draw on a wider menu of resin matrices, including PP, TPO, EVA, acetal, polycarbonate, biodegradable polylactic acid (PLA), and inherently conductive polyaniline. The nano-particles most widely used so far in these compounds are clays supplied by Nanocor and Southern Clay Products. But a new generation of emerging nanomaterials—including nanostructured silicas, carbon nano-tubes, and ceramic nanofibers—suggest that impressive gains in nanocomposite performance lie just a few years ahead.

"Progress has moved beyond nylon 6/clay composites to include products based on PP and PE," declares Nanocor president Peter Maul. Nanocor's data show up to 98 per cent stiffness improvement in PP and up to 52° F higher HDT. The company says the nanocomposite has virtually the same impact strength as unfilled PP homopolymer.

General Motors recently announced the first-ever automotive production part in an olefinic nanocomposite. It's an exterior step assist for 2002 vans, made of a nanoclay/TPO compound from Basell. William Windscheif, Basell's global business v.p. for advanced polyolefins, calls this application "a small step, but a giant one for nanocomposites," adding that it heralds a broader shift to nano-PP in automotive. Initially, the auto industry expressed most interest in nylon 6 nanocomposites for use under the hood, where higher HDT and lightweighting were the goals. Ingolf Buethe, senior v.p. for polymer research at BASF in Germany, says nylon nanocomposites show considerable promise in terms of enhanced stiffness, heat resistance, and gloss. But a serious downside for a 5 per cent nanoclay nylon compound tested by BASF was a loss of toughness more pronounced than with standard fillers. More recently, automotive OEMs and molders have turned their attention to PP and TPO nanocomposites. These polyolefin materials potentially offer engineering-thermoplastic properties at 20 per cent lower density and 50 per cent lower cost per pound. Consultant Kenneth Sinclair, head of STA Research in Snohomish, Wash., says the auto makers find that combination difficult to ignore. He estimates that about 30 per cent of nano-PP usage by 2004 will be in autos, mostly cannibalizing existing PP applications. But replacement of metals and engineering thermoplastics will follow. Sinclair says nano-PPs are stiffer and process better than standard PPs, so thin-walling of parts by around 40 per cent is feasible. In turn, thin-walling permits around 25 per cent reduction in cycle times, for 60-80 per cent total savings per part. Nonetheless, an official at one of GM's domestic competitors insists that car companies will balk at paying a premium for new materials. Nanocor, concluding that automotive OEMs want lighter parts at no extra cost, has focused its attention on non-automotive uses, including pallets, electronics, and appliance housings. Volvo Corp. of Sweden has studied 5 per cent-nanoclay composites based on a Basell TPO modified with maleic anhydride as a coupling agent. Volvo observed 32 per cent to 50 per cent higher stiffness than 20 per cent talc-filled PP. Impact strength was lower than unmodified TPO but higher than 20 per cent talc-filled TPO. Volvo found that nano-TPO still has 68 per cent lower stiffness than aluminum sheet. A long-term goal for Dow Plastics is in-reactor compounding of nano-PP by using nanoclays as the catalyst support for in-situ polymerization of PP homopolymer. Dow's effort is focused on highly loaded (up to 10 per cent clay) nano-PPs for semi-structural automotive uses. Dow sources say preliminary findings show "quite promising" performance of these composites.

Nanocor has developed a 40-50 per cent nanoclay concentrate in PP. One potential use is in heavy-duty electrical enclosures that must meet various fire ratings plus demanding specs for low-temperature toughness and weatherability. Switching to nano-PPs could bring 18 per cent weight savings and permit use of less halogenated FR additive to reach a given UL rating. In other polyolefins, Kabelwerk Eupen of Belgium says melt blending of nanoclay into EVA shows promise for wire and cable compounds. Calorimeter tests reveal a dramatic decline in heat release at relatively low (3-5 per cent) loadings. Nano-EVAs also exhibit superior mechanical properties, chemical resistance, and thermal stability. Meanwhile, not all automotive work in nanocomposites involves polyolefins. A role for nanocomposites in polycarbonate automotive glazing is being explored by Exatec of Wixom, Mich., the joint venture of Bayer and GE Plastics that is dedicated to PC auto-glazing development. Exatec marketing director Fritz Stein says nanotechnology is being considered for the exterior coating needed to achieve weatherability and abrasion-resistance without reducing clarity. A Bayer coating containing nano-particles is one of several promising approaches being pursued, Stein reports.

The major application focus for nylon 6 nanocomposites today is in high-barrier packaging. Much of the attention is on PET bottles, where nanocomposites demonstrate improved oxygen and carbon dioxide barrier. Honeywell offers a 2 per cent nanoclay nylon 6 for bottles and has a 4 per cent nanoclay version in development. An early commercial use is a pasteurizable beer bottle that will be introduced in China later this year. Nanoclays also enhance the oxygen barrier and stiffness of nylon 6 films. That could permit downgauging of packaging of oxygen-sensitive products—pet foods, boil-in bags, vacuum packs, and stand-up pouches. No modification of cast film equipment is needed. "Nanoclays significantly boost the barrier performance of nylon 6 while retaining most of its existing favorable characteristics," states Lance Altizer, Honeywell's market-development manager. He notes that nanocomposites retain nylon's toughness, clarity, hot-fill heat resistance, and oil/grease resistance. Honeywell claims that nylon 6 with 2 per cent nanoclay has three times the oxygen barrier of straight nylon 6, and 4 per cent nanoclay confers a six-fold improvement. That makes Honeywell's Aegis NC nano-nylon a candidate for medium-barrier bottles and films—those offering around 0.5 to 1 cc/mil/day O_2 transmission rate (OTR). Data also show a doubling of stiffness, higher HDT, and improved clarity for nano-nylon 6 packaging. Honeywell has turned its attention to creating nano-nylon materials that can beat the cost of high-barrier plastics or even glass. Its current contender is an active-passive barrier system called Aegis OX, which synergizes nanoclays as the passive barrier and a proprietary, nylon-specific oxygen scavenger as the active agent. Honeywell claims this one-two punch brings a 100-fold reduction in OTR versus nylon 6, taking oxygen ingress to near-zero levels. It also addresses a drawback of existing O_2 scavengers: In the Honeywell system, the passive barrier protects the scavenger from premature depletion. Efficiency of the system is also improved by uniform dispersion of the nano-platelets and by ensuring that the scavenger is positioned "to easily find the oxygen," as a Honeywell source puts it. Aegis materials are being tested by major PET bottle makers. Honeywell says nano-nylon 6 tends to stretch and orient in ways compatible with stretch-blow molding processes. Current barrier requirements for beer bottles (in which Aegis OX would be the core of a three-layer structure) set a maximum limit on oxygen ingress over 120 days, as well as a limit of 10 per cent CO_2 escape in that time. The beer industry appears headed toward a 180-day barrier standard for both gases. "We feel Aegis OX will compete against any existing barrier system for beer," says Honeywell's Altizer. He claims that Aegis OX meets both 120-day requirements and that the 180-day standard is achievable with processing refinements. Nanocor has come up with a different high-barrier option. Its Imperm compound supplements the inherent gas barrier (0.35 cc/mil/day) of amorphous MDX6 nylon from Mitsubishi Gas Chemical with the addition of a nanoclay. Used as the core of a three-layer PET bottle, Imperm is said to have 100-fold lower OTR than that of straight PET. It is being used in a 16-oz, non-pasteurized beer bottle in which the Imperm core (10 per cent of bottle thickness) reportedly ensures a 28.5-week shelf life. Imperm is said to adhere to PET without tie layers. Sufficient clarity is retained to meet requirements for the amber bottle. Bayer is aiming nylon 6 nanocomposites at cast film for multi-layer packaging, protective films for medical and corrosion-prone items, and more. Bayer's pre-commercial Durethan KU2-2601 compound uses Nanocor's clay to reduce OTR by around 50 per cent versus nucleated nylon 6. Stiffness of the nanocomposite is doubled, and its gloss and clarity rival those of a costly high-clarity copolyamide film, Bayer reports. Anti-blocking properties are also improved. Meanwhile, nanocomposites also limit emissions of gasoline, methanol, and organic solvents. Ube America is developing nanocomposite barriers for automotive fuel systems. It uses up to 5 per cent nanoclay in nylon 6 and 6/66 blends. Nylon 6 with 2 per cent nanoclay is said to be five times more resistant to gasoline permeation than unmodified nylon 6. Ube has developed a coextruded barrier fuel line, trade-named Ecobesta, using nylon 6/66 nanocomposite as the core layer.

Package Properties

Nanocomposite technology has been described by some as the next great frontier of material science in packaging. This technology was developed to improve barrier performance pertaining to gases such as oxygen and carbon dioxide. It also enhances the barrier performance to ultraviolet rays, as well as adding strength, stiffness, dimensional stability, and heat resistance. New plastics created with this technology demonstrate an increased shelf life and are less likely to shatter. Once perfected, these plastics will offer these improved characteristics at competitive prices. It will also make them attractive for use in food and beverage packaging and pharmaceutical packaging applications. In 2001, nearly 100,000 pounds of nanocomposite materials were used in the packaging industry; by 2011, the market in the United States is expected to be around 100 million pounds for rigid and flexible packaging. Polymer nanocomposites consist of resins [either thermoset or thermoplastics] that have fillers added with a least one dimension measured in nanometers. These fillers fall into three categories: nanotubes, nanoscale oxides and metals, and nanoclays. Nanotube-based nanocomposites are used for electrostatic dissipation applications; nanoscale oxides and metals are used for abrasion-resistant films; and nanoclay-based nanocomposites are used for barrier packaging applications. Nanocomposites have received considerable attention over the past decade because of their potential to achieve property enhancements significantly greater than that attainable using conventional fillers or polymer blends. The use of nanoclay-based nanocomposites in food and beverage packaging applications, and the properties that make them superior to conventional plastics, will be focused on. Nanocomposites were first referenced in the 1950s and polyamide nanocomposites were reported around 1976. In the early 1980s Toyota's Central Research and Development Labs began working with polymer-layered silicate-clay mineral composites and this is when the technology began to be studied more widely. The clay mineral that is generating the most interest for use in nanocomposites is montmorillonite, generically referred to as nanoclay, and sometimes referred to as bentonite. It is natural clay that is most commonly formed by the in situ alteration of volcanic ash or by the hydrothermal alteration of volcanic rocks. This clay is widely available and relatively inexpensive, thus becoming the most widely used clay in nanocomposite applications. The first step in the preparation of the nanoclay involves purifying approximately 99 per cent of the montmorillonite. The second step involves surface treatment of the clay. Montmorillonite is hydrophilic and relatively incompatible with most hydrophobic polymers, so it must be chemically modified to make its surface more receptive to dispersion. After the clays are chemically treated, they are dispersed in the polymer. The clays are incorporated into the polymer matrix by one of two approaches: during polymerization or by melt compounding. This is the difficult part of the technology and may limit the use of nanocomposites. The dispersion process requires a custom solution for each polymer used, so developing polymer nanocomposites becomes a capitol intensive research and development project. Very few compounding firms have this kind of capability, and this leaves resin producers or well-funded startup companies to develop these materials

A nanocomposite behavior is entirely dependent on the size of the clay fibers added. The smaller and thinner the fiber, the more surface area is available to interact with the polymer matrix and an improved plastic. The size of the clay fibers has grown smaller, which means that even tiny amounts of filler (two per cent of total volume) can be added to achieve the same effects. The nanoclays used in plastic composites are extremely small, irregular platelets. They are approximately one nanometer thick and 100 nanometers in diameter. A single human hair is 80,000 nanometers thick, so one can understand how a little goes a long way. Once these tiny flat platelets are dispersed into the plastic, they create a path that gases must follow to move through the material, greatly slowing their transmission.

Today's nanocomposites typically demonstrate unique improvements in materials properties, including rigidity, strength, and barrier characteristics, while maintaining a level of transparency and offering the potential for recyclability. Polymer-clay promise a veritable explosion of thermoplastic polymer applications because they are lightweight materials that rival metals in stiffness and strength while providing enhanced gas-barrier characteristics, superior dimensional stability, and high heat distortion temperatures—all with low mineral loading and virtually no loss in impact resistance. Making this technology even more impressive is the fact that the properties improve with few penalties (such as higher density, brittleness, or loss of clarity) than with conventional reinforcements such as talc or glass.

Nylon is often touted as one of the most promising resins for nanocomposite additives. In packaging, the major application focus for nylon 6 nanocomposites is in high-barrier packaging. Much of the attention is on PET bottles, where nanocomposites demonstrate an improved oxygen and carbon dioxide barrier.

Honeywell offers three grades of nanoclay nylon 6 resins; Aegis OX, HFX, and CSD. Aegis OX resin is an oxygen-scavenging nylon formulated for the high oxygen barrier demands of plastic beer bottles. Honeywell has turned its attention to creating nano-nylon materials that can beat the cost of higher-barrier plastics or even glass. The nanoclays act as the passive barrier and the nylon specific oxygen scavenger acts as the active agent Honeywell claims this one-two punch brings a 100 fold reduction in OTR versus nylon 6, taking oxygen ingress to near-zero levels. Aegis OX demonstrates excellent resistance to delamination, it can be processed easily, has excellent clarity, is recyclable, and cost competitive. Aegis OX is currently used in a three-layer PET bottle where it is the core layer for a 12-ounce premium beer bottle for Anchor Brewing of San Francisco

Aegis HFX is specifically formulated to address the growing requirement for improved delamination resistance in hot-fill bottle applications. It is a passive-active system that adds more oxygen resistance for foods. Customer trials have demonstrated that Aegis HFX meets shelf life requirements for hot-fill applications including juice, tea, and condiments.

For carbonated soft drinks, Aegis CSD provides delamination resistance during production. It also provides a high level of a passive carbon dioxide barrier compared with monolayer PET. At 3.5 per cent of bottle weight, Aegis CSD extends the shelf life of 0.5 liter bottles from 9 to 14 weeks. New data extends the shelf life to 16 weeks.

Films produced from Honeywell Specialty Polymers' provide excellent barrier properties to oxygen, flavors, and aromas. Each resin provides the extra performance needed for suitable films for a wide range of flexible packaging applications. They also provide toughness, strength, tear, and puncture resistance, and resistance to grease and gas penetration. Applications for this film include processed meat, fresh red meat, poultry, fish, cheese, dried food, and chilled fruit juices.

Nanocor has a different high barrier option. Through an alliance with the Mitsubishi Gas Chemical Company, Inc., Nanocor has rights to melt-compound its nanoclay additives with Mitsubishi's MXD6 nylon for use in barrier PET bottles and films under the trade name of "M9". Mitsubishi Gas Chemical Company, Inc. has supplied MXD6 plastics to the packaging market for 15 years. MXD6 features an excellent barrier for foods and beverages against harmful gases. It is preferred over competitive barrier plastics because it is easy to process and yields packages with high transparency. M9 boosts the CO_2 and O_2 barrier of standard MXD6 respectively by 50 per cent and 75 per cent. The material also retains high clarity and delamination resistance equal to standard MXD6. It is currently being used as the core of a three-layer 16-ounce non-pasteurized PET beer bottle and is said to have 100-fold lower OTR than that of straight PET. It also adheres to PET without tie layers and retains sufficient clarity to meet requirements for the amber bottle. M9's superior gas barrier provides flexibility in bottle design with a range of layer thickness to meet specific shelf life needs. Layer adhesion is virtually the same as standard N-MXD6, so it can be separated from PET by normal recycling methods. Because M9 is miscible with PET at layer thickness of five percent or lower, recyclers have the option of reprocessing thin-layer bottles in total, eliminating the separation step. Rancor-Mitsubishi is able to extend the shelf life of a 28 gram weight PET bottle from 14 weeks to 21 weeks. They use the standard N-MXD6 with a 5 per cent barrier layer of M9 due to its high carbon dioxide barrier. The cut-off is at 90 per cent CO_2 retention.

Triton Systems, Inc. also has a class of commodity polymers that have been enhanced with organically modified nanoclays. These materials are sold under the trade name, ORMLAS, and have demonstrated improved barrier performance, impact resistance, as well as flame resistance. ORMLAS have been incorporated into a long-life food tray that offers high barrier performance and impact resistance. They also include high clarity, ease of manufacture, and recyclability.

In addition to its alliance with Mitsubishi, Nanocor offers its nanoclay resins to other converters for use in various consumer applications. The dominant property improvement seen is a higher quality shelf life. This improvement can lead to lower weight packages because less material is needed to obtain the same, and if not better, barrier properties. This, in turn, can lead to reduced package cost. Improved shelf life and lower package cost are the reasons why nanotechnology is being pursued in consumer packaging.

Bayer's research team is aiming nylon 6 nanocomposites at cast film. It is referred to as a hybrid system that uses Nanocor's clay to reduce OTR by around 50 per cent versus nucleated nylon. Stiffness of the nanocomposite is doubled, and its gloss and clarity rival those of a costly high-clarity polyamide film. Anti-blocking properties are also improved. The new film material with nanoparticles unites the advantages of EVOH and polyamide 6. The embedded particles prevent gases from penetrating the film and moisture from escaping. Durethan KU 2-2601 is suitable for applications where conventional polyamides are too permeable and EVOH is too expensive. Bayer's Durethan KU 2-2601 makes an ideal candidate as a plastic coating for paperboard juice containers. It can protect highly oxygen-sensitive package contents, such as orange juice, at a much lower cost.

Real world applications for nanocomposites are coming slowly. There appears to be a reluctance to embrace this new technology due to cost and variability in the quality of some of the products. John Jones, a market development specialist at Honeywell, believes that manufacturers of these materials have to prove to the market that these new materials can meet their performance expectations. Another challenge nanocomposite producers face is the production of the nanocomposite itself. Both methods, pre-polymerization and post polymerization, for preparing nanocomposites have drawbacks. Pre-polymerization production can disrupt the polymerization process, which is often critical and requires much developmental time and expense to achieve good yields and controllability, and post polymerization often requires a lot of time to achieve a good dispersion of the nanoparticles in the composite. This then becomes an expensive and low cost-competitive initiative. Another concern deals with equipment conversion that accepts new material through recalibration; this is a big investment for converters to make. It's a complicated process to go from plastic pellets to a blown bottle. It requires heating and blowing that form to the shape of the bottle. This is expensive equipment, very high-speed equipment, designed for the material that you're going to run. You can't just take another material with different flow characteristics, crystallization rate, and those kinds of things, throw that in there and make it run.

Today, nanocomposite research is widespread and is conducted by companies and universities across the globe. Plastic suppliers who have already commercialized nanocomposite materials include Basell USA, Bayer, Dow Chemical, Eastman Chemical, Mitsubishi Gas Chemical, Nanocor, Triton Systems, Honeywell, and RTP Co. Most of these efforts are currently focused on either polyolefins or nylons, but, in theory, the clay nanoparticles could be used in any resin family. Optimism surrounding these novel materials has increased since they burst into industry consciousness two or three years ago, and exploratory effort has intensified as a growing body of data substantiates the potential established by nylon/clay nanocomposites, emerging polyolefin versions, and a range of other resin matrixes and nanofillers....the promise of nanocomposites is undiminished. It appears that the momentum is building. The success of Honeywell, Mitsubishi Gas and Chemical, Bayer, Triton Systems and

Nanocor will lead to other successes. As production reaches a sufficient scale, incorporating the clay into polymers will become more cost-effective as well. Equipment recalibration used for conventional plastic resins will also show success. The popularity of nanocomposites comes from the fact that a little goes a long way. It provides a marked increase in oxygen, carbon dioxide, moisture and odor barrier properties, increased stiffness, strength and heat resistance, and maintains film clarity and impact strength. For the packaging industry, these new materials and their commercial applications are coming into focus.

Suppliers of various nano-chemicals, nanofibers, and nanotubes argue that the potential of their technologies exceeds that of current nanoclay materials. But the prices are prohibitively high, and practical impacts probably lie five years ahead. Static-dissipative applications are emerging as one potentially large market. Applied Sciences has developed a vapor-grown carbon nano-tube made by pyrolysis of coal. Pyrograf-III comes in 100- and 200-nm diam. and has potential as an electrically conductive additive and modifier of plastics' coefficient of thermal expansion. Max Lake, president of Applied Sciences, says the nanotubes enhance electrical conductivity over a broad resistivity range and boost mechanical properties. Just 0.5 per cent loading provides volume resistivity in the 104 ohm-cm range. Another class of nanotube is a graphitic carbon type that is designed primarily to enhance electrical conductivity. These nano-tubes or "fibrils" are made from a hydrocarbon gas by Hyperion Catalysis International. They offer surface resistivity of 103 to 105 ohm/sq at 4 per cent to 7 per cent loadings. "Fibrils are more efficient at building electrical conductivity into plastics than carbon black or PAN carbon fibers," states John Hagerstrom, Hyperion's technical service manager. He says the fibrils' small diameter (10 nm average) and high aspect ratio (1000:1) mean a given level of conductivity is achieved at lower loadings than with conventional carbon particles or fibers. That reportedly means less sacrifice of matrix properties, lower warpage, and better surface smoothness. Fibrils have been used to enhance electrostatic painting of automotive mirrors, and as a static-dissipative additive in semiconductor components and disk drives, where the non-sloughing feature of nanotubes helps retain purity. Hybrid Plastics offers "POSS" nanochemicals, named for their polyhedral oligomeric silsesquioxane molecular building blocks.

These molecular silicas are hybrid organic-inorganic materials said to bridge conventional differences between minerals and monomers. POSS molecules are cage-like structures typically measuring 1.5 nm along each axis. The single molecule particles are truly dispersable and have no affinity for one another. Even loadings of 50 per cent or more by weight reportedly disperse without agglomeration. POSS molecules dissolve in a plastic melt, then recrystallize on cooling into a network that enhances mechanical and thermal properties, as well as flame-retardancy. These chemical tools allow the creation of superior reinforced, crosslinked, and chemically coupled alloys. A 10 per cent POSS loading elevates PP's flexural modulus by 12 per cent, HDT by 21 per cent, and impact strength by 36 per cent.

A 50 per cent POSS loading in crystal PS reportedly has no effect on optical clarity. Hybrid cites optical disks, micro-electronics, and medical products as target niches. At present, POSS costs around $200/lb. A larger plant for making POSS nanochemicals, due on-line in early 2002, could push down the cost of some POSS products to about $15/lb. Hybrid Plastics' sales and marketing arm is Divex, Inc. Argonide has a technology for producing alumina ceramic nano-whiskers by electro-explosion of metal wire. These NanoCeram whiskers offer potential for reinforcement because of their small size (2 nm diam.) and high aspect ratio (50:1 average). Applications for NanoCeram are being pursued as reinforcements and thermally conductive additives. Current price of pilot-plant materials is $280/lb, but that is likely to fall as usage develops.

For the first time, a PP-based nanocomposite has been adopted for use in an automotive production part. A step-assist for 2002 General Motors Safari and Astro vans is molded by Blackhawk Automotive Plastics, Troy, Mich., from a TPO matrix reinforced with 2.5 per cent nanoclay supplied by Southern Clay Products. The part is a low-volume, dealer-installed option. The nano-TPO replaces a 15 per cent talc-filled PP but is molded in the same tool. This is a beginning. Parties to this effort are investigating other exterior and interior automotive parts. Nano-TPO delivers stiffness equivalent to the talc-filled PP it replaced, translating into 7-8 per cent weight savings. Further, nano-TPO improves low-temperature impact strength and part surface quality. GM officials termed the project "cost-neutral," explaining that the higher price of Basell's nano-TPO is offset by lightweighting and other cost benefits. The nano-TPO processes at a somewhat lower temperature than talc-filled PP, decreasing cycle time, while also decreasing knit-line creation and other surface imperfections. Reports also show molding other prototype exterior claddings, including rocker and sail panels and body side moldings. Newer versions of Basell materials are being investigated, including some containing up to 5 per cent nanoclay and offering stiffness comparable to 30-40 per cent talc-filled PP. It is also noted that the much lower filler levels in nano-TPOs result in smoother surfaces while enhancing scratch resistance. When GM's R&D unit first announced the nano-TPO project in 1999, emphasis then was on vertical body panels and interior parts. Instead, the first commercial application is a relatively simple, low-volume part. But GM declares that there is no retreat from the applications initially targeted. Work is well under way on a rocker-panel prototype that is 20 per cent lighter than one of talc-filled PP. Interior trim is also being actively explored. A significant advance in chemistry by GM is critical to success of the van step. It allows more complete exfoliation (layer separation) and better dispersion of nanoclay particles into TPO. GM has granted Basell exclusive rights to use its advances in nano-PP technology. In turn, GM has been guaranteed exclusive rights to use Basell's nano-TPOs in the automotive field.

Future Application

Such mechanical property improvements have resulted in major interest in nanocomposite materials in numerous automotive and general/industrial applications. These include potential for utilisation as mirror housings on various vehicle types, door handles, engine covers and intake manifolds and timing belt covers. More general applications currently being considered include usage as impellers and blades for vacuum cleaners, power tool housings, mower hoods and covers for portable electronic equipment such as mobile phones, pagers etc.

The gaseous barrier property improvement that can result from incorporation of relatively small quantities of nanoclay materials is shown to be substantial. Data provided from various sources indicates oxygen transmission rates for polyamide-organoclay composites which are usually less than half that of the unmodified polymer. Further data reveals the extent to which both the amount of clay incorporated in the polymer, and the aspect ratio of the filler contributes to overall barrier performance. In particular, aspect ratio is shown to have a major effect, with high ratios (and hence tendencies towards filler incorporation at the nano-level) quite dramatically enhancing gaseous barrier properties. Such excellent barrier characteristics have resulted in considerable interest in nanoclay composites in food packaging applications, both flexible and rigid. Specific examples include packaging for processed meats, cheese, confectionery, cereals and boil-in-the-bag foods, also extrusion-coating applications in association with paperboard for fruit juice and dairy products, together with co-extrusion processes for the manufacture of beer and carbonated drinks bottles. The use of nanocomposite formulations would be expected to enhance considerably the shelf life of many types of food.

Honeywell have also been active in developing a combined active/passive oxygen barrier system for polyamide-6 materials. Passive barrier characteristics are provided by nanoclay particles incorporated via melt processing techniques whilst the active contribution comes from an oxygen scavenging ingredient (undisclosed). Oxygen transmission results reveal substantial benefits provided by nanoclay incorporation in comparison to the base polymer (rates approximately 15-20 per cent of the bulk polymer value, with further benefits provided by the combined active/passive system). Akkapeddi suggests that the increased tortuosity provided by the nanoclay particles essentially slows transmission of oxygen through the composite and drives molecules to the active scavenging species resulting in near zero oxygen transmission for a considerable period of time.

Triton Systems and the US Army are conducting further work on barrier performance in a joint investigation. The requirement here is for a non-refrigerated packaging system capable of maintaining food freshness for three years. Nanoclay polymer composites are currently showing considerable promise for this application. It is likely that excellent gaseous barrier properties exhibited by nanocomposite polymer systems will result in their substantial use as packaging materials in future years. A somewhat more esoteric possibility arising from enhanced barrier performance recently suggested has been blown–films for artificial intestines!

The ability of nanoclay incorporation to reduce solvent transmission through polymers such as polyamides has been demonstrated. Data provided by De Bievre and Nakamura of UBE Industries reveals significant reductions in fuel transmission through polyamide–6/66 polymers by incorporation of a nanoclay filler. As a result, considerable interest is now being shown in these materials as both fuel tank and fuel line components for cars. Of further interest for this type of application, the reduced fuel transmission characteristics are accompanied by significant material cost reductions.

The presence of filler incorporation at nano-levels has also been shown to have significant effects on the transparency and haze characteristics of films. In comparison to conventionally filled polymers, nanoclay incorporation has been shown to significantly enhance transparency and reduce haze. With polyamide based composites, this effect has been shown to be due to modifications in the crystallisation behaviour brought about by the nanoclay particles; spherilitic domain dimensions being considerably smaller. Similarly, nanomodified polymers have been shown, when employed to coat polymeric transparency materials, to enhance both toughness and hardness of these materials without interfering with light transmission characteristics. An ability to resist high velocity impact combined with substantially improved abrasion resistance was demonstrated by Haghighat of Triton Systems.

Water laden atmospheres have long been regarded as one of the most damaging environments which polymeric materials can encounter. Thus an ability to minimise the extent to which water is absorbed can be a major advantage. Data provided by Beall from Missouri Baptist College indicates the significant extent to which nanoclay incorporation can reduce the extent of water absorption in a polymer. Similar effects have been observed by van Es of DSM with polyamide based nanocomposites. In addition, van Es noted a significant effect of nanoclay aspect ratio on water diffusion characteristics in a polyimide nanocomposite. Specifically, increasing aspect ratio was found to diminish substantially the amount of water absorbed, thus indicating the beneficial effects likely from nanoparticle incorporation in comparison to conventional microparticle loading. Hydrophobic enhancement would clearly promote both improved nanocomposite properties and diminish the extent to which water would be transmitted through to an underlying substrate. Thus, applications in which contact with water or moist environments is likely could clearly benefit from materials incorporating nanoclay particles.

The ability of nanoclay incorporation to reduce the flammability of polymeric materials was a major theme of the paper presented by Gilman of the National Institute of Standards and Technology in the US. In his work Gilman demonstrated the extent to which flammability behaviour could be restricted in polymers such as polypropylene with as little as 2 per cent nanoclay loading. In particular heat release rates, as obtained from cone calorimetry experiments, were found to diminish substantially by nanoclay incorporation. Although conventional microparticle filler incorporation, together with the use of flame retardant and intumescent agents would also minimise flammability behaviour, this is usually accompanied by reductions in various other important properties. With the nanoclay approach, this is usually achieved whilst maintaining or enhancing other properties and characteristics.

Nanostructured Materials

Over the past decade nanostructured materials have been the subject of enormous interest, with the potential for wide-ranging industrial, biomedical, and electronic applications. These can include metals, ceramics, polymeric materials, or nanocomposites. The nanomaterials field has seen a huge increase in funding from private enterprises and government, and academic researchers within the field have formed many partnerships.

Nanomaterials have actually been produced and used by humans for hundreds of years - the ruby red colour of some stained glass is due to gold nanoparticles trapped in the glass matrix. The decorative glaze known as lustre, found on some medieval pottery, contains metallic spherical nanoparticles dispersed in a complex way in the glaze, which gives rise to its special optical properties.

The variety of nanomaterials is great, and their range of properties and possible applications enormous, from miniature batteries, to biomedical implants, as packaging films, super absorbants, components of army, and automobile parts.

Nanocrystalline materials are polycrystalline with grain sizes in the 1-100nm range that are found to exhibit improved hardness, increased ductility and toughness, superior magnetic and optoelectronic properties. Nanocrystals are aggregates of atoms from a few hundred to tens of thousands that combine into a crystalline form of matter known as a 'cluster'. Typically around 10nm in diameter, nanocrystals are larger than molecules but smaller than bulk solids and therefore frequently exhibit physical and chemical properties somewhere in-between.

Given that a nanocrystal is virtually all surface and no interior, its properties can vary considerably as the crystal grows in size. Nanoscale crystals are often harder, stronger and more wear resistant than the same materials in bulk form. Nanocrystals might be used to make super-strong and long-lasting metal parts. The crystals also might be added to plastics and other metals to make new types of composite structures for everything from cars to electronics.

These are particles that are less than 100nm in size. At this level the percentage of exposed area in proportion to its total volume changes dramatically when compared to bulk materials. In particles of around three to 5nm in diameter, about a third of the atoms are actually surface atoms; this means they behave a lot more like gas particles rather than solids. This gives them new properties, they float where bigger particles would sink, and dissolve more easily than the same material in bulk form. The extra exposed surface area also changes depending on how the elements and particles interact with each other.

Recently clay nanoparticles have made their way into composites in cars and packaging materials, where they offer transparency, reduction in permeability and increased strength, the latter being already attractive to the automotive industry, but potentially much more so to the aerospace industry. Sunscreens utilize nanoparticulate zinc oxide. Nanoparticles are also being used in antiseptics, as abrasives, in paints, in new coatings for spectacles (making them scratchproof and unbreakable), for tiles, and in electrochromic or self-cleaning coatings for windows.

Nanoparticles are the basis for anti-graffiti coatings for walls and improved ski waxes and ceramic coatings for solar cells. Glues containing nanoparticles have optical properties that give rise to uses in optoelectronics. Casings, containing nanoparticles used in for electronic devices, such as computers, offer improved shielding against electromagnetic interference.

Nanoscale Interfacial Material

In the development of nanometer-scale materials, ever since the early 1980s, scientists began discovering ultra-small crystal grains. The important point is that small is not only beautiful but also eminently useful. Therefore, the virtues of working in the nanodomain are increasingly recognized by the scientific community, the technological world, and the popular press. These small groups of atoms go by several different names: nanoparticles, nanocrystals, atomic clusters, and quantum dots. Much recent research has gone into finding ways to makesmall and uniform-size clusters that possess unusual optical, electrical, magnetic, and mechanical properties.

This effort has already paid several commercial dividends, for example, ceramics and chemical catalysts with increased efficiency due to high surface-to-volume ratios. Moreover, advances with semi- conducting nanoclusters are bringing several commercial applications into view as well. However, in order to develop new kinds of high-level functional materials, one should not only consider the proper- ties of a single nanoparticle (e.g., high surface ratios, quantum confinement effect), but also the cooperative effect among nanostructures and the macroscopic properties of their organized states. In this report, we attempt to introduce a new concept for the design and creation of new types of interfacial materials, i.e., binary cooperative complementary *nanoscale* materials.

In fact, from ancient times in China, the binary cooperative complementary concept has been used as "yin" and "yang" to explain the phenomenon and catalog the materials of nature and the universe. This initial philosophical principle has had a great influence on classical and modern science. Following the development of science, people have realized that binary cooperative complementary character is a universal feature of the material world. For example, the constitution of the chemical elements and the universe, as well as the physical and chemical properties of materials, are all involved in this binary concept.

In the development of materials science, people have also used the binary cooperative complementary effect to create unique functional materials. So far, binary atomic materials (e.g., binary alloys) and binary molecular materials have been developed that demonstrate tremendous potential in both basic research and practical applications. Well-known examples of binary atomic materials exist, such as the combination of Fe and Cr, which creates a type of stainless steel (also, cobalt is added to iron to make cobalt steels, among the hardest alloys known). Furthermore, many other binary alloys show a number of unique properties. Binary molecular, amphiphilic molecules, with both hydrophilic and hydrophobic groups on the same molecule, have been used in different fields, such as surfactants, in which a huge variety of industrial products have been developed. When electron donors and acceptors are incorporated into a single molecule, nonlinear optical materials can result. Also, binary molecular solids provide another good example. Through the synthesis of donor-acceptor complexes, organic conductors and even superconductors may be generated. The quintessential functional binary molecules (multi-binary molecules) are DNA and RNA, which precisely control the heredity of all living things, animal and plant. Still many unknowns remain, waiting to be elucidated in further research.

Above all, binary cooperative complementary atomic and molecular systems have shown us unique material words. Here we extend this concept from the molecular level to the nanostructural level and introduce a new material system, i.e., binary nanoscale materials. Before going into this topic, we would first like to start from the binary interfacial material, since understanding interfacial materials is key to the design of high-level functional materials. The motivation to develop binary cooperative complementary nanoscale interfacial materials is the desire to create unique functional macroscopic properties. By applying the binary coordinating complementary concept on the nanoscale on a solid surface, we expect to generate nanostructures with mutually compensating properties, e.g., hydrophilic and hydrophobic; conducting and insulating; convex and concave; p-type and n-type; oxidizing and reducing; ferromagnetic and antiferromagnetic, and so on. Whenever an interface (or a surface) is formed by the mixture of nanostructures with a pair of these mutually complementary properties, under certain cooperative conditions, unusual interfacial properties can be created, i.e., binary cooperative complementary nanoscale interfacial materials. Our recent results indicate that such a concept is useful for designing new types of interfacial materials. In the following sections we will introduce several examples for the unique character of binary cooperative complementary nanoscale interfacial materials.

These kinds of surfaces or interfaces can be generated by outside stimuli, e.g., light, electric fields, magnetic fields, heat, plasma etching, or chemical synthesis on the surface of the solid. The surface nanostructures were characterized by atomic force microscopy (AFM) in air (SPI300 Seiko Instruments) or in high vacuum (SPI300HV Seiko Instruments).

Recent research indicates that the binary cooperative complementary concept is truly useful for the design and creation of *nanoscale* functional interfacial materials. Here, we will simply mention these as follows.

Binary complementary hydrophilic and hydrophobic nanoscale structures on the surface of TiO thin films and single crystals appear to create super-amphiphilic properties, with contact angles of 0° for both water and oil. AFM images on the face of rutile TiO_2 single crystals and suggests the binary TiO_2 surface can be induced by UV light illumination, heating at 300 °C, or argon ion sputtering, respectively. The key transformation involves loss of oxygen atoms from Ti_{4+} sites to Ti_{3+} sites, which can adsorb hydroxyl groups and thus produce hydrophilic regions. In all of the images, hydrophilic domains appear as rectangular shapes, aligned along the direction of the face which was confirmed by friction force microscopy (FFM). This unique amphiphilic TiO surface character is ascribed to a composite of hydrophilic and hydrophobic (oleophilic) phases on the scale of several tens of nanometers, with the hydrophilic phases formed, for dissociative water adsorption. Microscopically,these areas may be hydrophilic and oleophilic; however macroscopically, the surface is highly amphiphilic. Furthermore, no visible water drops can form on the surface, and the material appears to have an antifogging property. The polycrystalline TiO_2 thin film, which has same character, can be coated on various substrates, e.g., glass, ceramics, plastics and metals. This highly amphiphilic surface has self-cleaning, antifogging properties. Furthermore, this basic idea can be also used to design and synthesize other types of super-amphiphilic materials, e.g., super-amphiphilic polymers.

Pt nanograins on the surface of TiO_2 thin films, i.e., a binary (with both oxidizing and reducing sites) coordinating complementary catalyst surface, can be created through synthesis processes. The effect is useful for the prevention of icing on windows and electrical cables in winter. The important advantage of this technique is that the anti-icing process is carried out through a photo-electrochemical reaction, but not a physical heating process. The necessary reaction for removing ice blocks on the substrate is just the decomposition of single molecular layer of H_2O contacting with the substrate. Furthermore, this kind of structure can be also applied to other catalysis systems for various industrial applications.

Through a phase-separation process, polyporous with hydrophobic convex and hydrophilic concave surfaces are generated in copolymer film which appears as an “amphiphobic” surface (i.e., the surface appears to have both hydrophobic and oileophobic properties). It was known that the roughness can enhance the wettability of a surface. However, a pure hydrophilic or hydrophobic surface of polyporous may only enhance the hydrophilic or hydrophobic property of a solid surface, but “amphiphobic” behavior can not be created. Interestingly, as-grown copolymer films show an "amphiphobic" property.

REFERENCES

Adeyeye, A., G. Lauhoff, J. Bland, C. Daboo, D. Hasko, and H. Ahmed. 1997. *Appl. Phys. Lett.* 70: 1046.

Adleman, L.M. 1994. Molecular computation to solutions of combinatorial problems. *Science* 266: 1021-1024.

Aizawa, M. 1994. Molecular interfacing for protein molecular devices and neurodevices. *IEEE Engineering in Medicine and Biology* (Feb./March): 94-102.

Ajayan, P.M., and S. Iijima. 1993. *Nature*. 361: 333. Arakawa, Y., H. Sakaki. 1982. *Appl. Phys. Lett.* 40: 939.

Akis, R., D. Ferry, C. Musgrave, "Kinetic Lattice Monte Carlo Simulations of Processes on the Silicon (100) Surface," *Physica E-Low Dimensional Systems and Nanostructures* 19, 183-187 (2003).

Aksay, I.A., et al., eds. 1992. Hierarchical structured materials. *MRS Proc*. 255. Angstrom Technology Partnership. 1995. *Challenging the angstrom*. Tokyo.

Aksay, I.A., M. Trau, S. Manne, I. Honma, N. Yao, L. Zhou, P. Fenter, P.M. Eisenberger, and S.M. Gruner. 1996. Biomimetic pathways for assembling inorganic thin films. *Science* 273: 892-898.

Aksay, Ilhan. 1998. Nanostructured ceramics through self-assembly. In *R&D status and trends*, ed. Siegel et al.

Al Globus, Charles W. Bauschlicher Jr., Jie Han, Richard L. Jaffe, Creon Levit, Deepack Srivastava, "Machine phase fullerene nanotechnology," *Nanotechnology* 9(1998): 192-199.

Albert Einstein, "Investigations on the Theory of Brownian Movement," with notes by R. Furth, translated into English from German by A.D. Cowper, Methuen, London (1926), Dover Publications (1956); original paper in *Ann. Phys.* 17(1905): 549.

Albrecht, T.R., Akamine, S., Carver, T.E., and Quate, C.F. (1990) Microfabrication of cantilever styli for the atomic force microscope. J. Vac. Sci. Technol. A 8(4), 3386-3396

Albrecht, T.R., Grütter, P., Horne, D., and Rugar, D. (1991) Frequency modulation detection using high-Q cantilevers for enhanced force microscope sensitivity. J. Appl. Phys. 69(2), 668-673

Alexander, S., Hellemans, L., Marti, O., Schneir, J., Elings, V., Hansma, P.K., Longmiro, M., and Gurley, J. (1989) An atomic-resolution atomic-force microscope implemented using an optical lever. J. Appl. Phys. 65(1), 164-167

Alivisatos, A.P. 1996. Perspectives on the physical chemistry of semiconductor nanocrystals. *Journal of Physical Chemistry* 100: 13226-13239.

Allara, D.L. 1996. Nanoscale structures engineered by molecular self-assembly of functionalized monolayers. In *Nanofabrication and biosystems*, ed. Hoch et al. Birge, R.R. 1995. Protein based computers. *Sci. Am.* (Mar.): 90-95.

Altmann, J. & Gubrud, M.A.: 2004, 'Military, Arms Control, and Security Aspects of Nanotechnology', in: D. Baird, A. Nordmann & J. Schummer (eds.), *Discovering the Nanoscale*, Amsterdam: IOS Press, pp. 269-277.

Altmann, J. Military Uses of Microsystem Technologies. Dangers and Preventive Arms Control. Münster: Agenda Verlag, 2001.

Altmann, J.: 2004, 'Military Uses of Nanotechnology: Perspectives and Concerns', *Security Dialogue*, 35 (1): 61-79.

Alves, H., M. Ferreira, U. Koster, and B. Muller. 1996. *Materials Science Forum* 225–227: 769.

Andrea G. Bodnar, Michel Ouellette, *et al.*, "Extension of life-span by introduction of telomerase into normal human cells," *Science* 279(16 January 1998): 349-352.

Andres, R.P., S. Datta, D.B. Janes, C.P. Kubiak, and R. Reifenberger. 1998. The design, fabrication and electronic properties of self-assembled molecular nanostructures. In *The handbook of nanostructured materials and nanotechnology*, ed. H.S. Nalwa. San Diego: Academic Press.

Arendt, H. The Human Condition, The University of Chicago Press, 1958.

Arriagada, F.J., and K. Osseo-Asave. 1995. *J. Colloid Interface Sci.* 170: 8.

Arribart, H.: 2004, 'Les nanomatériaux autres que ceux des technologies de l'information et des communications (TICS)', in: Académie des sciences, Académie des technologies (eds.), *Nanosciences, nanotechnologies*, éditions Tec & Doc, Paris, pp. 361-382.

Arthur C. Clarke, *Profiles of the Future*, Harper & Row, Publishers, New York, 1962.

Asada, M., Y. Mayamoto, and Y. Suematsu. 1986. *IEEE Journ. Quantum Electronics* QE- 22(9): 1915-1921.

Attard, G.S., et al. 1997. *Science* 278: 838. Averback, R.S., J. Bernholc, and D.L. Nelson. 1991. *MRS Symposium Proceedings* Vol. 206.

Attebery, B.: 2004, 'Dust, Lust, and Other Messages from the Quantum Wonderland', in: N.K. Hayles (ed.), *Nanoculture: Implications of the New Technoscience*, Bristol, UK: Intellect Books, pp. 161-172.

Averback, R.S., J. Bernholc, and D.L. Nelson. 1991. *MRS Symposium Proceedings* Vol. 206.

Axelbaum, R.L. 1997. Developments in sodium/halide flame technology for the synthesis of unagglomerated non-oxide nanoparticles. In *Proc. of the Joint NSF-NIST Conference on Nanoparticles: Synthesis, Processing into Functional Nanostructures and Characterization* (May 12-13, Arlington, VA).

B. Liu, N.B. Leontis and N.C. Seeman, Bulged 3-arm DNA Branched Junctions as Components for Nanoconstruction, Nanobiology 3, 177-188 (1994). N.C. Seeman, Molecular Craftwork with DNA, The Chemical Intelligencer 1(3), 38-47 (1995).

B. W. Matthews, H. Nicholson, and W. J. Becktel, "Enhanced protein thermostability from site-directed mutations that decrease the entropy of unfolding." *Proc. Nat. Acad. Sci., 84:* 6663-6667 (1987), and included references.

B. Yurke, A.J. Turberfield, A.P. Mills, Jr., F.C. Simmel, J.L. Neumann, "A DNA-fuelled molecular machine made of DNA," *Nature* 406(10 August 2000): 605-608.

B.H. Robinson and N.C. Seeman, The Design of a Biochip: A Self-Assembling Molecular-Scale Memory Device. Protein Engineering 1, 295-300 (1987).

Baibich, M..N., J.M. Broto, A. Fert, F. Nguyen Van Dau, F. Petroff, P. Etienne, G. Creuzet, A. Friederich, and J. Chazelas. 1998. *Phys. Rev. Lett.* 61: 2472.

Baird, D. & Shew, A.: 2004, 'Probing the History of Scanning Tunneling Microscopy', in: D. Baird, A. Nordmann & J. Schummer (eds.), *Discovering the Nanoscale*, Amsterdam: IOS Press, pp. 145-156.

Baird, D.: 2004, *Thing Knowledge: A Philosophy of Scientific Instruments*, University of California Press, Berkeley.

Baird, D.; Nordmann, A. & Schummer, J. (eds.): 2004, *Discovering the Nanoscale*, Amsterdam: IOS Press.

Baker, R.T.K. Synthesis, properties and applications of graphite nanofibers. In *R&D status and trends*, ed. Siegel et al.

Ball, P.: 2002, 'Natural Strategies for the molecular engineer', *Nanotechnology*, 13, 15-28.

Ball, P.: 2003, 'Nanotechnology in the firing line', *Nanotechweb.org,* 23 December [http: // www.nanotechweb.org/articles/society/2/12/1/1].

Bate, R., Frazier, G., Frensley, W., Reed., M., "An Overview of Nanoelectronics," *Texas Instruments Technical Journal*, July-August 1989, pp. 13-20.

Beck, J.S., J.C. Vartuli, W.J. Roth, M.E. Leonowicz, C.T. Kresge, K.D. Schmitt, C.T.-W. Chu, D.H. Olsen, E.W. Shepard, S.B. McCullen, J.B. Higgins, and J.L. Schlenker. 1992. *J. Am. Chem. Soc.* 114: 10834.

Becker, M.F., J.R. Brock, H. Cai, N. Chaudhary, D. Henneke, L. Hilsz, J.W. Keto, J. Lee, W.T. Nichols, and H.D. Glicksman. 1997. Nanoparticles generated by laser ablation. In *Proc. of the Joint NSF-NIST Conf. on Nanoparticles.*

Bensaude-Vincent, B.: 2001, 'The Construction of a Discipline: Materials Science in the United States', *Historical Studies in the Physical and Biological Sciences,* 31, 223-248.

Bensaude-Vincent, B.: 2004, 'Two Cultures of Nanotechnology?', *Hyle: International Journal for Philosophy of Chemistry*, 10(2), 65-82.

Bensaude-Vincent, B.; Arribart, B.Y.; Sanchez, C.: 2002, 'Chemists at the School of Nature', *New Journal of Chemistry*, 26, 1-5.

Berger, R., C. Gerber, H.P. Lang, and J.K. Gimzewski. 1996. Micromechanics: a toolbox for femtoscale science: "Towards a laboratory on a tip." *Microelectronic Engineering*, 35: 373-9. (International Conference on Micro- and Nanofabrication, Glasgow, U.K., 22- 25 Sept. 1996).

Berkowitz, A.E., J.R. Mitchell, M.J. Carey, A.P. Young, S. Zhang, F.E. Spada, F.T. Parker, A. Hutten, and G. Thomas. 1992. *Phys. Rev. Lett.* 68: 3745.

Berndt, C.C., J. Karthikeyan, T. Chraska, and A.H. King. 1997. Plasma spray synthesis of nanozirconia powder. In *Proc. of the Joint NSF-NIST Conf. on Nanoparticles.*

Berube, D.M.: 2004, 'The Rhetoric of Nanotechnology', in: D. Baird, A. Nordmann & J. Schummer (eds.), *Discovering the Nanoscale*, Amsterdam: IOS Press, pp. 173-192.

Bimberg, D., N. Kirstaedter, N.N. Ledentsov, Zh.I. Alferov, P.S. Kopev, and V.M. Ustinov. 1998. InGaAs-GaAs quantum-dot lasers. *IEEE J. of Selected Topics in Quant. Electronics* 3: 196-205.

Binnig, G., Quate, C.F., and Gerber, Ch. (1986) Atomic force microscope. Phys. Rev. Lett. 56(9), 930-933

Bishop, A.R., and R.G. Nuzzo. 1996. Self-assembled monolayers: Recent developments and applications. *Current Opinion in Colloid & Interface Sci.* 1: 127-136.

Bishop, J. et al. 1990. *Surface modified drug nanoparticles.* U.S. Patent application. Docket 61894 Filed 9/17/90.

Blackman, M. 1994. *An evaluation of the Link Nanotechnology Program and the National Initiative on Nanotechnology*. Cambridge.

Bohn, R., T. Haubold, R Birringer, and H. Gleiter. 1991. *Scripta Metall. Mater*. 25: 811.

Bowes, C.L., A. Malek, and G.A. Ozin. 1996. *Chem. Vap. Deposition* 2: 97.

Braun, P.V., P. Osenar, and S.I. Stupp. 1996. *Nature*. 368: 2. Brinker, C.J. 1996. *Curr. Opin. Solid State Mater. Sci*. 1: 798. Chen, C-Y., S.L. Burkett, H.-X. Li, and M.E. Davis. 1993. *Microporous Mater*. 2: 27.

Braun, T.; Schubert, A. & Zsindely, S.: 1997, 'Nanoscience and nanotechnology on the balance', *Scientometrics*, 38, 321-325.

Brinker, C.J. 1996. *Curr. Opin. Solid State Mater. Sci*. 1: 798.

Brotzman, R. 1998. Nanoparticle Dispersions. In *R&D status and trends*, ed. Siegel et al. Bu, X., P. Feng, and G.D. Stucky. 1998. Large-cage zeolite structures with multidimensional 12-ring channels. *Science* 278: 2080-2085.

Brown, N.: 2004, 'Needle on the Real: Technoscience and Poetry at the Limits of Fabrication', in: N.K. Hayles (ed.), *Nanoculture: Implications of the New Technoscience*, Bristol, UK: Intellect Books, pp. 173-190.

Brus, L. 1996. Semiconductor colloids: Individual nanocrystals, opals and porous silicon. *Current Opinion in Colloid & Interface Science* 1: 197-201. *Chemical Engineering News*. 1997. Particulate matter health studies to be reanalyzed (August 18): 33.

Bueno, O.: 2004, 'The Drexler-Smalley Debated on Nanotechnology: Incommensurability at Work?', *Hyle: International Journal for Philosophy of Chemistry*, 10(2), 83-98.

Bueno, O.: 2004, 'Von Neumann, Self-Reproduction and the Constitution of Nanophenomena', in: D. Baird, A. Nordmann & J. Schummer (eds.), *Discovering the Nanoscale*, Amsterdam: IOS Press, pp. 101-115.

C. Mao, T. LaBean, J.H. Reif and N.C. Seeman, Logical Computation Using Algorithmic Self-Assembly of DNA Triple Crossover Molecules, Nature 407, 493-496 (2000).

C. Mao, W. Sun and N.C. Seeman, Construction of Borromean Rings from DNA, Nature 386, 137-138 (1997).

C. Mao, W. Sun and N.C. Seeman, Designed Two-Dimensional DNA Holliday Junction Arrays Visualized by Atomic Force Microscopy, Journal of the American Chemical Society 121, 5437-5443 (1999).

C. Mao, W. Sun, Z. Shen and N.C. Seeman, A DNA Nanomechanical Device Based on the B-Z Transition, Nature 397, 144-146 (1999).

C. Mao, W. Sun, Z. Shen, N.C. Seeman, "A nanomechanical device based on the B-Z transition of DNA," *Nature* 397(14 January 1999): 144-146.

C.D. Montemagno, G.D. Bachand, "Constructing nanomechanical devices powered by biomolecular motors," *Nanotechnology* 10(1999): 225-231; G.D. Bachand, C.D. Montemagno, "Constructing organic/inorganic NEMS devices powered by biomolecular motors," *Biomedical Microdevices* 2(2000): 179-184.

C.T. Kuo, P.J. Bostick, R.F. Irie, D.L. Morton, A.J. Conrad, D.S. Hoon, "Assessment of messenger RNA of beta 1—>4-N-acetylgalactosaminyl-transferase as a molecular marker for metastatic melanoma," *Clin. Cancer Res*. 4(February 1998): 411-418.

Calcote, H.F., and D.G. Keil. 1997. Combustion synthesis of silicon carbide powder. In *Proc. of the Joint NSF-NIST Conf. on Nanoparticles*.

Canguilhem, G.: 1952, *Machine et organisme' in La connaissance de la vie*, Hachette, Paris [quoted from the fourth edition Vrin, Paris, 1971].

Carroll, J.S.: 2001, 'Social Science Research Methods for Assessing Societal Implications of Nanotechnology', in: M.C. Roco and W.S. Bainbridge (eds.), *Societal Implications of Nanoscience and Nanotechnology*, Dordrecht: Kluwer, pp. 188-192.

Carsley, J.E., R. Shaik, W.W. Milligan, and E.C. Aifantis. 1997. In *Chemistry and physics of nanostructures and related non-equilibrium materials*. ed. E. Ma, B. Fultz, R. Shull, J. Morral, and P. Nash. Warrendale, PA: TMS.

Chance, B., Mueller, P., DeVault, D. & Powers, L. (1980) *Phys. Today* 33 (10), 32-38.

Chang, K. "Smaller Computer Chips Built Using DNA as Template." New York Times, November 21, 2003.

Chantrell. 1994. *J. Appl. Phys.* 76: 6811. Erb, U., G. Palumbo, R. Zugic, and K.T. Aust. 1996. In *Processing and properties of nanocrystalline materials*, ed. C. Suryanarayana, J. Singh, and F.H. Froes. Warrendale, PA: TMS. Gertsman, V.Y., M. Hoffman, H. Gleiter and R. Birringer. 1994. *Acta Metall. Mater.* 42: 3539-3544.

Charles B. Musgrave, Jason K. Perry, Ralph C. Merkle, William A. Goddard III, "Theoretical studies of a hydrogen abstraction tool for nanotechnology," *Nanotechnology* 2(1991): 187-195; http: //www.zyvex.com/nanotech/Habs/Habs.html

Charles B. Musgrave, Jason K. Perry, Ralph C. Merkle, William A. Goddard III, "Theoretical studies of a hydrogen abstraction tool for nanotechnology," *Nanotechnology* 2(1991): 187-195; http: //www.zyvex.com/nanotech/Habs/Habs.html

Chen, J. and N.C. Seeman. 1991. The synthesis from DNA of a molecule with the connectivity of a cube. *Nature* 350: 631-633.

Chianelli, R.R. 1998. Synthesis, fundamental properties and applications of nanocrystals, sheets, and fullerenes based on layered transition metal chalcogenides. In *R&D status and trends*, ed. Siegel et al.

Chianelli, R.R. 1998. Synthesis, fundamental properties and applications of nanocrystals, sheets, and fullerenes based on layered transition metal chalcogenides. In *R&D status and trends*, ed. Siegel et al.

Chianelli, R.R., M. Daage, and M.J. Ledoux. 1994. *Advances in Catalysis* 40: 177.

Chidsey, E.E.D., and R.W. Murray. 1986. *Science*, 231: 25.

Chopra, N.G., R.J. Luyken, K. Cherrey, H. Crespi, M.L. Cohen, S.G. Louie, and A. Zettl. 1995. *Science* 269: 966.

Claeson, T., and Likharev, K.," Single Electronics," *Scientific American,* June 1992, pp. 80-85.

Clasen, R. 1990. *Int. Journal of Glass and Science Technology* 63: 291. Crandall, B.C. and J. Lewis, (eds.). 1992. *Nanotechnology: Research and perspectives*. Cambridge: MIT Press.

Coenen, C.: 2004, 'Nanofuturismus: Anmerkungen zu seiner Relevanz, Analyse und Bewertung', *Technikfolgenabschätzung ' Theorie und Praxis*, 13 (2), 78-85.

Crandall, B.C. and Lewis, J., *Nanotechnology: Research and Perspectives*, MIT Press, Cambridge, 1992.

Crow, M. & Sarewitz, D.: 2001, 'Nanotechnology and Societal Transformation', in: M.C. Roco and W.S. Bainbridge (eds.), *Societal Implications of Nanoscience and Nanotechnology*, Dordrecht: Kluwer, pp. 45-54.

Czekai, D., et al. 1994. *Use of smaller milling media to prepare nanoparticulate dispersions*. U.S. Patent application. Docket 69802(02) Filed 2/25/94.

D.B. Dusenbery, "Minimum size limit for useful locomotion by free-swimming microbes," *Proc. Natl. Acad. Sci. USA* 94(30 September 1997): 10949-10954.

D.T. Dennis et al, "Tularemia as a biological weapon: medical and public health management," *J. Amer. Med. Assoc.* 285(6 June 2001): 2763-2773.

D.W. Brenner, S.B. Sinnott, J.A. Harrison, O.A. Shenderova, "Simulated engineering of nanostructures," *Nanotechnology* 7(1996): 161-167; http: //www.zyvex.com/nanotech/nano4/brennerPaper.pdf

Dai, H., E.W. Wong, Y.Z. Lu, S. Fan, and C.M. Leiber. 1995. *Nature* 375: 769. deHeer, W.A., W.S. Bacsa, A. Chatelain, T. Gerfin, R. Humphrey-Baker, L. Forro, and D. Ugarte. 1995. *Science* 268: 845.

de Souza e Silva, A.: 2004, 'The Invisible Imaginary: Museum Spaces, Hybrid Reality and Nanotechnology', in: N.K. Hayles (ed.), *Nanoculture: Implications of the New Technoscience*, Bristol, UK: Intellect Books, pp. 27-82.

Decker, M.; Fiedeler, U.; Fleischer, Th.: 2004, 'Ich sehe was, was Du nicht siehst... zur Definition von Nanotechnologie', *Technikfolgenabschätzung ' Theorie und Praxis*, 13 (2), 10-16.

Dresselhaus, M., and G. Dresselhaus. 1995. *Ann. Rev. Mat. Sci.* 25: 487.

Dresselhaus, M.S. 1998. Carbon-based nanostructures. In *R&D status and trends*, ed. Siegel et al.

Dresselhaus, M.S., G. Dresselhaus, and P. Eklund. 1996. *Science of fullerenes and carbon nanotubes*. San Diego: Academic Press. Gleiter, H. 1989. *Prog. Mater. Sci.* 33: 223.

Drexler K.E.: 1995, 'Introduction to nanotechnology', in: Krummenacker, M. & Lewis, J. (eds.), *Prospects in Nanotechology. Proceedings of the 1st general conference on nanotechnology: developments, applications, and opportunities, November 11-14, 1992, Palo-Alto*, John Wiley & Sons, New York, pp. 1-20.

Drexler, E., Phoenix, C.: 2004, 'Self-Replication in Nanotechnology: Feasible, Potentially Safe, and Unnecessary' (unpublished paper, in preparation).

Drexler, E.: 1986, *Engines of Creation: The Coming Era of Nanotechnology*, Anchor Books, New York (expanded edition with a new afterword, 1990).

Drexler, E.: 1992, *Nanosystems: Molecular Machinery, Manufacturing, and Computation*, John Wiley & Sons, New York.

Drexler, E.: 2003a, 'Open Letter to Richard Smalley', *Chemical & Engineering News*, 81, 38-39.

Drexler, E.: 2003b, 'Drexler Counters', *Chemical & Engineering News*, 81, 40-41.

Drexler, K.E.: 1981, 'Molecular engineering: An approach to the development of general capabilities for molecular manipulation', *Proceedings of the National Academy of Sciences*, 78, no. 9, chemistry section, 5275-78.

Drexler, K.E.: 1986, *Engines of Creation*, Anchor Books, New York.

Drexler, K.E.: 1992, *Nanosystems. Molecular machinery, manufacturing and computation*, John Wiley & Sons, New York.

Drexler, K.E.: 2001, 'Machine-Phase nanotechnology', *Scientific American*, (Sept.), 66-67.

Drexler, K.E.: 2003a, 'Open Letter', *Chemical & Engineering News*, 81, no. 48, 37-42.

Drexler, K.E.: 2003b, 'An open letter to Richard Smalley', April 16, [published on KurzweilAI.net].

Drexler, K.E.: 2004, 'Nanotechnology: From Feynman to Funding', *Bulletin of Science, Technology & Society*, 24, 21-27.

Dupuy, J.P. "Common Knowledge, Common Sense." Theory and Decision 27 (1989): 37-62.

Dupuy, J.P. "Two Temporalities, Two Rationalities: A New Look at Newcomb's Paradox." in Bourgine, P.&Walliser, B. (eds.), Economics and Cognitive Science. New York: Pergamon. 1992, 191-220.

Dupuy, J.P.: 2004, 'Complexity and Uncertainty: A Prudential Approach To Nanotechnology', in: European Commission (Community Health and Consumer Protection): *Nanotechnologies: A Preliminary Risk Analysis on the Basis of a Workshop, Brussels, 1-2 March 2004*, pp. 71-93.

Dupuy, J.P.: 2004, 'Pour une évaluation normative du programme nanotechnologique', *Annales des Mines*, (February 2004), 27-32.

E. Francis J. Ring, Barbara Phillips, eds., *Recent Advances in Medical Thermology*, Plenum Press, New York, 1984.

E. Winfree, F. Liu, L. A. Wenzler, and N.C. Seeman, Design and Self-Assembly of Two-Dimensional DNA Crystals, Nature 394, 539-544 (1998).

E. Winfree, X. Yang and N.C. Seeman, Universal Computation via Self-assembly of DNA: Some Theory and Experiments, In: DNA Based Computers II, ed.by L.F. Landweber and E.B. Baum, Am. Math. Soc., Providence, pp. 191-213 (1998).

E.M. Purcell, "Life at low Reynolds number," *Am. J. Physics* 45(January 1977): 3-11.

E.T. Foley, A.F. Kam, J.W. Lyding, P.H. Avouris, P. H. (1998), "Cryogenic UHV-STM study of hydrogen and deuterium desorption from Si(100)," *Phys. Rev. Lett.* 80(1998): 1336-1339.

Einsiedel, E.F.; Goldenberg, L.: 2004, 'Dwarfing the Social? Nanotechnology Lessons from the Biotechnology Front', *Bulletin of Science, Technology & Society*, 24, 28-33.

Ellsberg, D. "Risk, Ambiguity and the Savage Axioms." Quarterly Journal of Economics 75 (1961): 643-669.

Entingh, D., Dunn, A., Glassman, E., Wilson, J. E., Hogan, E. & Damstra, T. (1975) in *Handbook of Psychobiology*, eds. Gazzinga, M. S. & Blakemore, C. (Academic, New York), pp. 201-238.

Estermann, M., L.B. McCusker, C. Baerlocher, A. Merrouche, and H. Kessler. 1991. *Nature* 331: 698.

Esumi, A., A. Suzuki, N. Aihara, K. Uswi, and K. Torigoe. 1998. *Langmuir*, 14: 3157.

ETC: 2003a, *The Big Down: Atom Tech ' Technologies Converging at the Atomic Scale*. Winnipeg, Canada: Action Group on Erosion, Technology and Concentration [www.etcgroup.org/documents/TheBigDown.pdf].

ETC: 2003b, 'No Small Matter II: The Case for a Global Moratorium ' Size Matters!' *Occasional Paper Series* 7(1) [www.etcgroup.org/documents/Occ.Paper_Nanosafety.pdf].

Etzkowitz, H., 2001, Nano-Science and Society: Finding a Social Basis for Science Policy', in: M.C. Roco and W.S. Bainbridge (eds.), *Societal Implications of Nanoscience and Nanotechnology*, Dordrecht: Kluwer, pp. 121-128.

European Comission: *European Workshop on Social and Economic Research on Nanotechnologies and Nanosciences, Brussels, 14-15 April 2004* [www.stage-research.net/STAGE/PAGES/Nano.html].

European Commission (Community Health and Consumer Protection): 2004, *Nanotechnologies: A*

Preliminary Risk Analysis on theBasis of a Workshop, Brussels, 1-2 March 2004 [www.europa.eu.int/comm/health/ ph_risk/documents/ev_20040301_en.pdf].

European Commission: 2004, *Converging Technologies: Shaping the Future of European Societies (Report of the High Level Expert Group "Foresighting the New Technology Wave")*, Brussels: European Commission Research [http: //europa.eu.int/comm/research/conferences/2004/ntw/index_en.html].

Evans, D.F.; Wennerstrom, H.: 1999, *The Colloidal Domain: Where Physics, Chemistry, Biology and Technology Meet*, John Wiley & Sons, New York.

F. Buot, "Mesoscopic Physics and Nanoelectronics: Nanoscience and Nanotechnology," *Physics Reports*, pp.73-174, 1993.

F. Capasso and S. Datta, "Quantum Electron Devices," *Physics Today*, pp.74-82, February 1990.

F. Liu, H. Wang and N.C. Seeman, Short Extensions to Sticky Ends for DNA Nanotechnology and DNA-Based Computation, Nanobiology 4, 257-262 (1999).

F. Liu, R. Sha and N.C. Seeman, Modifying the Surface Features of Two-Dimensional DNA Crystals, Journal of the American Chemical Society. 121, 917-922 (1999).

F. Mathieu, S. Liao, C. Mao, J. Kopatsch, T. Wang, and N.C. Seeman, Six-Helix Bundles Designed from DNA, NanoLetters 5, 661-665 (2005).

Faraday, M.: 1857, 'Experimental relations of gold (and other metals) to light', *Philosophical Transactions of the Royal Society London*, 147, 145-181.

Fedor N. Dzegilenko, Deepak Srivastava, Subhash Saini, "Simulations of carbon nanotube tip assisted mechano-chemical reactions on a diamond surface," *Nanotechnology* 9(December 1998): 325-330.

Fedor N. Dzegilenko, Deepak Srivastava, Subhash Saini, "Simulations of carbon nanotube tip assisted mechano-chemical reactions on a diamond surface," *Nanotechnology* 9(December 1998): 325-330.

Fennema, O. R. (1973) in *Low-Temperature Preservation of Foods and Living Matter*, eds. Fennema, O. R., Powrie, W. D. & Marth, E. H. (Dekker, New York), pp. 476-503.

Feyerabend, P.: 1981, *Realism, Rationalism and Scientific Method* (Philosophical Papers, volume 1), Cambridge University Press, Cambridge.

Feynman R. "There's Plenty of Room At the Bottom." Talk given on at the annual meeting of the American Physical Society at the California Institute of Technology, 1959.

Feynman, R. (1961) in *Miniaturization*, ed. Gilbert, H. D. (Reinhold, New York), pp. 282-296.

Feynman, R., "There's Plenty of Room at the Bottom: An invitation to Enter a New Field of Physics," Talk at the Annual Meeting of the American Physical Society, 29 December 1959.

Feynman, R.: 1960, 'There's Plenty of Room at the Bottom', *Engineering and Science*, 23, 22-36.

Fielder, F.A.; Reynolds, G.H.: 1994, 'Legal Problems of Nanotechnology: An Overview', *Southern California Interdisciplinary Law Journal*, 3, 593-629.

Filler, M., C. Mui, G. Wang, C. Musgrave, and S. Bent, "Competition and Selectivity in the Reaction of Nitriles on Ge (100)-2x1," *Journal of the American Chemical Society* 125, 4928-4936 (2003).

Fleischer, Th.; Decker, M. & Fiedeler, U. (eds.): 2004, *Große Aufmerksamkeit für kleine Welten Nanotechnologie und ihre Folgen*, special issue of *Technikfolgenabschätzung ' Theorie und Praxis*, 13 (2), 5-85 [www.itas.fzk.de/tatup/042/inhalt.htm].

Fleming, D. "The Economics of Taking Care: An Evaluation of the Precautionary Principle." in Freestone, D.&Hey, E. (eds.), The Precautionary Principle and International Law. La Haye: Kluwer Law International, 1996.

Foerster, H.v.: 1962, 'Bio-Logic', in: E.E. Bernard & M.R. Kare (eds.), *Biological prototypes and synthetic systems*, Plenum Press, New York, vol. 1., pp. 1-12

Fogelberg, H. & Glimell, H.: 2003, 'Molecular Matters: In Search of the Real Stuff', in: H. Fogelberg & H. Glimell, *Bringing Visibility to the Invisible: Towards A Social Understanding of Nanotechnology*, Göteborg: Göteborg University, pp. 5-32 [www.sts.gu.se/publications/STS_report_6.pdf].

Fogelberg, H.: 2003, 'The Material Culture of Nanotechnology', in: H. Fogelberg & H. Glimell, *Bringing Visibility to the Invisible: Towards A Social Understanding of Nanotechnology*, Göteborg: Göteborg University, pp. 99-114.

Frazier, G., "An Ideology For Nanoelectronics," in *Concurrent Computations: Algorithms, Architecture, and Technology*, Plenum Press, New York, 1988.

Friedlander, S.K. 1993. Controlled synthesis of nanosized particles by aerosol processes. *Aerosol Sci. Technol.* 19: 527.

Friedlander, S.K. 1998. Synthesis of nanoparticles and their agglomerates: aerosol reactors. In *R&D status and trends*, ed. Siegel et al.

Friedlander, S.K. 1998. Synthesis of nanoparticles and their agglomerates: Aerosol reactors. In *R&D status and trends*, ed. Siegel et al.

G. Serrano & N.C. Seeman, Nanotecnologia basada en ADN, Revista de Química 19, 11-20 (2004), in Spanish.

Gallego-Juárez, J.A. (1989) Piezoelectric ceramics and ultrasonic transducers. J. Phys. E: Sci. Instrum. 22, 804-816

George D. Skidmore, Eric Parker, Matthew Ellis, Neil Sarkar, Ralph Merkle, "Exponential assembly," paper presented at the 8th Foresight Conference, Washington DC, November 2000; http: //www.foresight.org/Conferences/MNT8/Papers/Skidmore/index.html

George M. Whitesides, "The Once and Future Nanomachine," *Scientific American* 285(September 2001): 78-83.

Gleiter, H. 1989. *Prog. Mater. Sci.* 33: 223. Goddard, W.A. 1998. Nanoscale theory and simulation. In *R&D status and trends*, ed. Siegel et al.

Gleiter, H. 1990. *Progress in Materials Science* 33: 4. Guinier, A. 1938. *Nature* 142: 569; Preston, G.D. 1938. *Nature* 142: 570.

Glimell, H.: 2001, 'Challenging Limits' Excerpts from an Emerging Ethnography of Nano Physicists', in: H. Glimell & O. Johlin (eds.), *The Social Production of Technology: On the Everyday Life with Things*, Göteburg, SE: BAS Publisher, chapter 7, pp. 111-131 (reprinted in: H. Fogelberg & H. Glimell, *Bringing Visibility to the Invisible: Towards A Social Understanding of Nanotechnology*, Göteborg: Göteborg University, pp. 115-139.

Glimell, H.: 2001, 'Dynamics of the Emerging Field of Nanoscience', in: M.C. Roco and W.S. Bainbridge (eds.), *Societal Implications of Nanoscience and Nanotechnology*, Dordrecht: Kluwer, pp. 156-160.

Glimell, H.: 2003, 'A Nano Narrative: The Mircopolitics of a New Generic Enabling Technology', in: H. Fogelberg & H. Glimell, *Bringing Visibility to the Invisible: Towards A Social Understanding of Nanotechnology*, Göteborg: Göteborg University, pp. 55-77 [www.sts.gu.se/publications/STS_report_6.pdf].

Glimell, H.: 2003, 'Dynamics of the Emerging Field of Nanoscience', in: H. Fogelberg & H. Glimell, *Bringing Visibility to the Invisible: Towards A Social Understanding of Nanotechnology*, Göteborg: Göteborg University, pp. 79-85 [www.sts.gu.se/publications/STS_report_6.pdf].

Glimell, H.: 2004, 'Grand Visions and Lilliput Politics: Staging the Exploration of the 'Endless Frontier'', in: D. Baird, A. Nordmann & J. Schummer (eds.), *Discovering the Nanoscale*, Amsterdam: IOS Press, pp. 231-246.

Goddard, W.A. 1998. Nanoscale theory and simulation. In *R&D status and trends*, ed. R. Siegel et al.

Godsell, D.: 2003, *Living Machinery. Bionanotechnology: Lessons from Nature*, Wiley-Liss, New-York.

Goldhaber-Gordon, D., Montemerlo, M.S., Love, J.C., Opiteck, G.J., and Ellenbogen, J.C., "Overview of Nanoelectronic Devices," submitted to the *Proceedings of the IEEE*, February 1997. For more information, please send e-mail to nanotech@mitre.org.

Gorman, M.E.; Groves, J.F. & Shrager, J.: 2004, 'Societal Dimensions of Nanotechnology as a Trading Zone: Results from a Pilot Project', in: D. Baird, A. Nordmann & J. Schummer (eds.), *Discovering the Nanoscale*, Amsterdam: IOS Press, pp. 63-73.

Grinbaum, A. & Dupuy, J.P.: 2004, 'Living with Uncertainty: Toward the Ongoing Normative Assessment of Nanotechnology', *Techne: Research in Philosophy and Technology*, 8(3) (forthcoming).

Grinbaum, A.: 2004, 'La condition de l'homme moderne et les nanotechnologies', in: G. Nivat (ed.), *Les limites de l´humain. 39èmes Rencontres Internationales de Genève*, L´Age d´Homme, Genève, p. 141.

Grunwald, A.: 2004, 'Ethische Aspekte der Nanotechnologie. Eine Felderkundung', *Technikfolgenabschätzung ' Theorie und Praxis*, 13 (2), 71-78.

Guarnieri, F., M. Fliss, and C. Bancroft. 1996. Making DNA Add. *Science* 273: 220-223. Hanes, J., J.L. Cleland, and R. Langer. 1997.

Gund, P., Andose, J. D., Rhodes, J. B. & Smith, G. M. (1980) *Science* 208, 1425-1431.

Günther, B., A. Baalmann, and H. Weiss. 1990. *Mater. Res. Soc. Symp. Proc.* 195: 611-615.

Gupta, V.K. & Pangannaya, N.B.: 2000, 'Carbon nanotubes: bibliometric analysis of patents', *World Patent Information*, 22, 185-189.

Gutte, B., Dannigen, M. & Wittschieber, E. (1979) *Nature (London)* 281, 650-655.

H. Qiu, J.C. Dewan and N.C. Seeman, A DNA Decamer with a Sticky End: The Crystal Structure of d-CGACGATCGT. J. Mol. Biol. 267, 881-898 (1997).

H. Yan and N.C. Seeman, Edge-Sharing Motifs in DNA Nanotechnology. Journal of Supramolecular Chemistry 1, 229-237 (2003).

H. Yan, X. Zhang, Z. Shen and N.C. Seeman, A Robust DNA Mechanical Device Controlled by Hybridization Topology, Nature 415, 62-65 (2002).

Hadjipanayis, C.G. 1998. Nanostructured magnetic materials. In *R&D Status and Trends*, ed. Siegel et al.

Han, W., S. Fam, O. Li, and Y. Hu. 1998. Synthesis of gallium nitride nanorods through a carbon nanotube-confined reaction. *Science* 277: 1287-1289.

Hansson, S.O. "The False Promises of Risk Analysis." Ratio 6 (1993): 16-26.

Hansson, S.O.: 2004, 'Great Uncertainty about Small Things', *Techne: Research in Philosophy and Technology*, 8(3) (forthcoming).

Harold J. Morowitz, "A Model of Reproduction," *American Scientist* 47(1959): 261-263.

Haruta, M. 1997. *Catalyst surveys of Japan* 1: 61 and references therein.

Hayles, N.K. (ed.): 2004, *Nanoculture: Implications of the New Technoscience*, Bristol, UK: Intellect Books.

Hayles, N.K.: 2004, 'Connecting the Quantum Dots: Nanotechscience and Culture', in: N.K. Hayles (ed.), *Nanoculture: Implications of the New Technoscience*, Bristol, UK: Intellect Books, pp. 11-26.

Held, R., T. Heinzel, A.P. Studerus, K. Ensslin, and M. Holland. 1997. Semiconductor quantum point contact fabricated by lithography with an atomic force microscope. *Appl. Phys. Lett.* 71: 2689-91.

Hellemans, A. 1998. X-rays find new ways to shine. *Science* 277: 1214-15.

Hemerén, G.: 2004, 'Nano Ethics Primer', in: European Commission (Community Health and Consumer Protection): *Nanotechnologies: A Preliminary Risk Analysis on theBasis of a Workshop, Brussels, 1-2 March 2004* (www.europa.eu.int/comm/health/ ph_risk/documents/ ev_20040301_en.pdf), pp. 95-102.

Hennig, J.: 2004, 'Changes in the Design of Scanning Tunneling Microscopic Images from 1980 to 1990', *Techne: Research in Philosophy and Technology*, 8(3) (forthcoming).

Hessenbruch, A.: 2004, 'Nanotechnology and the Negotiation of Novelty', in: D. Baird, A. Nordmann & J. Schummer (eds.), *Discovering the Nanoscale*, Amsterdam: IOS Press, pp. 135-144.

Hietpas, P.B., S.D. Gilman, R.A. Lee, M.R. Wood, N. Winograd, and A.G. Ewing. 1996. Development of votammetric methods, capillary electrophoresis and tof sims imaging for constituent analysis of single cells. In *Nanofabrication and biosystems*, ed.

Higgins, R.J. 1997. An economical process for manufacturing of nano-sized powders based on microemulsion-mediated synthesis. In *Proc. of the Joint NSF-NIST Conf. on Nanoparticles.*

Hiruma, K., M. Yazawa, T. Katsoyama, K. Ogawa, K. Haraguchi, M. Koguchi, and H. Kakibayashi. 1995. *J. Appl. Phys.* 77(2): 476.

Hoch et al. Ho, S.V., P.W. Sheridan, and E. Krupetsky. 1996. Supported polymeric liquid membranes for removing organics from aqueous solutions.

Hoch, H.C., L.W. Jelinski, and H.G. Craighead, eds. 1996. *Nanofabrication and biosystems*. New York: Cambridge University Press.

Hoh, J.H. and Hansma, P.K. (1992) Atomic force microscopy for high-resolution imaging in cell biology. Trends Cell Bio. 2, 208-213

Homer Jacobson, "On Models of Reproduction," *American Scientist* 46(1958): 255-284.

Hopwood, J., and S. Mann. 1997. *Chem. Mater.* 9: 1819.

Howland, H.: 1962, Structural, hydraulic and economic aspects of leaf venation and shape', in: E.E. Bernard & M.R. Kare (eds.), *Biological prototypes and synthetic systems*, Plenum Press, New York, vol. 1, pp. 183-192.

Hoyningen-Huene, P.: 1993, *Reconstructing Scientific Revolutions: Thomas S. Kuhn's Philosophy of Science* (trans. by A. Levin), University of Chicago Press, Chicago.

Huang, J.Y., Y.K. Wu, and H.Q. Ye. 1996. *Acta Mater*. 44: 1211. Inoue, A. 1997. Private communication. Inturi, R.B., and Z. Szklavska-Smialowska. 1992. *Corrosion* 48: 398. Karch, J., R. Birringer, and H. Gleiter. 1987. *Nature* 330: 556.

Hubbell, J.A., and R. Langer. 1995. Tissue engineering. *Chem. Eng. News* (March 13): 42- 54.

Hullmann, A. & Meyer, M.: 2003, 'Publications and Patents in Nanotechnology. An overview of previous studies and the state of the art', *Scientometrics*, 58 (3) 507-527.

Huo, Q., D.I. Margolese, U. Ciesla, P. Feng, T.E. Gier, P. Sieger, R. Leon, P.M. Petroff, F. Schuth, and G.D. Stucky. 1994. *Nature* 368: 317.

Imae, Y., and T. Atsumi. 1989. T. Na+-driven bacterial flagellar motors: A mini-review. *J. Bioenergetics and Biomembranes* 21: 705-716.

J. Chen and N.C. Seeman, The Synthesis from DNA of a Molecule with the Connectivity of a Cube, Nature 350, 631-633 (1991).

J. D. Le, Y. Pinto, N.C. Seeman, K. Musier-Forsyth, T. A. Taton and R.A. Kiehl, Self-Assembly of Nanoelectronic Component Arrays by In Situ Hybridization to 2D DNA Scaffolding, NanoLetters 4, 2343-2347 (2004).

J. Qi, X. Li, X.P. Yang and N.C. Seeman, The Ligation of Triangles Built from Bulged Three-Arm DNA Branched Junctions, Journal of the American Chemical Society, 118, 6121-6130 (1996).

J. Storrs Hall, "Architectural considerations for self-replicating manufacturing systems," *Nanotechnology* 10(1999): 323-330; http: //www.foresight.org/Conferences/MNT6/Papers/ Hall/index.html

J. Storrs Hall, "Architectural considerations for self-replicating manufacturing systems," *Nanotechnology* 10(1999): 323-330; http: //www.foresight.org/Conferences/MNT6/Papers/ Hall/index.html

J. Travis, "Tick, tock, enzyme rewinds cellular clock," *Science News* 153(17 January 1998): 37.

J.H. Chen, N.R. Kallenbach and N.C. Seeman, A Specific Quadrilateral Synthesized from DNA Branched Junctions. Journal of the American Chemical Society 111, 6402-6407 (1989).

J.H. Reif, T.H. LaBean & N.C. Seeman, Challenges and Applications for Self-Assembled DNA Nanostructures, Sixth International Workshop on DNA-Based Computers, DNA 2000, Leiden, The Netherlands, (June, 2000) ed. A. Condon, G. Rozenberg. Springer-Verlag, Berlin Heidelberg, Lecture Notes in Computer Science 2054, 173-198, (2001).

J.K. Gimzewski, C. Joachim, R.R. Schlittler, V. Langlais, H. Tang, I. Johannsen, "Rotation of a Single Molecule Within a Supramolecular Bearing," *Science* 281(24 July 1998): 531-533.

J.W. Lyding, K. Hess, G.C. Abeln, D.S. Thompson, J.S. Moore, M.C. Hersam, E.T. Foley, J. Lee, Z. Chen, S.T. Hwang, H. Choi, P.H. Avouris, I.C. Kizilyalli, "UHV-STM nanofabrication and hydrogen/deuterium desorption from silicon surfaces: implications før CMOS technology," *Appl. Surf. Sci.* 130(1998): 221-230.

Jaworek, T., D. Deher, G. Wegner, R.H. Wieringa, and A.J. Schouten. 1998. Electromechanical properties of an ultrathin layer of directionally aligned helical polypeptides. *Science* 279: 57-60.

Jena, P., S.N. Khanna, and B.K. Rao. 1996. In *Science and technology of atomically engineered materials*, ed. P. Jena. River Edge, NJ: World Scientific.

Jena, P., S.N. Khanna, and B.K. Rao. 1996. In *Science and technology of atomically engineered materials*, ed. P. Jena. River Edge, NJ: World Scientific.

Johansson, M.: 2003, 'Plenty of room at the bottom: Towards an anthropology of nanoscience', *Anthropology Today*, 19 (No. 6), 3-6.

John Cumings, A. Zettl, "Low-friction nanoscale linear bearing realized from multiwall carbon nanotubes," *Science* 289(28 July 2000): 602-604.

John von Neumann, *Theory of Self-Reproducing Automata*, A.W. Burks, ed., University of Illinois Press, Urbana IL, 1966.

Johnson, A.: 2004, 'The End of Pure Science: Science Policy from Bayh-Dole to the NNI', in: D. Baird, A. Nordmann & J. Schummer (eds.), *Discovering the Nanoscale*, Amsterdam: IOS Press, pp. 217-230.

Jonas, H. The Imperative of Responsibility. In Search of an Ethics for the Technological Age. Chicago: University of Chicago Press, 1985.

Jones, C.W., and W.J. Koros. 1994. *Carbon* 32: 1419.

Joshua Lederberg, "Infectious History," *Science* 288(14 April 2000): 287-293.

Jounet, C., W.K. Maser, P. Bernier, A. Loiseau, M. Lamy de la Chapelle, S. Lefrant, P. Deriard, R. Lee, and J.E. Fischer. 1997. *Nature* 388: 756.

Jounet, C., W.K. Maser, P. Bernier, A. Loiseau, M. Lamy de la Chapelle, S. Lefrant, P. Deniard, R. Lee, and J.E. Fischer. 1997. *Nature* 388: 756.

Joy, B.: 2000, 'Why the Future Doesn't Need US', *Wired*, 8, 1-11.

Junno, T., S.-B. Carlsson, H. Xu, L. Montelius, and L. Samuelson. 1998. Fabrication of quantum devices by angstrom-level manipulation of nanoparticles with an atomic force microscope. *Appl. Phys. Lett.* 72: 548-550.

Justus, B.L., R.J. Tonucci, and A.D. Berry. 1992. *Appl. Phys. Lett.* 61: 3151.

K. E. Drexler, "Molecular engineering: An approach to the development of general capabilities for molecular manipulation." *Proc. Nat. Acad. Sci.,* 78: 5275-5258 (1981).

K. E. Drexler, *Engines of Creation*, Anchor/Doubleday (New York, 1986).

K. Eric Drexler, "Molecular engineering: an approach to the development of general capabilities for molecular manipulation," *Proc. Natl. Acad. Sci. (USA)* 78(1981): 5275-5278.

K. Eric Drexler, "Molecular nanomachines: Physical principles and implementation strategies," *Annu. Rev. Biophys. Biomol. Struct.* 23(1994): 377-405.

K. Eric Drexler, "Nanomachinery: Atomically precise gears and bearings," *IEEE Micro Robots and Teleoperators Workshop*, Hyannis, Cape Cod, November 1987.

K. Eric Drexler, David Forrest, Robert A. Freitas Jr., J. Storrs Hall, Neil Jacobstein, Ralph Merkle, Christine Peterson, "On Physics, Fundamentals, and Nanorobots: A Rebuttal to Smalley's Assertion that Self-Replicating Mechanical Nanorobots Are Simply Not Possible"; http: //www.imm.org/SciAmDebate2/smalley.html

K. Eric Drexler, *Nanosystems: Molecular Machinery, Manufacturing, and Computation*, John Wiley & Sons, New York, 1992; http: //www.zyvex.com/nanotech/nanosystems.html; http: //www.foresight.org/NanoRev/Bookstore.html#anchor1025139

K. Eric Drexler, *Nanosystems: Molecular Machinery, Manufacturing, and Computation*, John Wiley & Sons, New York, 1992.

K.E. Drexler, *Nanosystems: Molecular Machinery, Manufacturing, and Computation,* Wiley, New York, 1992.

Kang, J. and C. Musgrave, "A Quantum Chemical Study of the Self-Directed Growth Mechanism of Styrene and Propylene Molecular Nanowires on the Silicon (100) 2x1 Surface," *Journal of Chemical Physics* 116, 9907-9913 (2002).

Kang, J. and C. Musgrave, "The Mechanism of Atomic Layer Deposition of SiO 2 on the Silicon (100)-2x1 Surface Using SiC1 4 and H 2 O as Precursors," *Journal of Applied Physics* 91, 3408-3414 (2002).

Karak, N., and S. Maiti. 1997. Dendritic polymers: A class of novel material. *J. Polym. Mater.*14: 105.

Karger, J., and D.M. Ruthven. 1992. *Diffusion in zeolites*. New York: J. Wiley.

Karplus, M. & Weaver, D. L. (1976) *Nature (London)* 260, 404-406.

Kasuga, T., M. Hiramatsu, A. Hoson, T. Sekino, and K. Niihara. 1998. *Langmuir* 14: 3160.

Katari, J.E.B., V.L. Colvin, and A.P. Alivisatos. 1994. *J. Phys. Chem*. 98: 4109.

Ke, M., S.A. Hackney, W.W. Milligan, and E.C. Aifantis. 1995. *Nanostructured Mater*. 5: 689.

Kear, B., and G. Skandan. 1998. Nanostructural bulk materials: Synthesis, processing, properties and performance. *In R&D status and trends*, ed.

Kear, B.H., R.K. Sadangi, and S.C. Liao. 1997. Synthesis of WC/Co/diamond nanocomposites. In *Proc. of the Joint NSF-NIST Conf. on Nanoparticles*.

Keiper, A.: 2003, 'The Nanotechnology Revolution', *The New Atlantis*, 2, 17-34.

Keller, D.J. and Chih-Chung, C. (1992) Imaging steep, high structures by scanning force microscopy with electron beam deposited tips. Surf. Sci. 268, 333-339

Khushf, G.: 2004, 'A Hierarchical Architecture for Nano-scale Science and Technology: Taking Stock of the Claims About Science Made By Advocates of NBIC Convergence', in: D. Baird, A. Nordmann & J. Schummer (eds.), *Discovering the Nanoscale*, Amsterdam: IOS Press, pp. 21-33.

Khushf, G.: 2004, 'Systems Theory and the Ethics of Human Enhancement: A Framework for NBIC Convergence', *Annals of the New York Academy of Sciences*, 1013, 124-149.

Kirschvink, J.L., A. Koyayashi-Kirschvink, and B.J. Woodford. 1992. Magnetite biomineralization in the human brain. *Proc. Nat'l. Acad. Sci.* USA 89: 7683-7687.

Kishida, M., T. Fujita, K. Umakoshi, J. Ishiyama, H. Nagata, and K. Wakabayashi. 1995. *Chem. Commun*. 763.

Klein, J.D., et al. 1993. *Chem. Mater*. 5: 902. Koch, C.C. 1989. Materials synthesis by mechanical alloying. *Annual Review of Mater. Sci.* 19: 121-143.

Knight, F.H. Risk, Uncertainty and Profit. London School of Economics and Political Science, London, 1933.

Koch, C.C. 1998. Bulk behavior. In *R&D status and trends*, ed. Siegel et al. 32 *Evelyn L. Hu and David T. Shaw.*

Kortan, A.R., R. Hull, R.L. Opila, M.G. Bawendi, M.L. Steigerwald, P.J. Carroll, and L.E. Brus. 1990. *J. Am. Chem. Soc.* 112: 1327.

Kourilsky, Ph.&Viney, G. Le Principe de précaution. Report to the Prime Minister, Paris, Éditions Odile Jacob, 2000.

Krätschmer, W., L.D. Lamb, K. Fostiropoulos, and D.R. Huffman. 1990. *Nature* 347: 354.

Krejchi, M.T., E.D.T. Atkins, A.J. Waddon, M.J. Fournier, T.L. Mason, and D.A. Tirrell. 1994. Chemical sequence control of beta-sheet assembly in macromolecular crystals of periodic proteins. *Science* 265: 1427-1432.

Krejchi, M.T., S.J. Cooper, Y. Deguchi, E.D.T. Atkins, M.J. Fournier, T.L. Mason, and D.A. Tirrell. 1997. Crystal structures of chain-folded antiparallel beta-sheet assemblies from sequence-designed periodic polypeptides. *Macromolecules* 30: 5012-5024.

Kresge, C.T., M.E. Leonowicz, W.J. Roth, J.C. Vartuli, and J.S. Beck. 1992. *Nature* 359: 710.

Kresge, C.T., M.E. Leonowicz, W.J. Roth, J.C. Vartuli, and J.S. Beck. 1992. *Nature* 359: 710. Kupperman, A., S. Nadimi, S. Oliver, G. Ozin, J. Garcés, and M. Olken. 1993. *Nature* 365: 239.

Kroes, P. & Meijers, A. "The Dual Nature of Technical Artifacts—Presentation of a New Research Programme." Techné 6, 2 (2002): 4 - 8.

Krstic, V., U. Erb, and G. Palumbo. 1993. *Scripta Metall. et Mater.* 29: 1501.

Krumhansl, J. A. & Pao, Y. H. (1979) *Phys. Today* 32 (11), 25-32.

Kuhn, T.: 1970, *The Structure of Scientific Revolutions* (second edition), University of Chicago Press, Chicago.

Kumar, A., and G.M. Whitesides. 1993. Features of gold having micrometer to centimeter dimensions can be formed through a combination of staming with an elastomeric stamp and an alkanethiol ink followed by chemical etching. *App. Phys. Lett.* 63: 2002-2004.

Kupperman, A., S. Nadimi, S. Oliver, G. Ozin, J. Garcés, and M. Olken. 1993. *Nature* 365: 239.

Kurzweil, R.: 1998, *The Age of Spiritual Machines, How We Will Live, Work and Think in the New Age of Intelligent Machines*, Phoenix, New York.

Kuusi, O.; Meyer, M.: 2002, 'Technological generalizations and leitbilder ' the anticipation of technological opportunities', *Technological Forecasting & Social Change*, 69, 625-639.

Kyprianidou-Leodidou, T., W. Caseri, and V. Suter. 1994. *J. Phys. Chem.* 98: 8992. Kung, H.H., and E.I. Ko, 1996. *Chem. Eng. J.* 64: 203.

L. Chong, B. van Steensel, D. Broccoli, H. Erdjument-Bromage, J. Hanish, P. Tempst, T. de Lange, "A Human Telomeric Protein," *Science* 270(8 December 1995): 1663-1667; D. Broccoli, A. Smogorzewska, L. Chong, T. de Lange, "Human telomeres contain two distinct Myb-related proteins, TRF1 and TRF2," *Nature Genet.* 17(October 1997): 231-235.

L. J. Perry and R. Wetzel, "Disulfide bond engineered into T4 lysozyme: stabilization of the protein toward thermal inactivation." *Science,* 226: 555-557 (1984).

L. Regan and W. F. DeGrado, "Characterization of a helical protein designed from first principles." *Science,* 241: 976-978 (1988).

L. Zhu, P.S. Lukeman, J. Canary & N.C. Seeman, Nylon/DNA: Single-Stranded DNA with Covalently Stitched Nylon Lining, J. Am. Chem. Soc. 125, 10178-10179 (2003).

L.J. Lauhon, W. Ho, "Inducing and observing the abstraction of a single hydrogen atom in bimolecular reaction with a scanning tunneling microscope," *J. Phys. Chem.* 105(2000): 3987-3992.

L.S. Penrose, "Self-Reproducing Machines," *Scientific American* 200(June 1959): 105-114.

Landon, B.: 2004, 'Less is More: Much Less is Much More: The Insistent Allure of Nanotechnology Narratives in Science Fiction', in: N.K. Hayles (ed.), *Nanoculture: Implications of the New Technoscience*, Bristol, UK: Intellect Books, pp. 131-146.

Larrère, R.: 2002, 'Agriculture: artificialisation ou manipulation de la nature?', *Cosmpolitiques*, 1 (June), 158-173.

Laszlo, P.: 2004, 'Is There Life After Partington?', *Hyle: International Journal for Philosophy of Chemistry*, 10(2), 169-178.

Laudan, L.: 1984, *Science and Values: The Aims of Science and their Role in Scientific Debate*, University of California Press, Berkeley.

Lehn, J.M., 2004: 'Une chimie supramoléculaire foisonnante', *La lettre de l'Académie des sciences*, 10, 12-13.

Lehn, J.M.: 1995, *Supramolecular Chemistry*, VCH, Weinheim.

Lenhard, J.: 2004, 'Nanoscience and the Janus-Faced Character of Simulations', in: D. Baird, A. Nordmann & J. Schummer (eds.), *Discovering the Nanoscale*, Amsterdam: IOS Press, pp. 93-100.

Lent, C.S., Tougaw, P.D., Porod, W., Bernstein, G.H., "Quantum Cellular Automata," *Nanotechnology*, Vol. 4, p. 49, 1993.

Leonard, D., M. Krishnamurthy, C.M. Reaves, S.P. Denbaars, and P.M. Petroff. 1993. Direct formation of quantum-sized dots from uniform coherent islands of InGaAs on GaAs surfaces. *Appl. Phys. Lett.* 63: 3203-5.

Lewak, S.E.: 2004, 'What's the Buzz? Tell Me What's A-Happening: Wonder, Nanotechnology, and Alice's Adventures in Wonderland', in: N.K. Hayles (ed.), *Nanoculture: Implications of the New Technoscience*, Bristol, UK: Intellect Books, pp. 201-.

Liang, G., Z. Li, and E. Wang. 1996. *J. Mater. Sci.* Makino, A., A. Inoue, T. Hatanai, and T. Bitoh. 1997. *Materials Science Forum* 235-238: 723.

Lide, D.R., ed. 1993-1994. *CRC Handbook of Chemistry and Physics*, 74th ed.

Lin-Easton, P.C.: 2001, 'It's Time for Environmentalists to Think Small ' Real Small: A Call for the Involvement of Environmental Lawyers in Developing Precautionary Policies Molecular Nanotechnology', *Georgetown International Law Review*, 14, 106-134.

López, J.: 2004, 'Bridging the Gaps: Science Fiction in Nanotechnology', *Hyle: International Journal for Philosophy of Chemistry*, 10(2), 129-152.

Lösch, A.: 2004, 'Nanomedicine and Space: Discursive Orders of Mediating Innovations', in: D. Baird, A. Nordmann & J. Schummer (eds.), *Discovering the Nanoscale*, Amsterdam: IOS Press, pp. 193-202.

Luce, R.D. & Raiffa, H. Games and Decisions, Wiley, New York, 1957.

Lukens, R.: 2004, 'Historic Nanotechnology', *Chemical Heritage*, 22, 29.

M.C. Hersam, G.C. Abeln, J.W. Lyding, "An approach for efficiently locating and electrically contacting nanostructures fabricated via UHV-STM lithography on Si(100)," *Microelectronic Engineering* 47(1999): 235-.

M.L. Petrillo, C.J. Newton, R.P. Cunningham, R.-I. Ma, N.R. Kallenbach and N.C. Seeman. The Ligation and Flexibility of 4-Arm DNA Junctions. Biopolymers, 27, 1337-1352 (1988).

Majetich, S.A. and A.C. Canter. 1993. *J. Phys. Chem.* 97: 8727.

Mao, C., W. Sun, and N.C. Seeman. 1997. Construction of Borromean rings from DNA. *Nature* 386: 137-138.

Mark A. Reed, James M. Tour, "Computing with molecules," *Scientific American* 282(June 2000): 86-93.

Marshall, K.: 2004, 'Atomizing the Risk Technology', in: N.K. Hayles (ed.), *Nanoculture: Implications of the New Technoscience*, Bristol, UK: Intellect Books, pp. 147-160.

Martin, T.P., N. Malinowski, U. Zimmerman, U. Naher, and H. Schaber. 1993. *J. Chem. Phys.* 99: 4210.

Martin, T.P., U. Naher, H. Schaber, U. Zimmerman. 1993. *Phys. Rev. Lett.* 70: 3079.

Mayer, S.: 2002, 'From genetic modification to nanotechnology: the dangers of 'sound science", in: T. Gilland (ed.), *Science: Can We Trust the Experts?*, London: Hodder and Stoughton, pp. 1-15.

McCammon, J. A., Gelin, B. R. & Karplus, M. (1977) *Nature (London)* 267, 585-590.

McConnell, H.M. 1996. Light-addressable potentiometric sensor: Applications to drug discovery. In *Nanofabrication and biosystems*, ed. Hoch et al.

McCulloch, W.S.: 1962, 'The imitation of one form of life by another – Biomimesis', in: E.E. Bernard & M.R. Kare (eds.), *Biological prototypes and synthetic systems*, Plenum Press, New York, vol. 1., p. 393-97.

Mehl, R.F., and R.W. Cahn. 1983. Historical development. In *Physical metallurgy*. North Holland.

Milligan, W.W., S.A. Hackney, M. Ke, and E.C. Aifantis. 1993. *Nanostructured Materials* 2: 267.

Mehta, M.D.: 2002, 'Nanoscience and Nanotechnology: Assessing the Nature of Innovation in These Fields', *Bulletin of Science, Technology & Society*, 22(4), 269-273.

Mehta, M.D.: 2004, 'From Biotechnology to Nanotechnology: What Can We Learn from Earlier Technologies?, *Bulletin of Science, Technology & Society*, 24, 34-39.

Meyer, G. and Amer, N.M. (1988) Novel optical approach to atomic force microscopy. Appl. Phys. Lett. 53(12), 1045-1047

Meyer, G. and Amer, N.M. (1990) Simultaneous measurement of lateral and normal forces with an optical-beam-deflection atomic force microscope. Appl. Phys. Lett. 57(20), 2089-2091

Meyer, M. & Kuusi, O.: 2004, 'Nanotechnology: Generalizations in an Interdisciplinary Field of Science and Technology', *Hyle: International Journal for Philosophy of Chemistry*, 10(2), 153-168.

Meyer, M.: 2000a, 'Does science push technology? Patents citing scientific literature', *Research Policy*, 29, 409-434.

Meyer, M.: 2000b, 'Patent citations in a novel field of technology: What can they tell about interactions of emerging communities of science and technology?', *Scientometrics*, 48, 151-178.

Meyer, M.: 2001a, 'Patent citations in a novel field of technology: An exploration of nano-science and nano-technology', *Scientometrics*, 51, 163-183.

Meyer, M.: 2001b, 'Socio-economic Research on Nanoscale Science and Technology: A European Overview and Illustration', in: M.C. Roco and W.S. Bainbridge (eds.), *Societal Implications of Nanoscience and Nanotechnology*, Dordrecht: Kluwer, pp. 217-241.

Meyer, M.; Persson, O.: 1998, 'Nanotechnology ' interdisciplinarity, patterns of collaboration and differences in application', *Scientometrics*, 47, 195'205.

Michael E.J. Holwill, Peter Satir, "II.4 Generation of Propulsive Forces by Cilia and Flagella," in J. Bereiter-Hahn, O.R. Anderson, W.-E. Reif, eds., *Cytomechanics: The Mechanical Basis of Cell Form and Structure*, Springer-Verlag, Berlin, 1987, pp. 120-130.

Michael Page, Donald W. Brenner, "Hydrogen abstraction from a diamond surface: *Ab initio* quantum chemical study using constrained isobutane as a model," *J. Am. Chem. Soc.* 113(1991): 3270-3274.

Milburn, C.: 2002, 'Nanotechnology in the age of post-human engineering: science fiction as science', *Configurations*, 10, 261-295 [reprinted in: N.K. Hayles (ed.), *Nanoculture: Implications of the New Technoscience*, Bristol, UK: Intellect Books, 2004, pp. 109-130].

Milburn, C.: 2004, 'Nano/Splatter: Disintegrating the Postbiological Body', *New Literary History*, 35 (in print).

Minsky, M.: 1995, 'Virtual Molecular Reality', in: Krummenacker, M. & Lewis, J. (eds.), *Prospects in Nanotechology. Proceedings of the 1st general conference on nanotechnology: developments, applications, and opportunities, November 11-14, 1992, Palo-Alto*, John Wiley & Sons, New York, pp. 187-205.

Mishra, R.S., and A.K. Mukherjee. 1997. Oral presentation at TMS meeting, Indianapolis, Indiana, 16-18 September 1997, to be published in proceedings of Symp. "Mechanical Behavior of Bulk Nano-Materials."

Mishra, R.S., R.Z. Valiev, and A.K. Mukherjee. 1997. *Nanostructured Materials* 9: 473. Morris, D.G., and M.A. Morris. 1991. *Acta Metall. Mater.* 39: 1763-1779.

Mnyusiwalla, A.; Abdallah, S.D.; Singer, P.A.: 2003, 'Mind the gap: science and ethics in nanotechnology', *Nanotechnology*, 14, R9-R13.

Mody, C.C.M.: 2004, 'How Probe Microscopists Became Nanotechnologists', in: D. Baird, A. Nordmann & J. Schummer (eds.), *Discovering the Nanoscale*, Amsterdam: IOS Press, pp. 119-133.

Mody, C.C.M.: 2004, 'Instruments in Training: The Growth of American Probe Microscopy in the 1980s', in: D. Kaiser (ed.), *Pedagogy and the Practice of Science: Producing Physical Scientists, 1800-2000*, Cambridge, MA: MIT Press (forthcoming).

Mody, C.C.M.: 2004, 'Small, but Determined: Technological Determinism in Nanoscience', *Hyle: International Journal for Philosophy of Chemistry*, 10(2), 99-128.

Mody, C.C.M.: 2004, *Crafting the Tools of Knowledge: The Invention, Spread, and Commercialization of Probe Microscopy, 1960-2000*, Ph.D. dissertation, Cornell University.

Montemerlo, M.S., Love, J.C., Opiteck, G.J., Goldhaber, D. J., and Ellenbogen, J.C., "Technologies and Designs for Electronic Nanocomputers," MITRE Technical Report 96W0000044, The MITRE Corporation, McLean, VA, July 1996. For more information, send e-mail to nanotech@mitre.org.

Moor, J.H. & Weckert, J.: 2004, 'Nanoethics: Assessing the Nanoscale From an Ethical Point of View', in: D. Baird, A. Nordmann & J. Schummer (eds.), *Discovering the Nanoscale*, Amsterdam: IOS Press, pp. 301-310.

Moore, J.C., H.M. Jin, O. Kuchner, and F.H. Arnold. 1997. Strategies for the in vitro evolution of protein function: Enzyme evolution by random recombination of improved sequences. *J. Mol. Biol.* 272: 336-347.

Morris, D.G., and M.A. Morris. 1997. *Materials Science Forum* 235-238: 861. Nagpal, P., and I. Baker. 1990. *Scripta Metall. Mater.* 24: 2381.

Moshe Sipper, "Fifty Years of Research on Self-Replication: An Overview," *Artificial Life 4*(Summer 1998): 237-257, in "Special Issue on Self-Replication." See also: Moshe Sipper, "The Artificial Self-Replication Page," http: //lslwww.epfl.ch/~moshes/selfrep/

Mui, C. and C. Musgrave, "The Hydroxylation and Oxidation of the Ge(100)-2x1 Surface by Hydrogen Peroxide and Water," *Langmuir*, 20, 7604-7609 (2004).

Mui, C., S. Bent and C. Musgrave, "A Quantum Chemistry Based Statistical Mechanical Model of Hydrogen Deposition from Si(100)-2x1, Ge(100)-2x1, and SiGe Alloy Surfaces," *Journal of Physical Chemistry B*, 108, 18243-18253, (2004).

Mui, C., Y. Widjaja, J. Kang, and C. Musgrave, "Surface Reaction Mechanisms for Atomic Layer Deposition of Silicon Nitride," *Surface Science,* 557, 159-170, (2004).

Munn Sanchez, E.: 2004, 'The Expert's Role in Nanoscience and Technology', in: D. Baird, A. Nordmann & J. Schummer (eds.), *Discovering the Nanoscale*, Amsterdam: IOS Press, pp. 257-266.

N. Hanai, K. Nakamura, K. Shitara, "Recombinant antibodies against gangliosides expressed on tumor cells," *Cancer Chemother. Pharmacol.* 46(2000): S13-S17.

N. Jonoska, P. Sa-Ardyen and N.C. Seeman, Computation by Self-Assembly of DNA Graphs, Journal of Genetic Programming and Evolvable Machines 4, 123-137 (2003).

N. Jonoska, S. Liao, & N.C. Seeman, Transducers with programmable input by DNA Self-assembly, Aspects of Molecular Computing, Lecture Notes in Computer Science 2340, Springer-Verlag, Berlin, 219-240 (2004).

N. Koumura, R.W. Zijlstra, R.A. van Delden, N. Harada, B.L. Feringa, "Light-driven monodirectional molecular rotor," *Nature* 401(9 September 1999): 152-155.

N.C. Seeman & A.M. Belcher, Emulating Biology: Nanotechnology from the Bottom Up, Proceedings of the National Academy of Sciences (USA) 99 (supp. 2), 6451-6455 (2002).

N.C. Seeman and N.R. Kallenbach, Design of Immobile Nucleic Acid Junctions. Biophysical Journal 44, 201-209 (1983).

N.C. Seeman and N.R. Kallenbach, Nucleic Acid Junctions, The Tensors of Life? In: Nucleic Acids: The Vectors of Life, ed. by B. Pullman and J. Jortner, D. Reidel, Dordrecht, 183-200 (1983).

N.C. Seeman and P.S. Lukeman, Nucleic Acid Nanostructures, Reports on Progress in Physics 68, 237-270 (2005).

N.C. Seeman, At the Crossroads of Chemistry, Biology and Materials: Structural DNA Nanotechnology, Chemistry & Biology 10, 1151-1159 (2003).

N.C. Seeman, B. Ding, S. Liao, T. Wang, W.B. Sherman, P.E. Constantinou, J. Kopatsch, C. Mao, R. Sha, F. Liu, H. Yan & P.S. Lukeman, Experiments in Structural DNA Nanotechnology: Arrays and Devices, Proc. SPIE; Nanofabrication: Technologies, Devices and Applications 5592, 71-81 (2005).

N.C. Seeman, Biochemistry and Structural DNA Nanotechnology: An Evolving Symbiotic Relationship, Biochemistry 42, 7259-7269 (2003).

N.C. Seeman, Branched DNA: A 3-D Structural Design System, Clinical Chemistry 39, 722-724 (1993).

N.C. Seeman, De Novo Design of Sequences for Nucleic Acid Structure Engineering, Journal of Biomolecular Structure and Dynamics 8, 573-581 (1990).

N.C. Seeman, Design and Engineering of Nucleic Acid Nanoscale Assemblies, Current Opinion in Structural Biology, 6, 519-526 (1996).

N.C. Seeman, Directing the Structure of Matter through Nanotechnology, Proceedings of IEEE International Joint Symposia on Intelligence and Systems IEEE Computer Society, Los Alamitos, CA, 146-150 (1998).

N.C. Seeman, DNA Components for Molecular Architecture. Accounts of Chemical Research 30, 357-363 (1997).

N.C. Seeman, DNA Engineering and its Application to Nanotechnology, Trends in Biotechnology 17, 437-443 (1999).

N.C. Seeman, DNA in a Material World, Nature 421, 33-37 (2003).

N.C. Seeman, DNA Nanostructures for Mechanics and Computing: Nonlinear Thinking With Life's Central Molecule. NanoBiotechnology. Editors, Chad Mirkin and Christof Niemeyer, Wiley-VCH Verlag GmbH & Co., KgaA, Weinheim, Chapter 20, pp 308-318 (2004).

N.C. Seeman, DNA Nanostructures for Mechanics and Computing: Nonlinear Thinking With Life's Central Molecule. NanoBiotechnology. Editors, Chad Mirkin and Christof Niemeyer, Wiley-VCH Verlag GmbH & Co., Weinheim, Chapter 20, pp 308-318 (2004).

N.C. Seeman, DNA Nanotechnology, *Encyclopedia of Supramolecular Chemistry*, Marcel Dekker, New York, 475-483 (2004).

N.C. Seeman, DNA Nanotechnology, In: WTEC Workshop Report on R & D Status and Trends in Nanoparticles, Nanostructured Materials, and Nanodevices in the United States, ed. by R.W. Siegel, E. Hu and M.C. Roco, International Technology Research Institute,Baltimore (1998).

N.C. Seeman, DNA Nanotechnology: From Topological Control to Structural Control, in Pattern Formation in Biology, Vision and Dynamics, ed. by A.Carbone, M.Gromov, P.Pruzinkiewicz, World Scientific Publishing Company, Singapore, 271-309 (2000).

N.C. Seeman, DNA Nanotechnology: Life's Central Performer in a New Role, Biological Physics Newsletter 2 (1) 2-6 (2002).

N.C. Seeman, DNA Nanotechnology: Novel DNA Constructions. Annual Review of Biophysics and Biomolecular Structure 27, 225-248 (1998).

N.C. Seeman, DNA Nicks and Nodes and Nanotechnology, NanoLetters 1, 22-26 (2001).

N.C. Seeman, DNA Structural Engineering Utilizing Immobile Junctions, Current Opinion in Structural Biology 1, 653-661 (1991).

N.C. Seeman, DNA: Beyond the Double Helix, Macromolecular Symposia 201, 237-244 (2003).

N.C. Seeman, From Genes to Machines: DNA Nanomechanical Devices, Trends in Biochemical Sciences 30, 119-235 (2005).

N.C. Seeman, H. Wang, B. Liu, J. Qi, X. Li, X. Yang, F. Liu, W. Sun, Z. Shen, Y. Wang, R. Sha, C. Mao, S. Zhang, T.-J. Fu, S. M. Du, J.E. Mueller, Y. Zhang and J. Chen, The Perils of Polynucleotides: The Experimental Gap Between the Design and Assembly of Unusual DNA Structures, In: DNA Based Computers II, ed.by L.F. Landweber and E.B. Baum, Am. Math. Soc., Providence, pp. 215-233 (1998).

N.C. Seeman, In the Nick of Space: Generalized Nucleic Acid Complementarity and the Development of DNA Nanotechnology, Synlett 2000, 1536-1548 (2000).

N.C. Seeman, It Started with Watson and Crick, But it Sure Didn't End There: Pitfalls and Possibilities beyond the Classic Double Helix, Natural Computing 1, 53-84 (2002).

N.C. Seeman, J. Qi, X. Li, X. Yang, N.B. Leontis, B. Liu, Y.Zhang, S.M. Du and J. Chen, The Control of DNA Structure: From Topological Modules to Geometrical Modules. In: Modular Chemistry, ed. by J. Michl, Kluwer, 95-104 (1997).

N.C. Seeman, Key Experimental Approaches in DNA Nanotechnology, Current Protocols in Nucleic Acid Chemistry, Unit 12.1, John Wiley & Sons, New York (2002).

N.C. Seeman, Macromolecular Design, Nucleic Acid Junctions and Crystal Formation. Journal of Biomolecular Structure and Dynamics 3, 11-34 (1985).

N.C. Seeman, Nanoengineering with DNA, Biomolecular Materials: Materials Research Society Symposium Proceedings 292, 123-134 (1993).

N.C. Seeman, Nanoscale Assembly and Manipulation of Branched DNA: A Biological Starting Point for Nanotechnology, NANOCON Proceedings, ed. by J. Lewis and J.L. Quel, NANOCON, P.O. Box 40176, Bellevue, WA 98004, pp. 101-123; transcript of oral presentation, pp. 30-36 (1989).

N.C. Seeman, Nucleic Acid Junctions and Lattices. Journal of Theoretical Biology 99, 237-247 (1982).

N.C. Seeman, Nucleic Acid Junctions: Building Blocks for Genetic Engineering in Three Dimensions. In: Biomolecular Stereodynamics, ed. by R.H. Sarma, Adenine Press, New York (1981), pp. 269-277.

N.C. Seeman, Nucleic Acid Nanostructures and Topology. Angewandte Chemie. 110, 3408-3428 (1998); Angewandte Chemie International Edition 37, 3220-3238 (1998).

N.C. Seeman, Structural Control and Engineering of Nucleic Acids, In: Concepts in Protein Engineering and Design, P. Wrede and G. Schneider, eds., Walter-de-Gruyter, Berlin, 319-343 (1994).

N.C. Seeman, Structural DNA Nanotechnology: A New Organizing Principle for Advanced Nanomaterials, Materials Today 6 (7), 24-29 (2003).

N.C. Seeman, Synthetic DNA Topology, Molecular Catenanes, Rotaxanes and Knots, ed. by J.-P. Sauvage and C. Dietrich-Buchecker, Wiley-VCH, Weinheim pp. 323-356 (1999).

N.C. Seeman. Nanotechnology and the Double Helix, Scientific American., 290, (6) 64-75 (2004).

N.C. Seeman. Structural DNA Nanotechnology: An Overview. Methods in Molecular Biology 303: Bionanotechnology Protocols, Editors, Sandra J. Rosenthal and David W. Wright, Humana Press, Totowa, NJ, pp. 143-166 (2005).

N.C.Seeman, J. Chen, Y. Zhang, T.-J. Fu, X. Li, X. Yang, Y. Wang, J. Qi, B. Liu and F. Liu, Control of Structure and Topology in DNA Nanotechnology. In: Molecular Nanotechnology: Biological Approaches and Novel Applications, ed. by S.A. Minden, IBC Libraries, Southborough, MA, Chapter 2.2 (31 pages) (1997).

N.R. Kallenbach, R.-I. Ma and N.C. Seeman, An Immobile Nucleic Acid Junction Constructed from Oligonucleotides. Nature 305, 829-831 (1983).

Nieman, G.W., J.R. Weertman, and R.W. Siegel. 1991a. *Mater. Res. Soc. Symp. Proc.* 206: 581-586.

Nomura, M. & Held, W. (1974) in *Ribosomes,* eds. Nomura, M., Tissiers, A. & Lengyel, P. (Cold Spring Harbor Laboratory, Cold Spring Harbor, NY), pp. 193-203.

Nordmann, A. (rapp.) Converging Technologies—Shaping the Future of European Societies, European Commission report, 2004.

Nordmann, A.: 2003, 'Shaping the World Atom by Atom: Eine nanowissenschaftliche WeltBildanalyse', in: A. Grunwald (ed.), *Technikgestaltung zwischen Wunsch und Wirklichkeit*, Berlin: Springer, pp. 191-199.

Nordmann, A.: 2004, 'Molecular Disjunctions: Staking Claims at the Nanoscale', in: D. Baird, A. Nordmann & J. Schummer (eds.), *Discovering the Nanoscale*, Amsterdam: IOS Press, pp. 51-62.

Nordmann, A.: 2004, 'Nanotechnology: Convergence and Integration', Presentation at *EuroNanoForum, Trieste, December 10, 2003* (Proceedings in preparation).

Nordmann, A.: 2004, 'Nanotechnology's WorldView: New Space for Old Cosmologies', *IEEE Technology and Society Magazine*, 23 (forthcoming).

Nordmann, A.: 2004, 'Social Imagination for Nanotechnology', in: European Commission (Community Health and Consumer Protection): *Nanotechnologies: A Preliminary Risk Analysis on the Basis of a Workshop, Brussels, 1-2 March 2004*, pp. 111-113 [www.europa.eu.int/comm/health/ph_risk/documents/ev_20040301_en.pdf].

Nordmann, A.: 2004, 'Was ist TechnoWissenschaft? ' Zum Wandel der Wissenschaftskultur am Beispiel von Nanoforschung und Bionik', in: T. Rossmann & C. Tropea (eds.), *Bionik ' Neue Forschungsergebnisse aus Natur-, Ingenieur- und Geisteswissenschaften*, Berlin: Springer, 2004 (forthcoming)

P. Livingston, "Ganglioside vaccines with emphasis on GM2," *Semin. Oncol.* 25(December 1998): 636-645.

P. Sa-Ardyen, N. Jonoska and N.C. Seeman, Self-Assembling DNA Graphs, DNA-Based Computers VIII, LNCS 2568, Springer-Verlag, Berlin, 1-9 (2003).

P. Sa-Ardyen, N. Jonoska and N.C. Seeman, The Construction of Graphs Whose Edges are DNA Helix Axes, J. Am. Chem. Soc., 126, 6648-6657 (2004).

P.S. Lukeman, A. Mittal, & N.C. Seeman, Two Dimensional PNA/DNA Arrays: Estimating the Helicity of Unusual Nucleic Acid Polymers, Chemical Communications 2004, 1694-1695 (2004).

Parker, J.C., et al. 1995. U.S. Patent 5,460,701. POST (Parliamentary Office of Science and Technology). 1995.

Paschen, H.; Coenen, C.; Fleischer, T.; Grünwald, R.; Oertel, D. & Revermann, C.: 2004, *Nanotechnologie: Forschung, Entwicklung, Anwendung*, Berlin: Springer (*Nanotechnologie, TAB-Arbeitsbericht 92*, Berlin: Büro für Technikfolgen-Abschätzung beim Deutschen Bundestag).

Paul A. Schulte, Frederica P. Perera, eds., *Molecular Epidemiology: Principles and Practices*, Academic Press, San Diego, CA, 1993.

Philip G. Collins, Michael S. Arnold, Phaedon Avouris, "Engineering carbon nanotubes and nanotube circuits using electrical breakdown," *Science* 292(27 April 2001): 706-709.

Philip G. Collins, Phaedon Avouris, "Nanotubes for electronics," *Scientific American* 283(December 2000): 62-69.

Phoenix, C.; Drexler, E.: 2004, 'Safe Exponential Manufacturing', *Nanotechnology*, 15, 869-72.

Pitt, J.C.: 2004, 'The Epistemology of the Very Small', in: D. Baird, A. Nordmann & J. Schummer (eds.), *Discovering the Nanoscale*, Amsterdam: IOS Press, pp. 157-163.

Pressman, J.: 2004, 'Nano Narrative: A Parable from Electronic Literature', in: N.K. Hayles (ed.), *Nanoculture: Implications of the New Technoscience*, Bristol, UK: Intellect Books, pp. 191-200.

Prigogine, I., and S. Rice. 1988. *Advances in chemical physics*, Vol. 70, Parts 1 & 2. New York: J. Wiley.

Putman, C.A.J., De Grooth, B.G., Van Hulst, N.F., and Greve, J. (1992) A detailed analysis of the optical beam deflection technique for use in atomic force microscopy. J. App. Phys. 72(1), 6-12

R. Sha, F. Liu and N.C. Seeman, Atomic Force Measurement of the Interdomain Angle in Symmetric Holliday Junctions, Biochemistry 41, 5950-5955 (2002).

R. Sha, F. Liu, D.P. Millar N.C. Seeman, Atomic Force Microscopy of Parallel DNA Branched Junction Arrays, Chemistry & Biology 7, 743-751 (2000).

R. Sha, F. Liu, M. Bruist and N.C. Seeman, Parallel Helical Domains in DNA Branched Junctions Containing 5', 5' and 3', 3' Linkages, Biochemistry 38, 2832-2841 (1999).

R. T. Bate, "Nanoelectronics," *Nanotechnology*, Vol. 1, pp. 1-7, 1990.

R.G. Postier, "Antibiotic-resistant organism infection," *Am. Surg.* 66(February 2000): 112-116.

R.I. Ma, N.R. Kallenbach, R.D. Sheardy, M.L. Petrillo and N.C. Seeman, 3-Arm Nucleic Acid Junctions Are Flexible. Nucleic Acids Research 14, 9745-9753 (1986).

R.J. Jackson, A.J. Ramsay, C.D. Christensen, S. Beaton, D.F. Hall, I.A. Ramshaw, "Expression of mouse interleukin-4 by a recombinant ectromelia virus suppresses cytolytic lymphocyte responses and overcomes genetic resistance to mousepox," *J. Virol.* 75(February 2001): 1205-1210.

R.N. Jones, "Resistance patterns among nosocomial pathogens: trends over the past few years," *Chest* 119(February 2001): 397S-404S.

Rademann, K., B. Kaiser, U. Even, F. Hensel. 1987. *Phys. Rev. Lett.* 59: 2319.

Ralph C. Merkle, "A new family of six degree of freedom positional devices," *Nanotechnology* 8(1997): 47-52; http: //www.zyvex.com/nanotech/6dof.html

Ralph C. Merkle, "A Proof About Molecular Bearings," *Nanotechnology* 4(1993): 86-90; http: // www.zyvex.com/nanotech/bearingProof.html

Ralph C. Merkle, "A proposed 'metabolism' for a hydrocarbon assembler," *Nanotechnology* 8(1997): 149-162; http: //www.zyvex.com/nanotech/hydroCarbonMetabolism.html

Ralph C. Merkle, "Binding sites for use in a simple assembler," *Nanotechnology* 8(1997): 23-28; http: //www.zyvex.com/nanotech/bindingSites.html

Ralph C. Merkle, "Casing an assembler," *Nanotechnology* 10(1999): 315-322; http: //www.zyvex.com/ nanotech/casing

Ralph C. Merkle, "Convergent assembly," *Nanotechnology* 8(1997): 18-22; http: //www.zyvex.com/ nanotech/convergent.html

Ralph C. Merkle, "Design considerations for an assembler," *Nanotechnology* 7(1996): 210-215; http: //www.zyvex.com/nanotech/nano4/merklePaper.html

Ralph C. Merkle, "Design-ahead for nanotechnology," in Markus Krummenacker, James Lewis, eds., *Prospects in Nanotechnology: Toward Molecular Manufacturing*, John Wiley & Sons, New York, 1995, pp. 23-52.

Ralph C. Merkle, "Molecular building blocks and development strategies for molecular nanotechnology," *Nanotechnology* 11(2000): 89-99; http: //www.zyvex.com/nanotech/mbb/mbb.html

Ralph C. Merkle, "Self replicating systems and molecular manufacturing," *J. Brit. Interplanet. Soc.* 45(1992): 407-413; http: //www.zyvex.com/nanotech/selfRepJBIS.html

Ramanan, V.R. 1998. Nanocrystalline soft magnetic alloys for applications in electrical and electronic devices. In *R&D Status and Trends*, ed. Siegel et al.

Rao, C.N.R., B.C. Satishkumar, and A. Govindaraj. 1997. *Chem. Commun*. 1581.

Rao, M.B., and S. Sircar. 1993. *Gas Separation and Purification* 7: 279.

Rechard, R.P. "Historical Relationship Between Performance Assessment for Radioactive Waste Disposal and Other Types of Risk Assessment." Risk Analysis 19, 5 (1999): 763-807.

Reed, M.A., "Quantum Dots," *Scientific American,* January 1993, pp. 118-123.

Richard A. Laing, "Automation introspection," *J. Comp. System Sci.* 13(1976): 172-183.

Richard A. Laing, "Automaton models of reproduction by self-inspection," *J. Theoret. Biol.* 66(1977): 437-456.

Richard A. Laing, "Some alternative reproductive strategies in artificial molecular machines," *J. Theoret. Biol.* 54(1975): 63-84.

Richard E. Smalley, "Of chemistry, love, and nanobots," *Scientific American* 285(September 2001): 76-77.

Richard P. Feynman, "There's Plenty of Room at the Bottom," *Engineering and Science* (California Institute of Technology), February 1960, pp. 22-36. Reprinted in B.C. Crandall, James Lewis, eds., *Nanotechnology: Research and Perspectives*, MIT Press, 1992. pp. 347-363, and in D.H. Gilbert, ed., *Miniaturization*, Reinhold, New York, 1961, pp. 282-296; http: // www.zyvex.com/nanotech/feynman.html

Rietman, E.A.: 2001, 'Drexler hypothesis of a universal assembler is supported not by theoretical arguments alone but by existence proof in the form of biological life', in: *Molecular Engineering of Nanosystems*, Springer, New York & Berlin.

Rip, A., Misa, Th. J.,&Schot, J. W. (eds.). Managing Technology in Society. The Approach of Constructive Technology Assessment. London: Pinter Publishers, 1995.

Robert A. Freitas Jr., "Clottocytes: Artificial Mechanical Platelets," Foresight Update No. 41, 30 June 2000, pp. 9-11; http: //www.imm.org/Reports/Rep018.html

Robert A. Freitas Jr., "Exploratory design in medical nanotechnology: A mechanical artificial red cell," *Artificial Cells, Blood Substitutes, and Immobil. Biotech.* 26(1998): 411-430; http: // www.foresight.org/Nanomedicine/Respirocytes.html

Robert A. Freitas Jr., "Microbivores: Artificial Mechanical Phagocytes using Digest and Discharge Protocol," Zyvex preprint, March 2001; http: //www.zyvex.com/Publications/articles/ Microbivores.html. See also: Robert A. Freitas Jr., "Microbivores: Artificial Mechanical Phagocytes," Foresight Update No. 44, 31 March 2001, pp. 11-13; http: //www.imm.org/ Reports/Rep025.html.

Robert A. Freitas Jr., "Nanodentistry," *J. Amer. Dent. Assoc.* 131(November 2000): 1559-1566.

Robert A. Freitas Jr., "Some Limits to Global Ecophagy by Biovorous Nanoreplicators, with Public Policy Recommendations," Zyvex preprint, April 2000; http: //www.foresight.org/NanoRev/ Ecophagy.html

Robert A. Freitas Jr., *Nanomedicine, Volume I: Basic Capabilities*, Landes Bioscience, Georgetown, TX, 1999; http: //www.nanomedicine.com/

Robert A. Freitas Jr., William P. Gilbreath, eds., *Advanced Automation for Space Missions, NASA Conference Publication CP-2255*, 1982; http: //www.islandone.org/MMSG/aasm/

Robert F. Service, "Is nanotechnology dangerous?" *Science* 290(24 November 2000): 1526-1527.

Roberts, J.A.: 2004, 'Deciding the Future of Nanotechnologies: Legal Perspectives on Issues of Democracy and Technology', in: D. Baird, A. Nordmann & J. Schummer (eds.), *Discovering the Nanoscale*, Amsterdam: IOS Press, pp. 247-255.

Robinson, C.: 2004, 'Images in NanoScience/Technology', in: D. Baird, A. Nordmann & J. Schummer (eds.), *Discovering the Nanoscale*, Amsterdam: IOS Press, pp. 165-169.

Robison, W.L.: 2004, 'Nano-Ethics', in: D. Baird, A. Nordmann & J. Schummer (eds.), *Discovering the Nanoscale*, Amsterdam: IOS Press, pp. 285-300.

Robot Factory, http: //www.fanuc.co.jp/eannai/servo.htm; citation courtesy of Tihamer Toth-Fejel, 2000.

Roco, M.&Bainbridge, W. (eds.). Converging Technologies for Improving Human Performances, National Science Foundation report, 2002.

Roco, M.C. & Bainbridge, W.S. (eds.): 2001, *Societal implications of nanoscience and nanotechnology*, (Proceedings of a workshop organized by the National Science Foundation, September 28-29, 2000), Kluwer: Dordrecht [available online at http: //itri.loyola.edu/nano/societalimpact/ nanosi.pdf]

Roco, M.C. & Tomellini, R. (eds.): 2002, *Nanotechnology: Revolutionary Opportunities and Societal Implications, Workshop, Lecce (Italy), 31 January - 1 February 2002*, Luxemburg: Office for Official Publications of the European Communities, [200 pp.]

Roco, M.C.; Bainbridge, W.S. (eds.): 2002, *Converging Technologies for Improving Human Performance: Nanotechnology, Biotechnology, Information Technology and the Cognitive Science*, Arlington, VA: National Science Foundation.

Rod Logic and Thermal Noise in the Mechanical Nanocomputer, by K. Eric Drexler, in *Proceedings of the 3rd International Symposium on Molecular Electronic Devices*, F.L. Carter, R.E. Siatkowski, H. Wohltjen, eds., North-Holland 1988.

Rofagha, R., R. Langer, A.M. El-Sherik, U. Erb, G. Palumbo, and K.T. Aust. 1991. *Scr. Metall. Mater*. 25: 2867.

Rohlfing, E.A., D.M. Cox, and A. Kaldor. 1984. *J. Chem. Phys*. 81: 3846.

Romanov, A.E., V.I. Vladimirov. 1992. In *Dislocations in solids*, ed. F.R.N. Nabarro, Vol. 9. Amsterdam: North-Holland. Salishekev, G.A., O.R. Valiakhmetov, V.A. Valitov, and S.K. Mukhtarov. 1994. *Materials Science Forum*. 170-172: 121.

Rosen, A. 1998. A periodic table in three dimensions: A sightseeing tour in the nanometer world. In *Advances in quantum chemistry*. In press.

Rosenberg, N. "Why Technology Forecasts Often Fail." Futurist July/August 1995, 16- 21.

Ruthven, D.M., S. Farooq, K.S. Knaebel. 1994. *Pressure swing adsorption*. New York: VCH Publishers.

Ruthven, D.M., S. Farooq, K.S. Knaebel. 1994. *Pressure swing adsorption*. New York: VCH Publishers. Sayari, A. 1996. *Chem. Mater*. 8: 1840.

S. Liao and N.C. Seeman, Translation of DNA Signals into Polymer Assembly Instructions, Science 306, 2072-2074 (2004).

S. P. Ho and W. F. DeGrado, "Design of a 4-Helix bundle protein: Synthesis of peptides which self-associate into a helical protein." *J. Am. Chem. Soc.*, 109: 6751-6758 (1987).

S. Sasaki, I. Karube, "The development of microfabricated biocatalytic fuel cells," *Trends Biotechnol.* 17(February 1999): 50-52.

S. Xiao, F. Liu, A. Rosen, J.F. Hainfeld, N.C. Seeman, K.M. Musier-Forsyth & R.A. Kiehl, Self-Assembly of Nanoparticle Arrays by DNA Scaffolding, J. Nanoparticle Research 4, 313-317 (2002).

S. Zhang, C. Cordon-Cardo, H.S. Zhang, V.E. Reuter, S. Adluri, W.B. Hamilton, K.O. Lloyd, P.O. Livingston, "Selection of tumor antigens as targets for immune attack using immunohistochemistry: I. Focus on gangliosides," *Int. J. Cancer* 73(26 September 1997): 42-49.

S.B. Levy, "The future of antibiotics: facing antibiotic resistance," *Clin. Microbiol. Infect.* 6(2000): 101-106 (Suppl. 3).

S.P. Walch, W.A. Goddard III, R.C. Merkle, "Theoretical studies of reactions on diamond surfaces," Fifth Foresight Conference on Molecular Nanotechnology, 1997; http: //www.foresight.org/Conferences/MNT05/Abstracts/Walcabst.html

Sanders, P.G., J.A. Eastman, and J.R. Weertman. 1996. In *Processing and properties of nanocrystalline materials*, ed. Suryanarayana et al.

Sanders, P.G., M. Rittner, E. Kiedaisch, J.R. Weertman, H. Kung, and Y.C. Lu. 1997. *Nanostructured Mater*. 9: 433.

Sankey, H.: 1994, *The Incommensurability Thesis*, Avebury, Aldershot.

Sarewitz, D.; Woodhouse, E.: 2003, 'Small is Powerful', in: A. Lightman, D. Sarewitz & Chr. Desser, (eds.), *Living with the Genie: Essays on Technology and the Quest for Human Mastery*, Washington, DC: Island Press, pp. 63-83.

Sarikaya, M.; Aksay, I. (eds.), 1995, *Biomimetics. Design and Processing of Materials*, AIP Press, Woodbury, New York.

Scheraga, H. A. (1978) in *Versatilty of Proteins*, ed. Li, C. H. (Academic, New York), pp. 119-132.

Schiefsky, M.J.: forthcoming, 'Art and Nature in Ancient Mechanics', in: W.R. Newman & B. Bensaude-Vincent (eds.), *The Artificial and the Natural: An Ancient Debate and its Modern Descendants*, MIT Press, Cambridge, MA.

Schiemann, G.: 2004, 'Dissolution of the Nature-Technology Dichotomy? Perspectives on Nanotechnology from an Everyday Understanding of Nature', in: D. Baird, A. Nordmann & J. Schummer (eds.), *Discovering the Nanoscale*, Amsterdam: IOS Press, pp. 209-213.

Schmidt, J.C.: 2004, 'Unbounded Technologies: Working Through Technological Reductionism of Nanotechnology', in: D. Baird, A. Nordmann & J. Schummer (eds.), *Discovering the Nanoscale*, Amsterdam: IOS Press, pp. 35-50.

Schuler, E.: 2004, 'Perception of Risks and Nanotechnology', in: D. Baird, A. Nordmann & J. Schummer (eds.), *Discovering the Nanoscale*, Amsterdam: IOS Press, pp. 279-284.

Schultz, L., J. Wecker, and E. Hellstern. 1987. *J. Appl. Phys.* 61: 3583.

Schultz, L., K. Schnitzke, and J. Wecker. 1989. *J. Magn. Mater.* 80: 115.

Schummer, J.: 2001, 'Ethics of Chemical Synthesis', *Hyle: International Journal for Philosophy of Chemistry*, 7, 103-124 [online: www.hyle.org/journal/issues/7/schummer.htm].

Schummer, J.: 2003, 'Aesthetics of Chemical Products: Materials, Molecules, and Molecular Models', *Hyle: International Journal for Philosophy of Chemistry*, 9, 73-104 [online: www.hyle.org/journal/issues/9-1/schummer.htm].

Schummer, J.: 2003, 'The Notion of Nature in Chemistry', *Studies in History and Philosophy of Science*, 34 (2003), 705-736.

Schummer, J.: 2004, 'Interdisciplinary Issues in Nanoscale Research', in: D. Baird, A. Nordmann & J. Schummer (eds.), *Discovering the Nanoscale*, Amsterdam: IOS Press, pp. 9-20.

Schummer, J.: 2004, 'Multidisciplinarity, Interdisciplinarity, and Patterns of Research Collaboration in Nanoscience and Nanotechnology', *Scientometrics*, 59, 425-465.

Schummer, J.: 2004, 'Naturverhältnisse in der modernen Wirkstoff-Forschung', in: K. Kornwachs (ed.), *Technik ' System ' Verantwortung*, Münster: LIT, pp. 629-638.

Schummer, J.: 2004, 'Why do Chemists Perform Experiments?', in: D. Sobczynska, P. Zeidler, E. Zielonacka-Lis (eds.), *Chemistry in the Philosophical Melting Pot*, Peter Lang (Frankfurt/ M.), 2004, pp. 395-410.

Schummer, J.: 2005, 'Reading Nano: The Public Interest in Nanotechnology as Reflected in Book Purchase Patterns", *Public Understanding of Science*, 14 (2) (forthcoming).

Schummer, J.; Baird, D. (eds.): 2004-5, *Nanotech Challenges*, joint special issue of *Hyle: International Journal for Philosophy of Chemistry & Techne: Research in Philosophy and Technology* (forthcoming).

Schwarz, A.E.: 2004, 'Shrinking the 'Ecological Footprint' with NanoTechnoScience?', in: D. Baird, A. Nordmann & J. Schummer (eds.), *Discovering the Nanoscale*, Amsterdam: IOS Press, pp. 203-208.

Schwarz, R.B. 1998. Storage of hydrogen in powders with nanosized crystalline domains. In *R&D status and trends*, ed. Siegel et al.

Senosiain, J., C. Musgrave and D. Golden, "Temperature and Pressure Dependence of the Reaction of OH and CO: Master Equation Modeling on a High Level Potential Energy Surface," *International Journal of Chemical Kinetics* 35, 464-474 (2003).

Sha, R., Zhang, X., Liao, S., Constantinou, P.E., Ding, B., Wang, T., Garibotti, A.V., Zhong, H., Israel, L.B., Wang, X., Wu, G., Chakraborty, B., Chen, J., Zhang, Y., Mao, C., Yan, H., Kopatsch, J., Zheng, J., Lukeman, P.S., Sherman, W.B. and Seeman, N.C., Motifs and Methods in Structural DNA Nanotechnology, Proc. Intl. Conf. Nanomaterials, NANO 2005, July 13-15, 2005, Mepco Schlenk Engineering College, Srivakasi, India, V. Rajendran, ed., pp. 3-10 (2005).

Shen, T.D., C.C. Koch, T.Y. Tsui, and G.M. Pharr. 1995. *J. Mater. Res.* 10: 2892.

Shull, R.D. 1998. NIST activities in nanotechnology. In *R&D Status and Trends*, ed.

Shull, R.D., R.D. McMichael, and J.J. Ritter. 1993. *Nanostructured Mater.* 2: 205.

Siegel et al. Ogunnaike, B and W. Ray. 1994. *Process dynamics, modeling and control*. Oxford University Press, pp 5-21; 1033-48.

Siegel et al. Siegel, R.W. 1997. *Materials Science Forum* 235-238: 851.

Siegel R.W. and G.E. Fougere. 1994. In *Nanophase materials*, ed. G.C. Hadjipanayis and R.W. Siegel. Netherlands: Kluwer Acad. Publ.

Siegel, H.: 1980. 'Objectivity, Rationality, Incommensurability, and More', *British Journal for the Philosophy of Science*, 31, 359-384.

Siegel, R.W., E. Hu, and M.C. Roco, eds. 1998. *R&D status and trends in nanoparticles, nanostructured materials, and nanodevices in the United States*. Baltimore: Loyola College, International Technology Research Institute. NTIS #PB98-117914.

Smalley, R.: 2001, 'Of Chemistry, Love and Nanobots', *Scientific American*, 285, 76-77.

Smalley, R.: 2003a, 'Smalley Responds', *Chemical & Engineering News*, 81, 39-40.

Smalley, R.: 2003b, 'Smalley Concludes', *Chemical & Engineering News*, 81, 41-42.

Smalley, R.E.: 1999, 'Prepared written statement and supplemental material', Rice University, 22 June [http//www.house.gov/science/smalley_062299.htm].

Smalley, R.E.: 2001, 'Of Chemistry, Love and Nanobots', *Scientific American*, (Sept.), 76-77.

Smith, C.S. 1981. *A search for structure*. Cambridge, Mass.: MIT Press.

Snow, E.S., P.M. Campbell, and F.K. Perkins. 1997. Nanofabrication with proximal probes. *Proceedings of the IEEE* 85: 601-11.

Soffer, A., J.E. Koresh, S. Saggy. 1987. United States Patent 4,685,940.

Stephen P. Walch, Ralph C. Merkle, "Theoretical studies of diamond mechanosynthesis reactions," *Nanotechnology* 9(1998): 285-296.

Stix Gary., "Toward 'Point One'," *Scientific American*, February 1995, pp. 90-95.

Stuckless, J.T, D.E. Starr, D.J. Bald, C.T. Campbell. 1997. *J. Chem. Phys*. 107: 5547.

Stupp, S., and P.V. Braun. 1997. Molecular manipulation of microstructures: biomaterials, ceramics, and semiconductors. *Science* 277: 1242-1248.

Suchman, M.: 2001, 'Envisioning Life on the Nano-Frontier', in: M.C. Roco and W.S. Bainbridge (eds.), *Societal Implications of Nanoscience and Nanotechnology*, Dordrecht: Kluwer, pp. 211-216.

Suchman, M.: 2002, 'Social Science and Nanotechnologies', M. Roco & R. Tomellini (eds.), *Nanotechnology ' Revolutionary Opportunities and Societal Implications*, Luxembourg: European Communities, pp. 95-99.

Sunstrom, J.E., IV, W.R. Moser, and B. Marshik-Guerts. 1996. *Chem. Mater*. 8: 2061.

Surowiecki, J.: 2004, 'Bring on the Nanobubble', *The New Yorker*, March 15, p. 68.

Susan B. Sinnott, Richard J. Colton, Carter T. White, Donald W. Brenner, "Surface patterning by atomically-controlled chemical forces: molecular dynamics simulations," *Surf. Sci*. 316(1994): L1055-L1060.

Susan Smith, Izabela Giriat, Anja Schmitt, Titia de Lange, "Tankyrase, a poly(ADP-ribose) polymerase at human telomeres," *Science* 282(20 November 1998): 1484-1487.

Suslick, K.S., T. Hyeon, and F. Fang. 1996. *Chem. Mater*. 8: 2172.

Sweeney, A.E.; Seal, S. & Vaidyanathan, P.: 2003, 'The promises and perils of nanoscience and nanotechnology: Exploring emerging social and ethical issues', *Bulletin of Science, Technology & Society*, 23 (4), 236-245.

T. E. Creighton, "The protein-folding problem." *Science,* 240: 267, 344 (1988).

T. LaBean, H. Yan, J. Kopatsch, F. Liu, E. Winfree, J.H. Reif and N.C. Seeman, The Construction of DNA Triple Crossover Molecules, Journal of the American Chemical Society 122, 1848-1860 (2000).

T.R. Kelly, H. De Silva, R.A. Silva, "Unidirectional rotary motion in a molecular system," *Nature* 401(9 September 1999): 150-152.

Tanev, P.T., M. Chibwe, and T.J. Pinnavaia. 1994. *Nature* 368: 321.

Taylor, K.J., C.L. Pettiette-Hall, O. Cheshnovsky, and R.J. Smalley. 1992. *J. Chem. Phys.* 96: 3319.

Thompson, D.: 1992 [1942], *On Growth and Forms*, Cambridge University Press, Cambridge.

Tighe, T.S., J.M. Worlock, and M.L. Roukes. 1997. Direct thermal conductance measurements on suspended monocrystalline nanostructures. *Appl. Phys. Lett.* 70: 2687-9.

Tihamer Toth-Fejel, "Modeling kinematic cellular automata: An approach to self-replication," Environmental Research Institute of Michigan, Ann Arbor, MI, 4 October 2000.

Tohru Watanabe, Watanabefv@aol.com, Department of Robotics, Ritsumeikan University; personal communication to Tihamer Toth-Fejel, 21 March 2000.

Toumey, C.: 2004, 'Anticipating Public Reactions to Nanotechnology', *Techne: Research in Philosophy and Technology*, 8(3) (forthcoming).

Trivelli, A., and W.F. Smith. 1939. *Photog. J.* 79: 330,463,609. As referred to in T.H. James. 1977. *The theory of the Photographic process.* New York: MacMillan (p. 100).

Trudeau, M.L., and J.Y. Ying. 1996. *Nanostructured Materials* 7: 245.

Trudeau, M.L., and J.Y. Ying. 1996. *Nanostructured Materials* 7: 245. Tschope, A.S., W. Liu, M. Flyzani-Stephanopoulos, and J.Y. Ying 1995. *J. Catal.* 157: 42.

Tschope, A.S., and J.Y. Ying. 1994. *Nanostructured Materials* 4: 617. Yeo, Y.Y., C.E. Wartnaby, and D.A. King. 1995. *Science* 268: 1731.

Tschope, A.S., W. Liu, M. Flyzani-Stephanopoulos, and J.Y. Ying 1995. *J. Catal.* 157: 42.

Turton, R., *The Quantum Dot*, Oxford University Press, 1995.

Uyeda, R. 1991. *Prog. in Mater. Sci.* 35: 1. Van de Zande, B.M.I., M.R. Bohmer, L.G.J. Fokkink, and C. Shonenberger. 1997. *J. Phys. Chem.* 101: 852.

Verbeek, A. & Callaert, J.: 2002, *Detailed analysis of the science-technology interaction in the field of Nanotechnology*, Luxemburg: Office for Official Publications of the European Communities (Linking Science to Technology. Bibliographic References in Patents, Vol. 9) [79 pp.].

Vermaas, P.E.: 2004, 'Nanoscale Technology: A Two-Sided Challenge for Interpretations of Quantum Mechanics', in: D. Baird, A. Nordmann & J. Schummer (eds.), *Discovering the Nanoscale*, Amsterdam: IOS Press, pp. 77-91.

Vicki Glaser, "Carbohydrate-Based Drugs Move Closer to Market," *Genetic Engineering News*, 15 April 1998, pp. 1, 12, 32, 34.

Vidyasankar, S., M. Ru, and F.H. Arnold. 1997. Molecularly imprinted ligand-exchange adsorbents for the chiral separation of underivatized amino acids. *J. Chromatography A* 775: 51-63.

Vogel, S.: 1998, *Cats' paws and catapults*, Norton & Cy, New York, London.

W. Shen, M. Bruist, S. Goodman & N.C. Seeman, A Nanomechanical Device for Measuring the Excess Binding Energy of Proteins that Distort DNA, Angew. Chem. Int. Ed. 43, 4750-4752 (2004); Angew.Chem. 116, 4854-4856 (2004).

W.B. Sherman and N.C. Seeman, A Precisely Controlled DNA Bipedal Walking Device, NanoLetters 4, 1203-1207 (2004); Erratum, 4, 1801-1801.

W.C. Albrich, M. Angstwurm, L. Bader, R. Gartner, "Drug resistance in intensive care units," *Infection* 27(1999): S19-S23.

Wald, C.A.: 2004, 'Working Boundaries on the NANO Exhibition', in: N.K. Hayles (ed.), *Nanoculture: Implications of the New Technoscience*, Bristol, UK: Intellect Books, pp. 83-108.

Walsh, C. (1979) *Enzymatic Reaction Mechanisms* (Freeman, San Francisco), pp. 33, 38.

Wang, C.C., Z. Zhang, and J.Y. Ying. 1997. *Nanostructured Mater*. 9: 583.

Weckert, J.: 2001, 'The Control of Scientific Research: The Case of Nanotechnology', *Australian Journal of Professional and Applied Ethics*, 3, 29-44.

Weil, V.: 2001, 'Ethical Issues in Nanotechnology', in: M.C. Roco and W.S. Bainbridge (eds.), *Societal Implications of Nanoscience and Nanotechnology*, Dordrecht: Kluwer, pp. 193-198.

Weisenhorn, A.L., Hansma, P.K., Albrecht, T.R., and Quate, C.F. (1989) Forces in atomic force microscopy in air and water. Appl. Phys. Lett. 54(26), 2651-2653

Whitesides G. 2001 'The Once and Future nanomachines', *Scientific American*, (Sept.), 78-83.

Whitesides G.: 1995, 'Self-Assembling Materials,' *Scientific American*, (Sept.), 146-9.

Whitesides, G.: 1998, 'Nanotechnology: Art of the Possible', *Technology, MIT Magazine of Innovation*, (Nov.-Dec.), pp. 8-13.

Whitesides, G.; Love, J.C.: 2001, 'The Art of Building Small', *Scientific American*, (Sept.), 38-47.

Whitesides, G.M.; Ismagilov, R.F.: 1999, 'Complexity in Chemistry', *Science*, 284, 89-92.

Widjaja, Y. and C. Musgrave, "Atomic Layer Deposition of Hafnium Oxide: A Detailed Reaction Mechanism from First Principles," *Journal of Chemical Physics* 117, 1931-1934 (2002).

Widjaja, Y. and C. Musgrave, "Indirect Adsorbate-Adsorbate Interactions Mediated Through the Surface Electronic Structure of the Si(100)-(2x1) Surface," *Journal of Chemical Physics* 120, 1555-1559 (2004).

Widjaja, Y. and C. Musgrave, "Quantum Chemical Study of the Elementary Reactions in Zirconium Oxide Atomic Layer Deposition," *Applied Physics Letters* 81, 304-306 (2002).

Wilder, K., H.T. Soh, T. Soh, A. Atalar, and C.F. Quate. 1997. Hybrid atomic force/scanning tunneling lithography. *J. of Vacuum Science & Technol. B.* 15: 1811-17.

Wildoer, J.W.G., L.C. Venema, A.G. Rinzler, R.E. Smalley, and C. Dekker. 1998. Electronic structure of atomically resolved carbon nanotubes. *Nature* 391: 59-62.

Williams, R.S. 1998. Functional nanostructures. In *R&D status and trends*, ed. Siegel et al. Wiltzius, P. 1998.

Wilson Ho, Hyojune Lee, "Single bond formation and characterization with a scanning tunneling microscope," *Science* 286 (26 November 1999): 1719-1722; http: //www.physics.uci.edu/~wilsonho/stm-iets.html

Winningham, M.J. and D.Y. Sogah. 1997. A modular approach to polymer architecture control via catenation of prefabricated biomolecular segments: Polymers containing parallel beta-sheets templated by a phenoxathiin-based reverse turn mimic. *Macromolecules* 30: 862 - 876.

Wolfson, J.R.: 2003, 'Social and Ethical Issues in Nanotechnology: Lessons from Biotechnology and Other High Technologies', *Biotechnology Law Report*, 22 (4), 376-96.

Woods, S.; Jones, R. & Geldart, A.: 2003, *The Social and Economic Challenges of Nanotechnology*, Report to the Economic and Social Research Council (ESRC), Swindon, UK [www.esrc.ac.uk/esrccontent/DownloadDocs/Nanotechnology.pdf].

Wu, M.K., R.S. Windeler, C.K. Steiner, T. Bors, and S.K. Friedlander. 1993. Controlled synthesis of nanosized particles by aerosol processes. *Aerosol Sci. Technol.* 19: 527.

X. Yang, B. Liu, A. Vologodskii, B. Kemper and N.C. Seeman, Torsional Control of Double Stranded DNA Branch Migration. Biopolymers 45, 69-83 (1998).

X. Yang, L.A. Wenzler, J. Qi, X. Li and N.C. Seeman, Ligation of DNA Triangles Containing Double Crossover Molecules, Journal of the American Chemical Society 120, 9779-9786 (1998).

X. Zhang, H. Yan, Z. Shen and N.C. Seeman, Paranemic Cohesion of Topologically-Closed DNA Molecules, J Am. Chem. Soc.124, 12940-12941 (2002).

X.J. Li, X.P. Yang, J. Qi, and N.C. Seeman, Antiparallel DNA Double Crossover Molecules as Components for Nanoconstruction, Journal of the American Chemical Society, 118, 6131-6140 (1996).

Xu, Y. and C. Musgrave, "A DFT Study of the Atomic Layer Deposition of Al 2 O 3 on SAMs: The Effect of SAM Termination," *Chemistry of Materials,* 16, 646 (2004).

Y. Wang, J.E. Mueller, B. Kemper, and N.C. Seeman, The Assembly and Characterization of 5-Arm and 6-Arm DNA Junctions, Biochemistry 30, 5667-5674 (1991).

Y. Zhang and N.C. Seeman, A Solid-Support Methodology for the Construction of Geometrical Objects from DNA, Journal of the American Chemical Society 114, 2656-2663 (1992).

Y. Zhang and N.C. Seeman, The Construction of a DNA Truncated Octahedron, Journal of the American Chemical Society 116, 1661-1669 (1994).

Yeo, Y.Y., C.E. Wartnaby, and D.A. King. 1995. *Science* 268: 1731.

Yin, H., M.D. Wang, K. Svoboda, R. Landick, J. Gelles, and S.M. Block. 1995. Transcription against an applied force. *Science* 270: 1653-1657.

Ying, J.Y. 1998. Nanostructure processing of advanced catalytic materials. In *R&D status and trends*, ed. Siegel et al.

Ying, J.Y., and T. Sun. 1997. Research needs assessment on nanostructured catalysts. *J. of Electroceramics* 1: 219-238.

Yuan, Y., J. Fendler, and I. Cabasso. 1992. *Chem. Mater.* 4: 312.

Z. Shen, H. Yan, T. Wang and N.C. Seeman, Paranemic Crossover DNA: A Generalized Holliday Structure with Applications in Nanotechnology J. Am. Chem. Soc. 126, 1666-1674 (2004).

Zachariah, M.R. 1994. Flame processing, in-situ characterization, and atomistic modeling of nanoparticles in the reacting flow group at NIST. In *Proc. of the Joint NSF-NIST Conf. on Ultrafine Particle Engineering* (May 25-27, Arlington, VA).

Zener, C. 1948. *Elasticity and anelasticity of metals*. Chicago: University of Chicago Press. Zhang, Z., M.L. Roukes, and P.C. Hammel. 1996.

Zhang, L., and A. Manthiram. 1997. *Appl. Phys. Lett.* 70(18): 2469. 34 *Evelyn L. Hu and David T. Shaw.*

Zhang, Shuguang (2003) 'Fabrication of novel biomaterials through molecular self-assembly', *Nature Biotechnology*, 21, no. 10, 1171-78.

Glossary

Anisotropic: Exhibiting different values of a property in different crystallographic directions.

At the Core: A Non-Heavy Metal Quantum Dot. Research at Evident, which is focused on advancing product development, led to the first commercially available nanocrystals that comprise a III-V semiconductor material: Indium, Gallium and Phosphide core quantum dots, with a unique molecular metallic platting compound to form the shell or InGaP/ZnS EviDots-MP™.

Avogadro's number: The number of atoms in exactly 12 grams of pure ^{12}C, equal to 6.022×10^{23}.

Brownian Assembly: Brownian motion in a fluid brings molecules together in various position and orientations. If molecules have suitable complementary surfaces, they can bind, assembling to form a specific structure. Brownian assembly is a less paradoxical name for self-assembly (how can a structure assemble itself, or do anything, when it does not yet exist?).

Dissolution: Deterioration in an organism such that its original structure cannot be determined from its current state.

Distributed Intelligence: An intelligent entity which is distributed over a large volume (or inside another system, like a computer network) with no distinct center. This is the opposite to the strategy of Concentrated intelligences. Distributed intelligences have much longer communications lags, but are more flexible in their structure and can survive damage to their parts.

Fertilization: Fusion of a male and a female gamete (both haploid) to form a diploid zygote, which develops into a new individual.

FET: Field-Effect Transistor — semiconductor device whose insulated gate electrode controls current flow.

Fiber-optic: Relates to transmission of information as modulated light in tiny transparent fibers instead of copper wires.

Field-Effect Transistor (FET): Field-Effect Transistor — semiconductor device whose insulated gate electrode controls current flow.

Fluorescent dye: Molecule that absorbs light at one wavelength and responds by emitting light at another wavelength; the emitted light is of longer wavelength (and hence of lower energy) than the light absorbed.

Free-energy change (DG): Change in the free energy during a reaction: the free energy of the product molecules minus the free energy of the starting molecules. A large negative value of DG indicates that the reaction has a strong tendency to occur.

Highest Occupied Molecular Orbital (HOMO): The highest energy molecular orbital of an atom or molecule that contains an electron.

Histonatation: In medical nanorobotics, locomotion (swimming) through tissues by a nanorobot.

Histone: A protein present in the eukaryotic nucleus that is bound to the DNA at regularly spaced intervals.

Homogeneous and Heterogeneous Assays: A homogeneous assay does not require a separation step to remove free antigen from bound antigen and relies upon the fact that the function of the label is modified upon binding, leading to a change in signal intensity. Because of high background signal a heterogeneous approach incorporating a separation step of bound and unbound makes the detection limit lower, approaching the values obtained by RIA. The homogeneous assay is less technically demanding.

Induction (embryonic): Change in the developmental fate of one tissue caused by an interaction with another tissue.

Inductor: Energy storage circuit component consisting of a coil of wire and possibly a magnetic material.

Inflammatory response: Local response of a tissue to injury or infection. Caused by invasion of white blood cells, which release various local mediators such as histamine.

Infrared: Invisible electromagnetic radiation having a longer wavelength, and lower frequency, than visible red light.

Inhibitor: A chemical substance that, when added in relatively low concentrations, slows down a chemical reaction.

Inline Universities: (As opposed to online universities), nanocomputer implants serving to increase intelligence and education of their owners, essentially turning them into walking universities.

Interface: In physical terms, an interface is the boundry between two phases, for instance between a solid and liquid or between a liquid and gas.

Intermolecular: Between more than one molecule. Describes an interaction (e.g., a chemical reaction) between different molecules.

Ion: An atom or molecule with a net charge. **(2)** An atom with more or fewer electrons than those needed to cancel the electronic charge of the nucleus. An ion is an atom with a net electric charge.

Ionic bond: A chemical bond resulting chiefly from the electrostatic attraction between positive and negative ions. A coulombic interatomic bond existing between two adjacent and oppositely charged ions.

Ionophore: A macro-organic molecule capable of specifically solubilizing an inorganic ion of suitable size in organic mediums.

Karyotype: Full set of chromosomes of a cell arranged with respect to size, shape, and number.

Keratin (cytokeratin): Member of the family of proteins that form keratin intermediate filaments, mainly in epithelial cells. Some specialized keratins are found in hair, nails, and feathers.

Ketone: Organic molecule containing a carbonyl group linked to two alkyl groups.

Kevlar (TM): A synthetic fiber made by E. I. du Pont de Nemours & Co., Inc. Stronger than most steels, Kevlar is among the strongest commercially available materials and is used in aerospace construction, bulletproof vests, and other applications requiring a high strength-to-weight ratio.

Kilobyte (kB): 2^{10} (= 1024, or about one thousand) bytes of information.

Leucine zipper: Structural motif seen in many DNA-binding proteins in which two a helices from separate proteins are joined together in a coiled-coil, forming a protein dimer.

Life (lifetime): The length of time the sensor can be used before its performance changes.

Limit of detection: The smallest measurable input. This differs from resolution, which defines the smallest measurable change in input. For a temperature measurement, this would provide an indication of the lowest temperature a sensor could generate an output in response to.

Limited Assembler: An assembler with built-in limits that constrain its use (for example, to make hazardous uses difficult or impossible, or to build just one thing). Assembler capable of making only certain products; faster, more efficient, and less liable to abuse than a general-purpose assembler.

Machine-phase chemistry: The chemistry of systems in which all potentially reactive moieties follow controlled trajectories (e.g., guided by molecular machines working in vacuum).

Macroscale: Larger than nanoscale; often implies a design that humans can directly interact with; too large to be built by a single assembler (one cubic micron of diamond contains 176 billion atoms).

Macrosensing: In medical nanorobotics, the detection of global somatic states (inside the human body) and extrasomatic states (sensory data originating outside of the human body) by in vivo nanorobots.

Magnetic field strength (designated by H) [A/m]: Magnetic field produced by a current, independent of the presence of magnetic material. The units of H are ampere-turns per meter, or just amperes per meter.

Magnetic Force Microscopy (MFM): A method for observing local magnetic fields near a surface by scanning the surface with a magnetic probe.

Magnetic susceptibility ($\div_m$): The proportionality constant between the magnetization M (see under"magnetization") and the magnetic field strength H. The magnetic susceptibility is unitless.

Magnetization (M): The total magnetic moment per unit volume of material. Also, a measure of the contribution to the magnetic flux by some material within an H field. The magnitude of M is proportional to the applied field as: $M = \div_m \times H$, with $\div_m$ the magnetic susceptibility.

Magnetostrictive material: A material that changes dimension in the presence of a magnetic field or generates a magnetic field when mechanically deformed.

Manufacturing: The ability to make products, in this case ranging from clothing, to electronics, to medical devices, to books, to building materials, and much more.

Martensite: A metastable iron phase supersaturated in carbon that is the product of a diffusionless (athermal) transformation from austenite.

Mask: Pattern on glass, like a photographic negative, for producing integrated-circuit elements on semiconductor wafer.

Mass to Charge Ratio: A number defining how a particle will respond to an electric or magnetic field that can be calculated by dividing the mass of a particle by its charge.

Massometer: In medical nanorobotics, a nanosensor device for measuring the mass of individual molecules or small physical objects to single-proton resolution.

Materials science: Science of ceramics, glass, metals, plastics, semiconductors.

MatML Materials Markup Language: Materials property data distributed on the World Wide Web in documents using hypertext markup language.

Matter as Software: "Autonomous, motile microdevices clearly are on the horizon. They may be regarded as the first step in the evolution of a technology for "programming" the structure and properties of material objects at the microscopic and the submicroscopic levels. As this evolution progresses, *the physical and economic properties of such programmable matter are likely to become much like those of present day software."* [MITRE Corporation]

MCM: MultiChip Module; the interconnection of two or more semiconductor chips in a semiconductor-type package.

Measurand: A physical quantity, condition, or property that is to be measured.

Meat Machine: AKA Cabinet Beast. A box containing assemblers and raw material, within which is formed meat [or whatever else it was programmed to make].

Mechatronics: The study of the melding of AI and electromechanical machines to make machines that are greater than the sum of their parts. The synergistic combination of precision mechanical engineering with electronic control.

Megabyte (MB): 2^{20} (= 1,048,576, or about one million) bytes of information.

Messenger molecule: A chemically recognizable molecule which can convey information after it is received and decoded by an appropriate chemical sensor.

Molecular surgery (molecular repair): In medical nanorobotics, the analysis and physical correction of molecular structures in the body using medical nanomachines. Analysis and physical correction of molecular structures in the body using medical nanomachines.

Molecular Systems Engineering: Design, analysis, and construction of systems of molecular parts working together to carry out a useful purpose.

Molecular Wire: A molecular wire - the simplest electronic component - is a quasi-one-dimensional molecule that can transport charge carriers (electrons or holes) between its ends. [Michael D Ward]

Molecularly imprinted polymers MIPs: A new class of materials that have artificially created receptor structures. Since their discovery in 1972, MIPs have attracted considerable interest from scientists and 3engineers involved with the development of chromatographic absorbents, membranes, sensors and enzyme and receptor mimics.

Monkeywrenching: In medical nanorobotics, the mechanical or chemical jamming of cellular equilibrium processes, with a cytocidal objective.

Op-amp: Operational Amplifier — semiconductor amplifier characterized by high gain and high internal resistance.

Operator: Short region of DNA in a bacterial chromosome that controls the transcription of an adjacent gene.

Operon: In a bacterial chromosome, a group of contiguous genes that are transcribed into a single mRNA molecule.

Patch-clamp recording: Electrophysiological technique in which a tiny electrode tip is sealed onto a patch of cell membrane, thereby making it possible to record the flow of current through individual ion channels in the patch.

Pathogen (adjective pathogenic): An organism or other agent that causes diseases.

PCR (polymerase chain reaction): Technique for amplifying specific regions of DNA by multiple cycles of DNA polymerization, each followed by a brief heat treatment to separate complementary strands.

Peptide bond: Chemical bond between the carbonyl group of one amino acid and the amino group of a second amino acid - a special form of amide linkage.

Peptide map: Characteristic two-dimensional pattern (on paper or gel) formed by the separation of the mixture of peptides produced by the partial digestion of a protein.

Peptide nanotubes: NANOTUBES formed from cyclic peptides (PEPTIDES, CYCLIC). Alternating D and L linkages create planar rings that self assemble by stacking into nanotubes. They can form pores through CELL MEMBRANE causing damage. MeSH 2004

Permeability [µ]: From the relation between magnetic induction and magnetic field (B = µH); for free space, $\mu_0 = 1.26 \times 10^{-6}$ H/m.

Permitivity [å]: From the relation between polarization charge and electric field; for free space, $å_0 = 8.85 \times 10^{-12}$ F/m.

Peroxisome: Small membrane-bounded organelle that uses molecular oxygen to oxidize organic molecules. Contains some enzymes that produce and others that degrade hydrogen peroxide (H2O2)

Phagocyte: General term for a professional phagocytic cell - that is, a cell such as a macrophage or neutrophil that is specialized to take up particles and microorganisms by phagocytosis.

Phagocytosis: Process by which particulate material is endocytosed ("eaten") by a cell. Prominent in carnivorous cells, such as *Amoeba proteus*, and in vertebrate macrophages and neutrophils. (From Greek *phagein*, to eat.)

Pharmacyte: In medical nanorobotics, a theorized (nanorobotic) device capable of delivering precise doses of biologically active chemicals to individually-addressed human body tissue cells (e.g. cell-by-cell drug delivery).

Phase shift: A time difference between the input and output signals.

Phase space: A classical system of *N* particles can be described by its 3*N* position and 3*N* momentum coordinates. The phase space associated with the system is the 6*N* dimensional space defined by these coordinates.

Phase transformation: A change in the number and/or character of the phases that make up the microstructure of an alloy.

Photolithography: Writing or creating patterns by means of light.

Photon: A particle of light. The energy *E* of a photon is given by the Einstein relation, $E = hf$. Here *f* is the frequency of the light and *h* is Planck's constant. The momentum *p* of a photon is given by the de Broglie relation, $p = h/ë$. Here ë is the wavelength of the light.

Photonic Band Gap (PBG): An energy range (and corresponding wavelength range) for which a material neither absorbs light nor allows light propagation.

Phototransistor: Transistor that, when powered, produces amplified voltage (current) in response to illumination.

Pico Technology: (trillionth of a meter) — The next step smaller, after Nano-technology. The art of manipulating materials on a quantum scale.

Piezoelectric material: A ferroelectric material in which an electrical potential difference is created due to mechanical deformation, or conversely, in which the application of a voltage causes dimensional changes in the material.

Piezoelectric: A piezoelectric material expands or contracts when a voltage is applied across it. Piezoelectric tubes are often used to move the tip in an AFM or STM. If a piezoelectric crystal is compressed, a voltage appears across it. This effect is sometimes used to measure displacements. A commonly used piezoelectric is PZT $(Pb,Zr)TiO_3$.

Pinhole: The term pinhole embraces a wide variety of oxide defects and is used in a broad sense today. Listed in this category are cracks caused by thermal contraction after oxidation or by handling, and regions of oxide with low dielectric strength caused by dust particles, inadequate masking, contamination, or poor resist adhesion.

Pin-out: Diagram showing for electronic components the relations between connecting pins and internal components.

Pitting: A form of very localized corrosion wherein small pits or holes form, usually in a vertical direction.

Pixel: Picture Element — smallest element of an image, such as a dot on a computer monitor screen.

pK value: A measure of the strength of an acid on a logarithmic scale. The pK value is given by log10 (1/Ka), where Ka is the acid dissociation constant pK values often are used to compare the strengths of different acids.

Plastic deformation: Permanent or nonrecoverable deformation, accompanied by permanent atomic displacement.

Plasticizer: A low molecular weight polymer additive that enhances flexibility and workability and reduces stiffness and brittleness.

Point defect: A crystalline defect associated with one or several atomic sites.

Poisson's ratio (í): For elastic deformation, the negative ratio of lateral and axial strains that result from an applied axial stress.

Poisson's ratio: A bar of an isotropic, elastic material ordinarily shrinks laterally when it is stretched longitudinally. The lateral contracting strain divided by the applied tensile strain is Poisson's ratio, which varies from material to material.

Polarization (P): The total electric dipole moment per unit volume of dielectric material.

Polyclonal antibodies: Antibodies produced by an animal's white blood cells (lymphocytes, specifically) in response to an antigen. This response occurs naturally or can purposely be created by injecting an animal, such as a rabbit or goat, with a specific antigen. More than one kind of anti-body is produced since more than one lymphocyte is producing antibodies. This is referred to as"polyclonal". The polyclonal antibodies are isolated from the animal and can be used for detection purposes. Because the antibodies are actually a mixture with different affinities (binding capability) for the antigen of interest, some variability in performance can occur from one test to another or one batch of antibodies to another.

Polycyclic: A cyclic structure contains rings of bonds; a structure having many such rings is termed polycyclic. In the polycyclic structures of interest in this volume, a large fraction of the atoms are members of multiple small rings, resulting in considerable rigidity.

Polymer: A molecule consisting of many smaller sub-units covalently linked together.

Polymerase Chain Reaction (PCR): A chemical reaction that uses the polymerase enzyme to carry out *in vitro* replication of DNA.

Potentiometric Device: Monitors the voltage between a sensing electrode and a reference electrode. A high input impedance voltmeter is used to minimize current flow. The voltage typically is proportional to the logarithm of the analyte concentration.

Power [W]: Product of voltage and current in a component; also, refers to the field of electric energy supply.

Precision: The degree of reproducibility among several independent measurements of the same true value under specified conditions.

Presentation semaphore: In medical nanorobotics, a mechanical device used to display specific antigens, chemical ligands, or other molecular objects to the external environment, with the purpose of selectively modifying the chemical or other surface characteristics of a nanorobot exterior.

Printed circuit board: PCB — selectively metallized insulating sheet for supporting and interconnecting circuit components.

Probability density function: Consider an uncertain physical property and a corresponding space describing the range of values that the property can have (e.g., the configuration of a thermally excited N particle system and the corresponding $3N$ dimensional configuration space). The probability density function associated with a property is defined over the corresponding space; its value at a particular point is the probability per unit volume that the property has a value in an infinitesimal region around that point.

Protein Design, Protein Engineering: The design and construction of new proteins; an enabling technology for nanotechnology.

Protein Folding: "The process by which proteins acquire their functional, preordained, three-dimensional structure after they emerge, as linear polymers of amino acids, from the ribosome." [The Scientist]

Protein: Living cells contain many molecules that consist of amino acid polymers folded to form more-or-less definite three-dimensional structures; these are termed proteins. Short polymers lacking definite three-dimensional structures are termed peptides. Many proteins incorporate structures other than amino acids, either as covalently attached side chains or as bound ligands. Molecular objects made of protein form much of the molecular machinery of living cells.

Proteomics: The study of the full expression of proteins in an organism The term proteome refers to all the proteins expressed by a genome, and thus proteomics involves the identification of proteins in the body and the determination of their role in physiological and pathophysiological functions. ... Ultimately it is believed that through proteomics new disease markers and drug targets can be identified that will help design products to prevent, diagnose and treat disease. [e-proteomics.net]

Proximal probes: A family of devices capable of fine positional control and sensing, including scanning tunneling and atomic force microscopes; an enabling technology for nanotechnology.

p-Type semiconductor: A semiconductor for which the predominant charge carriers responsible for electrical conduction are holes.

Pyroelectricity: The property of certain crystals, such as tourmaline, of acquiring opposite electrical charges on opposite faces when heated.

Q factor: A rating, applied to coils, capacitors, and resonant circuits, equal to the reactance divided by the resistance. The ratio of energy stored to energy dissipated per cycle in an electrical or mechanical system.

Quantization: The concept that energy can occur only in discrete units called quanta.

Quantum Computer: A computer that exploits quantum mechanical phenomena such as superposition and entanglement.

Quantum Confined Atoms (QCA): atoms caged inside nanocrystals. May find uses in clear-glass sunglasses, bio-sensors, and optical computing.

Quantum corral: Structures where electrons are corralled (confined) by atoms placed on the surface of a metal.

Quantum Cryptography: A system based on quantum- mechanical principles. Eavesdroppers alter the quantum state of the system and so are detected. Developed by Brassard and Bennett, only small laboratory demonstrations have been made.

Quaternary Structure: Three-dimensional relationship of the different polypeptide chains in a protein complex.

Qubit: The quantum computing analog to a bit. Qubits exhibit superposition. Thus, unlike normal bits, qubits can be both 1 and 0 at the same time.

Resistor: Energy dissipative element consisting of a poor conductor in series with connecting wires.

Resolution: The minimum distance between two objects that can be distinguished in microscopy or the minimum spacing between two features that can be fabricated with lithography. The smallest measurable change in input that will produce a small but noticeable change in the output. In the context of chemical separations, defines the completeness of separation.

Resonant frequency: The frequency at which a moving member or a circuit has a maximum output for a given input.

Reverse bias: The insulating bias for a p-n junction rectifier; electrons flow into the p side of the junction.

RF: Radio Frequency — refers to alternating voltages and currents having frequencies between 9 kHz and 3 MHz.

Ribonuclease: An enzyme that cuts RNA molecules into smaller pieces.

ROM: Read only memory; memory used for permanent, storage of unalterable data; nonvolatile memory.

***Saccharomyces*:** Genus of yeasts that reproduce asexually by budding or sexually by conjugation. Economically important in brewing and baking, they are also widely used in genetic engineering and as simple model organisms in the study of eucaryotic cell biology.

Sacrificial anode: An active metal or alloy that corrodes and protects another metal or alloy to which it is electrically coupled.

Sacrificial layer: A thin film that is later removed to release a microstructure from its substrate.

Salmonella: Rod-shaped, motile, aerobic genus of bacteria. Includes species that cause food poisoning.

Salt bridge: An ionic bond between charged groups that are part of larger covalent structures; salt bridges occur in many proteins.

SAMFET: (self assembled monolayer field effect transistor). Where a few molecules act as FETs, exhibiting both very strong gain, and extraordinarily rapid response. [Mark Ratner & MT 5(2) p. 20]

Sanguinatation: In medical nanorobotics, locomotion (especially swimming by a nanórobot) through the bloodstream.

Sarcomere: Repeating unit of a myofibril in a muscle cell, composed of an array of overlapping thick (myosin) and thin (actin) filaments between two adjacent Z discs.

Sarcoplasmic reticulum: Network of internal membranes in the cytoplasm of a muscle cell that contains high concentrations of sequestered Ca2+ : that is released into the cytosol during muscle excitation.

Satellite DNA: Regions of highly repetitive DNA from a eucaryotic chromosome, usually identifiable by its unusual nucleotide composition. Satellite DNA is not transcribed and has no known function.

Saturated molecule: Molecule containing carbon-carbon bonds that has only single covalent bonds.

Saturated: An organic molecule is described as saturated if it is a closed shell species lacking double or triple bonds; forming a new bond to a saturated molecule requires the cleavage of an existing bond.

Scanning Capacitance Microscopy (SCM): A variation of scanning force microscopy that maps the localized capacitance of a surface. A method for mapping the local capacitance of a surface. [NTN]

Scanning electron microscope (SEM): A microscope producing an image by using reflected electron beams that scan the surface of a specimen.

Scanning Electron Microscopy (SEM): A method of producing images of a surface by scanning an electron beam over the sample and measuring the electronic interactions with the interface.

Scanning Force Microscope (SFM): An instrument able to image surfaces to molecular accuracy by mechanically probing their surface contours. A kind of proximal probe. .... A device in which the deflection of a sharp stylus mounted on a soft spring is monitored as the stylus is moved across a surface. If the deflection is kept constant by moving the surface up and down by measured increments, the result (under favorable conditions) is an atomic-resolution topographic map of the surface. Also termed an atomic force microscope.

Scanning Near Field Optical Microscopy: A method for observing local optical properties of a surface that can be smaller than the wavelength of the light used. A type of scanning probe microscopy that maps the near-field optical properties of an interface.

Scanning Probe Microscopy (SPM): A method for imaging nanoscale features of surfaces by scanning a sensor (probe) over a surface. Near-field effects such as tunneling, van der Waals forces, local fields and more are serially detected at localized points on the surface and used to create an SPM image.

Scanning Thermal Microscopy (SThM): A type of scanning probe microscopy that maps the local temperature and thermal conductivity of an interface. A method for observing local temperatures and temperature gradients on a surface. [NTN]

Scanning Tunneling Microscope (STM): An instrument able to image conducting surfaces to atomic accuracy; has been used to pin: molecules to a surface. A device in which a sharp conductive tip is moved across a conductive surface close enough to permit a substantial tunneling current (typically a nanometer or less). In a common mode of operation, the voltage is kept constant and the current is monitored and kept constant by controlling the height of the tip above the surface; the result, under favorable conditions, is an atomic-resolution map of the surface reflecting a combination of topography and electronic properties. The STM has been used to manipulate atoms and molecules on surfaces. A type of scanning probe microscopy that maps the local electrical properties of an interface.

Schwann cell: Glial cell responsible for forming myelin sheaths in the peripheral nervous system.

Science Court: A name (originally applied by the media) for a government-conducted fact forum.

Science: The process of developing a systematized knowledge of the world through the variation and testing of hypotheses. The pursuit of knowledge and understanding, from the Latin term *scientia*, which means 'knowledge'.

TATA box: Consensus sequence in the promoter region of many eucaryotic genes that binds a general transcription factor and hence specifies the position where transcription is initiated.

Technocyte: A nanoscale artificial device (especially a nanite) in the human bloodstream used for repairs, cancer protection, as an artificial immune system or for other uses. [AS 1995]

Tight junction: Cell-cell junction that seals adjacent epithelial cells together, preventing the passage of most dissolved molecules from one side of the epithelial sheet to the other.

Tight-receptor structures: A receptor structure in which a bound ligand of a particular kind is confined on all sides by repulsive interactions (note that favorable binding energies are compatible with repulsive forces). A tight-receptor structure discriminates strongly against all molecules larger than the target.

Time constant: The time it takes for the output change to reach 63% of its final value.

Triple bond: A double bond is formed when a pi bond is superimposed on a single bond; adding a second pi bond results in a triple bond. The two pi bonds have perpendicular nodal planes, and their sum has roughly cylindrical symmetry, permitting rotation in much the same manner as a single bond.

Virtual reality system: A combination of computer and interface devices (goggles, gloves, etc.) that presents a user with the illusion of being in a three dimensional world of computer-generated objects.

Viscoelasticity: A type of deformation exhibiting the mechanical characteristics of viscous flow and elastic deformation.

Viscosity: The ratio of the magnitude of an applied shear stress to the velocity gradient that it produces; in other words: a measure of a noncrystalline material's resistance to permanent deformation.

Volitional normative model of disease: In medical nanorobotics, disease is said to be present in a human being upon either (1) the failure of optimal physical (e.g. biological) functioning, or (2) the failure of desired (by the patient) functioning.

Voltage [V]: Potential difference between two points: energy to move a 1-C charge through a 1-V potential difference is 1-J.

Wave function: In quantum mechanics, a complex function extending over the configuration space of a system; its complex conjugate yields the probability density function, and other mathematical operations yield other physical quantities.

Weber: Unit of magnetic flux. One weber is a magnetic flux that, linking a circuit of 1 turn, would produce in it an electromotive force of 1 volt if it were reduced to zero at a uniform rate in 1 second.

Weight percent (wt%): Concentration specification on the basis of weight (or mass) of a particular element relative to the total alloy weight (or mass).

Western blotting: Technique by which proteins are separated and immobilized on a paper sheet and then analyzed, usually by means of a labeled antibody.

Bibliography

Berube, D.M.: 2004, 'The Rhetoric of Nanotechnology', in: D. Baird, A. Nordmann & J. Schummer (eds.), *Discovering the Nanoscale*, Amsterdam: IOS Press, pp. 173-192.

Bowes, C.L., A. Malek, and G.A. Ozin. 1996. *Chem. Vap. Deposition* 2:97.

Dresselhaus, M.S., G. Dresselhaus, and P. Eklund. 1996. *Science of fullerenes and carbon nanotubes*. San Diego: Academic Press.

Drexler, K.E.: 2004, 'Nanotechnology: From Feynman to Funding', *Bulletin of Science, Technology &.Society*, 24, 21-27.

Dupuy, J.-P.: 2004, 'Complexity and Uncertainty: A Prudential Approach To Nanotechnology', in: European Commission (Community Health and Consumer Protection): *Nanotechnologies: A Preliminary Risk Analysis on the Basis of a Workshop, Brussels, 1-2 March 2004*, pp. 71-93.

Dupuy, J.-P.: 2004, 'Pour une évaluation normative du programme nanotechnologique', *Annales des Mines*, (February 2004), 27-32.

Einsiedel, E.F.; Goldenberg, L.: 2004, 'Dwarfing the Social? Nanotechnology Lessons from the Biotechnology Front', *Bulletin of Science, Technology & Society*, 24, 28-33.

ETC: 2003a, *The Big Down: Atom Tech ' Technologies Converging at the Atomic Scale*. Winnipeg, Canada: Action Group on Erosion, Technology and Concentration [www.etcgroup.org/documents/TheBigDown.pdf].

Etzkowitz, H., 2001, 'Nano-Science and Society: Finding a Social Basis for Science Policy', in: M.C. Roco and W.S. Bainbridge (eds.), *Societal Implications of Nanoscience and Nanotechnology*, Dordrecht: Kluwer, pp. 121-128.

European Comission: *European Workshop on Social and Economic Research on Nanotechnologies and Nanosciences, Brussels, 14-15 April 2004.*

European Commission: 2004, *Converging Technologies: Shaping the Future of European Societies (Report of the High Level Expert Group "Foresighting the New Technology Wave")*, Brussels: European Commission Research.

F. Buot, "Mesoscopic Physics and Nanoelectronics: Nanoscience and Nanotechnology," *Physics Reports*, pp.73-174, 1993.

F. Capasso and S. Datta, "Quantum Electron Devices," *Physics Today*, pp.74-82, February 1990.

Feynman, R., "There's Plenty of Room at the Bottom: An invitation to Enter a New Field of Physics," Talk at the Annual Meeting of the American Physical Society, 29 December 1959. Reprinted in Appendix B of Crandall and Lewis. See citation above.

Fielder, F.A.; Reynolds, G.H.: 1994, 'Legal Problems of Nanotechnology: An Overview', *Southern California Interdisciplinary Law Journal*, 3, 593-629.

Fleischer, Th.; Decker, M. & Fiedeler, U. (eds.): 2004, *Große Aufmerksamkeit für kleine Welten ' Nanotechnologie und ihre Folgen*, special issue of *Technikfolgenabschätzung ' Theorie und Praxis*, 13 (2), 5-85 [www.itas.fzk.de/tatup/042/inhalt.htm].

Fogelberg, H. & Glimell, H.: 2003, 'Molecular Matters: In Search of the Real Stuff', in: H. Fogelberg & H. Glimell, *Bringing Visibility to the Invisible: Towards A Social Understanding of Nanotechnology*, Göteborg: Göteborg University, pp. 5-32 [www.sts.gu.se/publications/STS_report_6.pdf].

Fogelberg, H. & Glimell, H.: 2003, *Bringing Visibility to the Invisible: Towards A Social Understanding of Nanotechnology*, Göteborg: Göteborg University [www.sts.gu.se/publications/STS_report_6.pdf].

Fogelberg, H.: 2003, 'The Grand Politics of Technoscience: Contextualizing Nanotechnology', in: H. Fogelberg & H. Glimell, *Bringing Visibility to the Invisible: Towards A Social Understanding of Nanotechnology*, Göteborg: Göteborg University, pp. 33-54 [www.sts.gu.se/publications/STS_report_6.pdf].

Fogelberg, H.: 2003, 'The Material Culture of Nanotechnology', in: H. Fogelberg & H. Glimell, *Bringing Visibility to the Invisible: Towards A Social Understanding of Nanotechnology*, Göteborg: Göteborg University, pp. 99-114.

Frazier, G., "An Ideology For Nanoelectronics," in *Concurrent Computations: Algorithms, Architecture, and Technology*, Plenum Press, New York, 1988.

Glimell, H.: 2001, 'Challenging Limits ' Excerpts from an Emerging Ethnography of Nano Physicists', in: H. Glimell & O. Johlin (eds.), *The Social Production of Technology: On the Everyday Life with Things*, Göteburg, SE: BAS Publisher, chapter 7, pp. 111-131 (reprinted in: H. Fogelberg & H. Glimell, *Bringing Visibility to the Invisible: Towards A Social Understanding of Nanotechnology*, Göteborg: Göteborg University, pp. 115-139.

Glimell, H.: 2001, 'Dynamics of the Emerging Field of Nanoscience', in: M.C. Roco and W.S. Bainbridge (eds.), *Societal Implications of Nanoscience and Nanotechnology*, Dordrecht: Kluwer, pp. 156-160.

Glimell, H.: 2003, 'A Nano Narrative: The Mircopolitics of a New Generic Enabling Technology', in: H. Fogelberg & H. Glimell, *Bringing Visibility to the Invisible: Towards A Social Understanding of Nanotechnology*, Göteborg: Göteborg University, pp. 55-77 [www.sts.gu.se/publications/STS_report_6.pdf].

Gupta, V.K. & Pangannaya, N.B.: 2000, 'Carbon nanotubes: bibliometric analysis of patents', *World Patent Information*, 22, 185-189.

Hansson, S.O.: 2004, 'Great Uncertainty about Small Things', *Techne: Research in Philosophy and Technology*, 8(3) (forthcoming).

Haruta, M. 1997. *Catalyst surveys of Japan* 1:61 and references therein.

Hennig, J.: 2004, 'Changes in the Design of Scanning Tunneling Microscopic Images from 1980 to 1990', *Techne: Research in Philosophy and Technology*, 8(3) (forthcoming).

Hessenbruch, A.: 2004, 'Nanotechnology and the Negotiation of Novelty', in: D. Baird, A. Nordmann & J. Schummer (eds.), *Discovering the Nanoscale*, Amsterdam: IOS Press, pp. 135-144.

Jounet, C., W.K. Maser, P. Bernier, A. Loiseau, M. Lamy de la Chapelle, S. Lefrant, P. Deniard, R. Lee, and J.E. Fischer. 1997. *Nature* 388:756.

K.E. Drexler, *Nanosystems: Molecular Machinery, Manufacturing, and Computation,* Wiley, New York, 1992.

Keiper, A.: 2003, 'The Nanotechnology Revolution', *The New Atlantis*, 2, 17-34.

Khushf, G.: 2004, 'A Hierarchical Architecture for Nano-scale Science and Technology: Taking Stock of the Claims About Science Made By Advocates of NBIC Convergence', in: D. Baird, A. Nordmann & J. Schummer (eds.), *Discovering the Nanoscale*, Amsterdam: IOS Press, pp. 21-33.

Khushf, G.: 2004, 'Systems Theory and the Ethics of Human Enhancement: A Framework for NBIC Convergence', *Annals of the New York Academy of Sciences*, 1013, 124-149.

Krätschmer, W., L.D. Lamb, K. Fostiropoulos, and D.R. Huffman. 1990. *Nature* 347:354.

Kresge, C.T., M.E. Leonowicz, W.J. Roth, J.C. Vartuli, and J.S. Beck. 1992. *Nature* 359:710.

Kupperman, A., S. Nadimi, S. Oliver, G. Ozin, J. Garcés, and M. Olken. 1993. *Nature* 365:239.

Kuusi, O.; Meyer, M.: 2002, 'Technological generalizations and leitbilder ' the anticipation of technological opportunities', *Technological Forecasting & Social Change*, **69**, 625-639.

Landon, B.: 2004, 'Less is More: Much Less is Much More: The Insistent Allure of Nanotechnology Narratives in Science Fiction', in: N.K. Hayles (ed.), *Nanoculture: Implications of the New Technoscience*, Bristol, UK: Intellect Books, pp. 131-146.

Laszlo, P.: 2004, 'Is There Life After Partington?', *Hyle: International Journal for Philosophy of Chemistry*, 10(2), 169-178.

Lenhard, J.: 2004, 'Nanoscience and the Janus-Faced Character of Simulations', in: D. Baird, A. Nordmann & J. Schummer (eds.), *Discovering the Nanoscale*, Amsterdam: IOS Press, pp. 93-100.

Lent, C.S., Tougaw, P.D., Porod, W., Bernstein, G.H., "Quantum Cellular Automata," *Nanotechnology*, Vol. 4, p. 49, 1993.

Lewak, S.E.: 2004, 'What's the Buzz? Tell Me What's A-Happening: Wonder, Nanotechnology, and Alice's Adventures in Wonderland', in: N.K. Hayles (ed.), *Nanoculture: Implications of the New Technoscience*, Bristol, UK: Intellect Books, pp. 201-.

Lide, D.R., ed. 1993-1994. *CRC Handbook of Chemistry and Physics*, 74th ed.

Lin-Easton, P.C.: 2001, 'It's Time for Environmentalists to Think Small ' Real Small: A Call for the Involvement of Environmental Lawyers in Developing Precautionary Policies Molecular Nanotechnology', *Georgetown International Law Review*, 14, 106-134.

López, J.: 2004, 'Bridging the Gaps: Science Fiction in Nanotechnology', *Hyle: International Journal for Philosophy of Chemistry*, 10(2), 129-152.

Lösch, A.: 2004, 'Nanomedicine and Space: Discursive Orders of Mediating Innovations', in: D. Baird, A. Nordmann & J. Schummer (eds.), *Discovering the Nanoscale*, Amsterdam: IOS Press, pp. 193-202.

Marshall, K.: 2004, 'Atomizing the Risk Technology', in: N.K. Hayles (ed.), *Nanoculture: Implications of the New Technoscience*, Bristol, UK: Intellect Books, pp. 147-160.

Martin, T.P., N. Malinowski, U. Zimmerman, U. Naher, and H. Schaber. 1993. *J. Chem. Phys.* 99:4210.

Martin, T.P., U. Naher, H. Schaber, U. Zimmerman. 1993. *Phys. Rev. Lett.* 70:3079.

Mayer, S.: 2002, 'From genetic modification to nanotechnology: the dangers of 'sound science", in: T. Gilland (ed.), *Science: Can We Trust the Experts?*, London: Hodder and Stoughton, pp. 1-15.

Mehta, M.D.: 2002, 'Nanoscience and Nanotechnology: Assessing the Nature of Innovation in These Fields', *Bulletin of Science, Technology & Society*, 22(4), 269-273.

Mehta, M.D.: 2004, 'From Biotechnology to Nanotechnology: What Can We Learn from Earlier Technologies?, *Bulletin of Science, Technology & Society*, 24, 34-39.

Meyer, M. & Kuusi, O.: 2004, 'Nanotechnology: Generalizations in an Interdisciplinary Field of Science and Technology', *Hyle: International Journal for Philosophy of Chemistry*, 10(2), 153-168.

Meyer, M.: 2000a, 'Does science push technology? Patents citing scientific literature', *Research Policy*, 29, 409-434.

Meyer, M.: 2000b, 'Patent citations in a novel field of technology: What can they tell about interactions of emerging communities of science and technology?', *Scientometrics*, 48, 151-178.

Meyer, M.: 2001a, 'Patent citations in a novel field of technology: An exploration of nano-science and nano-technology', *Scientometrics*, 51, 163-183.

Meyer, M.: 2001b, 'Socio-economic Research on Nanoscale Science and Technology: A European Overview and Illustration', in: M.C. Roco and W.S. Bainbridge (eds.), *Societal Implications of Nanoscience and Nanotechnology*, Dordrecht: Kluwer, pp. 217-241.

Milburn, C.: 2002, 'Nanotechnology in the age of post-human engineering: science fiction as science', *Configurations*, 10, 261-295 [reprinted in: N.K. Hayles (ed.), *Nanoculture: Implications of the New Technoscience*, Bristol, UK: Intellect Books, 2004, pp. 109-130].

Milburn, C.: 2004, 'Nano/Splatter: Disintegrating the Postbiological Body', *New Literary History*, 35 (in print).

Mnyusiwalla, A.; Abdallah, S.D.; Singer, P.A.: 2003, 'Mind the gap: science and ethics in nanotechnology', *Nanotechnology*, 14, R9-R13.

Mody, C.C.M.: 2004, 'How Probe Microscopists Became Nanotechnologists', in: D. Baird, A. Nordmann & J. Schummer (eds.), *Discovering the Nanoscale*, Amsterdam: IOS Press, pp. 119-133.

Mody, C.C.M.: 2004, 'Instruments in Training: The Growth of American Probe Microscopy in the 1980s', in: D. Kaiser (ed.), *Pedagogy and the Practice of Science: Producing Physical Scientists, 1800-2000*, Cambridge, MA: MIT Press (forthcoming).

Mody, C.C.M.: 2004, 'Small, but Determined: Technological Determinism in Nanoscience', *Hyle: International Journal for Philosophy of Chemistry*, 10(2), 99-128.

Mody, C.C.M.: 2004, *Crafting the Tools of Knowledge: The Invention, Spread, and Commercialization of Probe Microscopy, 1960-2000*, Ph.D. dissertation, Cornell University.

Montemerlo, M.S., Love, J.C., Opiteck, G.J., Goldhaber, D. J., and Ellenbogen, J.C., "Technologies and Designs for Electronic Nanocomputers," MITRE Technical Report 96W0000044, The MITRE Corporation, McLean, VA, July 1996. For more information, send e-mail to nanotech@mitre.org.

Moor, J.H. & Weckert, J.: 2004, 'Nanoethics: Assessing the Nanoscale From an Ethical Point of View', in: D. Baird, A. Nordmann & J. Schummer (eds.), *Discovering the Nanoscale*, Amsterdam: IOS Press, pp. [illegible]10.

Pitt, J.C.: 2004, 'The Epistemology of the Very Small', in: D. Baird, A. Nordmann & J. Schummer (eds.), *Discovering the Nanoscale*, Amsterdam: IOS Press, pp. 157-163.

Pressman, J.: 2004, 'Nano Narrative: A Parable from Electronic Literature', in: N.K. Hayles (ed.), *Nanoculture: Implications of the New Technoscience*, Bristol, UK: Intellect Books, pp. 191-200.

Prigogine, I., and S. Rice. 1988. *Advances in chemical physics*, Vol. 70, Parts 1 & 2. New York: J. Wiley.

Roco, M.C. & Bainbridge, W.S. (eds.): 2001, *Societal implications of nanoscience and nanotechnology*, (Proceedings of a workshop organized by the National Science Foundation, September 28-29, 2000), Kluwer: Dordrecht [available online at http://itri.loyola.edu/nano/societalimpact/nanosi.pdf]

Roco, M.C. & Tomellini, R. (eds.): 2002, *Nanotechnology: Revolutionary Opportunities and Societal Implications, Workshop, Lecce (Italy), 31 January - 1 February 2002*, Luxemburg: Office for Official Publications of the European Communities, [200 pp.]

Roco, M.C.; Bainbridge, W.S. (eds.): 2002, *Converging Technologies for Improving Human Performance: Nanotechnology, Biotechnology, Information Technology and the Cognitive Science*, Arlington, VA: National Science Foundation.

Roher, H. 1993. *Jpn. J. Appl. Phys.* 32:1335.

Rohlfing, E.A., D.M. Cox, and A. Kaldor. 1984. *J. Chem. Phys.* 81:3846.

Rosen, A. 1998. A periodic table in three dimensions: A sightseeing tour in the nanometer world. In *Advances in quantum chemistry*. In press.

Ruthven, D.M., S. Farooq, K.S. Knaebel. 1994. *Pressure swing adsorption*. New York: VCH Publishers.

Sarewitz, D.; Woodhouse, E.: 2003, 'Small is Powerful', in: A. Lightman, D. Sarewitz & Chr. Desser, (eds.), *Living with the Genie: Essays on Technology and the Quest for Human Mastery*, Washington, DC: Island Press, pp. 63-83.

Schiemann, G.: 2004, 'Dissolution of the Nature-Technology Dichotomy? Perspectives on Nanotechnology from an Everyday Understanding of Nature', in: D. Baird, A.

Nordmann & J. Schummer (eds.), *Discovering the Nanoscale*, Amsterdam: IOS Press, pp. 209-213.

Schmidt, J.C.: 2004, 'Unbounded Technologies: Working Through Technological Reductionism of Nanotechnology', in: D. Baird, A. Nordmann & J. Schummer (eds.), *Discovering the Nanoscale*, Amsterdam: IOS Press, pp. 35-50.

Schummer, J.: 2004, 'Why do Chemists Perform Experiments?', in: D. Sobczynska, P. Zeidler, E. Zielonacka-Lis (eds.), *Chemistry in the Philosophical Melting Pot*, Peter Lang (Frankfurt/M.), 2004, pp. 395-410.

Schummer, J.: 2005, 'Reading Nano: The Public Interest in Nanotechnology as Reflected in Book Purchase Patterns", *Public Understanding of Science*, 14 (2) (forthcoming).

Schummer, J.; Baird, D. (eds.): 2004-5, *Nanotech Challenges*, joint special issue of *Hyle: International Journal for Philosophy of Chemistry & Techne: Research in Philosophy and Technology* (forthcoming).

Stuckless, J.T, D.E. Starr, D.J. Bald, C.T. Campbell. 1997. *J. Chem. Phys*. 107:5547.

Suchman, M.: 2001, 'Envisioning Life on the Nano-Frontier', in: M.C. Roco and W.S. Bainbridge (eds.), *Societal Implications of Nanoscience and Nanotechnology*, Dordrecht: Kluwer, pp. 211-216.

Index